Earth Science:
Earth Materials and Resources

Earth Science: Earth Materials and Resources

Volume 2

Editor

Stephen I. Dutch, Ph.D.
University of Wisconsin-Green Bay

Salem Press
A Division of EBSCO Publishing
Ipswich, Massachusetts

Library of Congress Cataloging-in-Publication Data
Earth science. Earth materials and resources / editor, Steven I. Dutch, Ph. D., University of Wisconsin-Green Bay.
  pages cm
Includes bibliographical references and index.
ISBN 978-1-58765-989-8 (set) – ISBN 978-1-58765-981-2 (set 3 of 4) – ISBN 978-1-58765-982-9 (volume 1) – ISBN 978-1-58765-983-6 (volume 2) 1. Geology, Economic. 2. Mines and mineral resources. I. Dutch, Steven I. II. Title: Earth materials and resources.
 TN260.E25 2013
 553–dc23

                                                                          2012027580

PRINTED IN THE UNTED STATES OF AMERICA

CONTENTS

COMMON UNITS OF MEASURE

Notes: Common prefixes for metric units—which may apply in more cases than shown below—include giga- (1 billion times the unit), mega- (1 million times), kilo- (1,000 times), hecto- (100 times), deka- (10 times), deci- (0.1 times, or one tenth), centi- (0.01, or one hundredth), milli- (0.001, or one thousandth), and micro- (0.0001, or one millionth).

UNIT	QUANTITY	SYMBOL	EQUIVALENTS
Acre	Area	ac	43,560 square feet 4,840 square yards 0.405 hectare
Ampere	Electric current	A *or* amp	1.00016502722949 international ampere 0.1 biot *or* abampere
Angstrom	Length	Å	0.1 nanometer 0.0000001 millimeter 0.000000004 inch
Astronomical unit	Length	AU	92,955,807 miles 149,597,871 kilometers (mean Earth-sun distance)
Barn	Area	b	10^{-28} meters squared (approximate cross-sectional area of 1 uranium nucleus)
Barrel (dry, for most produce)	Volume/capacity	bbl	7,056 cubic inches; 105 dry quarts; 3.281 bushels, struck measure
Barrel (liquid)	Volume/capacity	bbl	31 to 42 gallons
British thermal unit	Energy	Btu	1055.05585262 joule
Bushel (U.S., heaped)	Volume/capacity	bsh *or* bu	2,747.715 cubic inches 1.278 bushels, struck measure
Bushel (U.S., struck measure)	Volume/capacity	bsh *or* bu	2,150.42 cubic inches 35.238 liters
Candela	Luminous intensity	cd	1.09 hefner candle
Celsius	Temperature	C	1° centigrade
Centigram	Mass/weight	cg	0.15 grain
Centimeter	Length	cm	0.3937 inch
Centimeter, cubic	Volume/capacity	cm³	0.061 cubic inch
Centimeter, square	Area	cm²	0.155 square inch
Coulomb	Electric charge	C	1 ampere second

UNIT	QUANTITY	SYMBOL	EQUIVALENTS
Cup	Volume/capacity	C	250 milliliters 8 fluid ounces 0.5 liquid pint
Deciliter	Volume/capacity	dl	0.21 pint
Decimeter	Length	dm	3.937 inches
Decimeter, cubic	Volume/capacity	dm^3	61.024 cubic inches
Decimeter, square	Area	dm^2	15.5 square inches
Dekaliter	Volume/capacity	dal	2.642 gallons 1.135 pecks
Dekameter	Length	dam	32.808 feet
Dram	Mass/weight	dr *or* dr avdp	0.0625 ounce 27.344 grains 1.772 grams
Electron volt	Energy	eV	$1.5185847232839 \times 10^{-22}$ Btu $1.6021917 \times 10^{-19}$ joule
Fermi	Length	fm	1 femtometer 1.0×10^{-15} meter
Foot	Length	ft *or* '	12 inches 0.3048 meter 30.48 centimeters
Foot, cubic	Volume/capacity	ft^3	0.028 cubic meter 0.0370 cubic yard 1,728 cubic inches
Foot, square	Area	ft^2	929.030 square centimeters
Gallon (British Imperial)	Volume/capacity	gal	277.42 cubic inches 1.201 U.S. gallons 4.546 liters 160 British fluid ounces
Gallon (U.S.)	Volume/capacity	gal	231 cubic inches 3.785 liters 0.833 British gallon 128 U.S. fluid ounces
Giga-electron volt	Energy	GeV	$1.6021917 \times 10^{-10}$ joule
Gigahertz	Frequency	GHz	—
Gill	Volume/capacity	gi	7.219 cubic inches 4 fluid ounces 0.118 liter

UNIT	QUANTITY	SYMBOL	EQUIVALENTS
Grain	Mass/weight	gr	0.037 dram 0.002083 ounce 0.0648 gram
Gram	Mass/weight	g	15.432 grains 0.035 avoirdupois ounce
Hectare	Area	ha	2.471 acres
Hectoliter	Volume/capacity	hl	26.418 gallons 2.838 bushels
Hertz	Frequency	Hz	$1.08782775707767 \times 10^{-10}$ cesium atom frequency
Hour	Time	h	60 minutes 3,600 seconds
Inch	Length	in *or* "	2.54 centimeters
Inch, cubic	Volume/capacity	in^3	0.554 fluid ounce 4.433 fluid drams 16.387 cubic centimeters
Inch, square	Area	in^2	6.4516 square centimeters
Joule	Energy	J	$6.2414503832469 \times 10^{18}$ electron volt
Joule per kelvin	Heat capacity	J/K	$7.24311216248908 \times 10^{22}$ Boltzmann constant
Joule per second	Power	J/s	1 watt
Kelvin	Temperature	K	-272.15 degrees Celsius
Kilo-electron volt	Energy	keV	$1.5185847232839 \times 10^{-19}$ joule
Kilogram	Mass/weight	kg	2.205 pounds
Kilogram per cubic meter	Mass/weight density	kg/m^3	$5.78036672001339 \times 10^{-4}$ ounces per cubic inch
Kilohertz	Frequency	kHz	—
Kiloliter	Volume/capacity	kl	—
Kilometer	Length	km	0.621 mile
Kilometer, square	Area	km^2	0.386 square mile 247.105 acres
Light-year (distance traveled by light in one Earth year)	Length/distance	lt-yr	5,878,499,814,275.88 miles 9.46×1012 kilometers
Liter	Volume/capacity	L	1.057 liquid quarts 0.908 dry quart 61.024 cubic inches

UNIT	QUANTITY	SYMBOL	EQUIVALENTS
Mega-electron volt	Energy	MeV	—
Megahertz	Frequency	MHz	—
Meter	Length	m	39.37 inches
Meter, cubic	Volume/capacity	m³	1.308 cubic yards
Meter per second	Velocity	m/s	2.24 miles per hour 3.60 kilometers per hour
Meter per second per second	Acceleration	m/s²	12,960.00 kilometers per hour per hour 8,052.97 miles per hour per hour
Meter, square	Area	m²	1.196 square yards 10.764 square feet
Metric. See unit name			
Microgram	Mass/weight	mcg *or* μg	0.000001 gram
Microliter	Volume/capacity	μL	0.00027 fluid ounce
Micrometer	Length	μm	0.001 millimeter 0.00003937 inch
Mile (nautical international)	Length	mi	1.852 kilometers 1.151 statute miles 0.999 U.S. nautical mile
Mile (statute or land)	Length	mi	5,280 feet 1.609 kilometers
Mile, square	Area	mi²	258.999 hectares
Milligram	Mass/weight	mg	0.015 grain
Milliliter	Volume/capacity	mL	0.271 fluid dram 16.231 minims 0.061 cubic inch
Millimeter	Length	mm	0.03937 inch
Millimeter, square	Area	mm²	0.002 square inch
Minute	Time	m	60 seconds
Mole	Amount of substance	mol	6.02×10^{23} atoms or molecules of a given substance
Nanometer	Length	nm	1,000,000 fermis 10 angstroms 0.001 micrometer 0.00000003937 inch

Unit	Quantity	Symbol	Equivalents
Newton	Force	N	0.224808943099711 pound force 0.101971621297793 kilogram force 100,000 dynes
Newton-meter	Torque	N·m	0.7375621 foot-pound
Ounce (avoirdupois)	Mass/weight	oz	28.350 grams 437.5 grains 0.911 troy or apothecaries' ounce
Ounce (troy)	Mass/weight	oz	31.103 grams 480 grains 1.097 avoirdupois ounces
Ounce (U.S., fluid or liquid)	Mass/weight	oz	1.805 cubic inch 29.574 milliliters 1.041 British fluid ounces
Parsec	Length	pc	30,856,775,876,793 kilometers 19,173,511,615,163 miles
Peck	Volume/capacity	pk	8.810 liters
Pint (dry)	Volume/capacity	pt	33.600 cubic inches 0.551 liter
Pint (liquid)	Volume/capacity	pt	28.875 cubic inches 0.473 liter
Pound (avoirdupois)	Mass/weight	lb	7,000 grains 1.215 troy or apothecaries' pounds 453.59237 grams
Pound (troy)	Mass/weight	lb	5,760 grains 0.823 avoirdupois pound 373.242 grams
Quart (British)	Volume/capacity	qt	69.354 cubic inches 1.032 U.S. dry quarts 1.201 U.S. liquid quarts
Quart (U.S., dry)	Volume/capacity	qt	67.201 cubic inches 1.101 liters 0.969 British quart
Quart (U.S., liquid)	Volume/capacity	qt	57.75 cubic inches 0.946 liter 0.833 British quart
Rod	Length	rd	5.029 meters 5.50 yards

Unit	Quantity	Symbol	Equivalents
Rod, square	Area	rd²	25.293 square meters 30.25 square yards 0.00625 acre
Second	Time	s or sec	$\frac{1}{60}$ minute $\frac{1}{3,600}$ hour
Tablespoon	Volume/capacity	T or tb	3 teaspoons 4 fluid drams
Teaspoon	Volume/capacity	t or tsp	0.33 tablespoon 1.33 fluid drams
Ton (gross or long)	Mass/weight	t	2,240 pounds 1.12 net tons 1.016 metric tons
Ton (metric)	Mass/weight	t	1,000 kilograms 2,204.62 pounds 0.984 gross ton 1.102 net tons
Ton (net or short)	Mass/weight	t	2,000 pounds 0.893 gross ton 0.907 metric ton
Volt	Electric potential	V	1 joule per coulomb
Watt	Power	W	1 joule per second 0.001 kilowatt $2.84345136093995 \times 10^{-4}$ ton of refrigeration
Yard	Length	yd	0.9144 meter
Yard, cubic	Volume/capacity	yd³	0.765 cubic meter
Yard, square	Area	yd²	0.836 square meter

COMPLETE LIST OF CONTENTS

Volume 2

CATEGORY LIST OF CONTENTS

Earth Science:
Earth Materials and Resources

NUCLEAR POWER

Nuclear power obtained from the fission of uranium and plutonium nuclei represents a significant percentage of world energy resources. Its production in appropriately designed nuclear fission reactors is especially important as a low-pollution supplement to fossil fuels.

PRINCIPAL TERMS

- **half-life:** the time required for half of the atoms in a given amount of a radioactive isotope to disintegrate
- **isotopes:** an element's variant forms, whose atoms have the same number of protons but different numbers of neutrons
- **moderator:** a material used in a nuclear reactor for slowing neutrons to increase their probability of causing fission
- **millisievert (mSv):** a measure of the biological effect of radiation; 1000 mSv in a short time will result in radiation sickness, and 4,000 mSv will kill half of those exposed
- **nuclear fission:** the splitting of an atomic nucleus into two lighter nuclei, resulting in the release of neutrons and some of the binding energy that held the nucleus together
- **nuclear fusion:** the collision and combining of two nuclei to form a single nucleus with less mass than the original nuclei, with a release of energy equivalent to the mass reduction
- **radioactivity:** the spontaneous emission from unstable atomic nuclei of alpha particles (helium nuclei), beta particles (electrons), and gamma rays (electromagnetic radiation)

NUCLEAR FISSION

The idea of the atom as a source of energy developed near the beginning of the twentieth century following the discovery of radioactivity by Antoine-Henri Becquerel. The energy of this spontaneous emission, first measured by Pierre and Marie Curie, was found to be far greater than ordinary chemical energies. Nuclear fission was discovered in 1939 after Otto Hahn and Fritz Strassmann had bombarded uranium with neutrons at their laboratory in Berlin, leaving traces of radioactive barium. Their former colleague Lise Meitner and her nephew Otto Frisch calculated the enormous energy—about 200 million electron volts—that would be released in reactions of this type. These results were reported to Niels Bohr and quickly verified in several laboratories in 1939. Soon Bohr developed a theory of fission showing that the rare isotope uranium-235 (uranium with 235 nucleons: 92 protons and 143 neutrons) is far more likely to produce fission, especially with slow neutrons, than the common isotope uranium-238, which makes up 99.3 percent of natural uranium. It also was recognized that if a sufficient number of neutrons were emitted in fission, they could produce new fissions with even more neutrons, resulting in a self-sustaining chain reaction. In this process, the fissioning of one gram of uranium-235 would release energy equivalent to burning about three tons of coal.

The first nuclear reactor to achieve a controlled, self-sustaining chain reaction was developed under the leadership of the Italian physicist Enrico Fermi in 1942 at the University of Chicago. To increase the probability of fission in natural uranium, only 0.7 percent of which is uranium-235, and to prevent any chance of explosion, the neutrons were slowed down by collisions with carbon atoms in a graphite "moderator." It was necessary to assemble a large enough lattice of graphite (385 tons) and uranium (40 tons) to achieve a "critical mass" of fissile material, in which the number of neutrons not escaping from the "pile" would be sufficient to sustain a chain reaction. Cadmium "control rods" were inserted to absorb neutrons during construction so that the chain reaction would not begin the instant the critical size was reached.

The uranium-235 isotope is the only natural material that can be used to produce nuclear energy directly. By early 1941, however, it was known that uranium-238 captures fast neutrons to produce the new element plutonium. Plutonium has a 24,000-year half-life and is fissile, so it can be used as a nuclear fuel. Plutonium can be "bred" in a uranium reactor from uranium-238 with excess neutrons from the fissioning of uranium-235. One other fissionable isotope, uranium-233, can be obtained by neutron capture from the thorium isotope thorium-232. Uranium-233 is a possible future nuclear fuel. Uranium from which uranium-235 has been removed is almost entirely uranium-238, and is called "depleted uranium." Depleted uranium has few uses, mostly exploiting its low cost but high specific gravity of 19 (almost twice the density of lead). It is used where massive weights are needed in small spaces, like counterweights for aircraft control surfaces, and for armor-piercing artillery shells.

THERMAL REACTORS

The two basic types of reactors in use are thermal reactors, which use slow neutrons, and fast breeder reactors, which use fast neutrons to breed plutonium. Plutonium can be separated by chemical methods, but very expensive physical methods are necessary to separate uranium-235 from uranium-238; these methods involve many stages of gaseous diffusion or centrifuge processes to distinguish between their slightly different masses. Most nuclear reactors use 3.5 percent enriched uranium, but there are also reactor designs that use natural uranium. Weapons-grade uranium is usually enriched to 93.5 percent. Fast breeder reactor fuel is generally enriched with plutonium.

Thermal reactors for generating useful power consist of a core that contains a critical assembly of fissionable fuel elements surrounded by a moderator to slow the neutrons, a coolant to transfer heat, and movable control rods to absorb neutrons and establish the desired fission rate. Reactor fuel elements are made of natural or enriched uranium metal or oxide in the form of thin rods clad with a corrosion-resistant alloy of magnesium, zirconium, or stainless steel. Moderator materials must be low neutron absorbers, with small atomic masses close to the mass of neutrons so that they can slow them down by repeated collisions. Most reactors use moderators made of graphite, water, or heavy water, which contains the hydrogen isotope deuterium. Ordinary water is low in cost and doubles as a coolant, but it absorbs neutrons about one hundred times more than graphite and about one thousand times more than heavy water. Coolants such as water, carbon dioxide, and helium transfer heat liberated by fission from the core, producing steam or hot gas to drive a turbine for generating electricity in the conventional manner. Control rods are made of high neutron absorbers, such as cadmium or boron, and can be adjusted for any desired power output.

LIGHT VERSUS HEAVY WATER REACTORS

Most reactors in the United States are "light water reactors" (LWRs); they use ordinary water as both moderator and coolant and require some fuel enrichment. Some are "pressurized water reactors" (PWRs), and some are "boiling water reactors" (BWRs). Most LWRs use 2 to 3 percent enriched uranium dioxide fuel elements clad in a zirconium alloy, although the PWR was first developed with much higher fuel enrichments for compact shipboard use. In a PWR, water is circulated at high pressure through the reactor core at above 300 degrees Celsius and then through a heat exchanger, where steam is produced in a secondary loop. In a BWR, the water is boiled in the core at about 280 degrees Celsius, eliminating the high cost of an external heat exchanger and highly pressurized containment vessel. LWRs have the fail-safe feature: If the temperature increases fast enough to expel water from the core, neutrons will be slowed less effectively, and the fission rate will decrease.

Canada has specialized in heavy water reactors, since Canada has access to natural uranium with no need for fuel enrichment. In the Canadian deuterium-uranium system (CANDU), the heavy water coolant is circulated past fuel elements inside pressure tubes, which are surrounded by a heavy water moderator in a low-pressure tank. The coolant is pumped through a heat exchanger to boil ordinary water for driving steam turbines. Since 1968, several CANDU plants in the 200 to 700 megawatt range have been built in Canada, Argentina, India, Pakistan, and South Korea. Variants of this system employ light water or gas as a coolant to reduce the high cost of heavy water, but they may require enriched fuel.

FAST BREEDER REACTORS

The main alternatives to thermal reactors are fast breeder reactors, which can obtain about fifty times as much energy from natural uranium by producing more plutonium from uranium-238 than the uranium-235 they use. Because neutron capture by uranium-238 requires fast neutrons (about one thousand times faster than thermal neutrons), no moderator can be used, and a 15 to 30 percent fuel enrichment is needed to sustain the chain reaction. A typical breeder core consists of a compact assembly of fuel rods with 20 percent plutonium and 80 percent depleted uranium (most uranium-235 is removed) oxides surrounded by a "blanket" of depleted uranium carbide to absorb neutrons and yield more plutonium. The "liquid metal fast breeder reactor" (LMFBR) uses sodium in liquid form (above 99 degrees Celsius) as a coolant, since water would slow the neutrons. Loss or interruption of sodium can lead to meltdown of the core, so some designs seal the core in a pool of sodium.

The first commercial fast breeder reactor began operating in the Soviet Union in 1972, producing 350 megawatts. France had the most advanced fast breeder reactor program, with its 1,200-megawatt Super Phénix breeder reactor. India has successfully converted thorium into uranium-233 and used it as fuel in the ICGAR fast breeder test reactor. In its pure form, thorium is a silver-white metal similar to uranium. Thorium metal is more stable in air, retaining its luster for months, whereas uranium quickly tarnishes. Since thorium is three to four times more abundant than uranium, the ability to use it in commercial reactors would greatly extend the nuclear fuel supply.

NUCLEAR FUSION

Most of the problems associated with fission power could be eliminated with nuclear fusion reactors. These problems include the handling, storage, and reprocessing of highly radioactive materials such as plutonium; the possible theft of such materials by terrorists; the disposal of radioactive waste products; the dangers of a reactor accident; and the limited availability of fission fuels. The fusion of hydrogen isotopes to produce helium releases energy comparable to fission but requires no critical mass of fuel that might cause meltdown. It has many fewer radioactive products with no storage or disposal problems and uses a fuel of almost unlimited supply. About 0.01 percent of the hydrogen in ocean water is in the form of deuterium. To overcome electrical repulsion and bring deuterium atoms close enough to cause a fusion reaction, an ignition temperature of about 100 million degrees is required. Ignition and isolation of such reactions require some kind of magnetic confinement of a plasma (ionized gas) or inertial confinement of deuterium pellets. Energy would be extracted by nuclear reactions in a surrounding lithium blanket caused by neutrons emitted during fusion. Some progress has been made in achieving these requirements, but a practical source of fusion power is many years away.

Experiments at several laboratories in the 1980's claimed to have found evidence for room-temperature fusion by using electrolysis to draw deuterium ions into the crystal lattice of hydrogen-absorbing materials such as palladium or titanium. Because electrical repulsion between charged nuclei increases greatly as they approach each other, it is difficult to understand how this process could bring deuterium ions close enough for fusion to occur. Even if such experiments are confirmed and explained, the development of a reactor to produce electrical power with this technique may be difficult if not impossible. Hydrogen absorption declines sharply with increasing temperature, decreasing by a factor of at least ten as the temperature approaches the boiling point of water. Much higher temperatures would have to be produced for an efficient steam-driven electrical generator. Most physicists believe the initial results were the result of poor experimental design.

SAFETY OF NUCLEAR POWER

The study of reactor safety involves estimating the biological effects of radiation and analyzing the risk factors in possible reactor accidents. Information on the effects of large doses of radiation is based on medical X rays, animal experiments, and studies of Japanese atomic-bomb survivors. Radiation doses are monitored by photographic film dosimeters and simple ionization chambers. Normal background radiation from radioactivity in the earth, radon gas, and cosmic radiation is about double the average dose received by a person for medical purposes annually. The radioactivity from normal reactor operation is considerably less than background radiation. The major public concern focuses on accidental releases of large amounts of radioactivity. The Nuclear Regulatory Commission estimates the risk from a reactor accident at less than one death over its service lifetime.

The risks of a nuclear accident can be studied only when one such accident occurs. The most serious commercial reactor accident in the United States occurred in 1979 at the Three Mile Island power station in Pennsylvania; the loss of some coolant led to the shutdown of one reactor, as designed, but resulted in costly damage to the core. Because of containment structures, including a thick steel vessel around the core and a reinforced concrete building with walls several feet thick, the highest average dose released was about one-tenth the annual background radiation. A much more serious accident occurred at Chernobyl in the Soviet Union in 1986; a loss of coolant in a graphite reactor led to increased power followed by explosions and fire, killing thirty-one men. Approximately 270,000 people were evacuated from areas in the former Soviet Union where the average radiation from Chernobyl fallout ranged from 6 to 60 millisieverts (mSv). While these levels are above the average of 2.2 millisieverts for natural background radiation, they are no higher than levels that occur naturally in some regions of Brazil, India, and China. In 2011, a large earthquake in Japan resulted in serious damage and radiation leaks at the nuclear plant at Fukushima. This disaster was eventually given a hazard rating equal to that of Chernobyl. According to the National Research Council report on the health effects of low levels of ionizing radiation, although an increased frequency of chromosome aberrations is found, no increase in the frequency of cancer has been documented in populations residing in areas of high natural background radiation.

The disposal of radioactive wastes is another area of concern and continuing study. Methods of solidifying such waste in glass or other materials for confinement in metal canisters and burial are being studied. The solid waste projected through the year 2010 would cover about 40,000 square meters (10 acres). Of the several disposal sites and methods under study, the most likely is deep underground burial in formations of salt or rock. The principal problem is plutonium-239, which has a half-life of 24,000 years, meaning waste would have to remain isolated for a million years or more. Eventual reprocessing of the waste in breeder reactors might alleviate this problem.

FUTURE OF NUCLEAR POWER

Nuclear power is an important source of low-pollution energy in spite of serious problems that have emerged since 1980. By the end of the 1990's, 19 percent of electrical energy in the United States was generated by 104 nuclear reactors producing nearly 100,000 megawatts of power, while 16 percent of electrical energy worldwide was generated by about 440 nuclear reactors. Using more than 63,000 metric tons of uranium in 1998, the reactors produced more than

350,000 megawatts of power. At this rate, known uranium reserves will last about fifty years. Reprocessing spent fuel and using the military's excess of highly enriched uranium (HEU) may extend the supply for many years, and new reserves are still being discovered. The use of fast breeder reactors rather than thermal reactors would extend nuclear fuel supplies by a factor of fifty or sixty, while the use of thorium could add an additional factor of three or four.

The future of conventional nuclear power is uncertain. Since the Three Mile Island accident in 1979, public distrust of nuclear power has increased. The concern about reactor safety has led to new requirements that have raised the cost of nuclear power plants by a factor of five above inflation. Bankers and investors have become increasingly cautious about financing new plant construction. The high rate of government subsidies has complicated the evaluation of real dollar costs of nuclear power, but estimates indicate that the profit potential of conventional nuclear power is about half that of coal power. The total U.S. nuclear power capacity, operational and planned, slipped from 236 reactors in 1976 to 104 in 1999.

One promising approach to the safer production of nuclear fission power is the development of small-scale modular reactors that use tiny ceramic-coated fuel pellets in small enough quantities in their cores that meltdown is impossible. Although initially expensive, such units would not require expensive safety systems and could be built on an assembly line, producing one module at a time to match operating capacity with the demand for power. Using advanced technology, India, Japan, South Korea, Russia, and other nations are expanding their use of nuclear energy; in 1999, thirty-one new reactors were under construction, and sixty-seven more were on order or planned worldwide.

Joseph L. Spradley

FURTHER READING

Andrews, John, and Nick Jelley. *Energy Science: Principles, Technologies, and Impact.* New York: Oxford University Press, 2007. Discusses various forms of energy and their environmental and socioeconomic impacts. Covers principles of energy consumption as an entire system, including information on the generation, storage, and transmission of energy. Requires a strong mathematics background in certain chapters. Appropriate for undergraduate and professionals with interest in energy resources.

Blair, Ian. *Taming the Atom: Facing the Future with Nuclear Power*. Bristol, England: Adam Hilger, 1983. An account of the development and use of nuclear power from a British perspective. Covers the basic physical principles of nuclear power, along with a good survey of the nuclear industry. An appendix on world nuclear reactors includes diagrams and data.

Byrne, John, and Steven M. Hoffman, eds. *Governing the Atom: The Politics of Risk*. New Brunswick, N.J.: Transaction, 1996. Offers research into the environmental and social aspects of the nuclear industry. Focuses on government policies that have been implemented to monitor safety and accident prevention measures.

Cameron, I. R. *Nuclear Fission Reactors*. New York: Plenum Press, 1983. Provides a technical treatment of nuclear fission and reactor theory, as well as a more readable survey of reactor types and a discussion of the safety and environmental aspects of reactors. The bibliography lists about 150 references to books and technical articles.

Cohen, Bernard L. *Before It's Too Late: A Scientist's Case for Nuclear Energy*. New York: Plenum Press, 1982. Discusses problems in the public understanding of nuclear power. Covers the danger of radiation, the possibility and results of a meltdown accident, the problem of radioactive waste, and risk assessments. Each chapter lists many references.

Craig, J. R., D. J. Vaughan, and B. J. Skinner. *Resources of the Earth*. 3d ed. Upper Saddle River, N.J.: Prentice Hall, 2001. Contains a chapter on nuclear power, the nuclear fuel cycle, reactor safety, and uranium mining. Other chapters on fossil fuels and environmental problems give useful comparative information.

Inglis, David R. *Nuclear Energy: Its Physics and Its Social Challenge*. Reading, Mass.: Addison-Wesley, 1973. Based on an introductory college course for general students. Explains the basic principles of nuclear energy and describes reactors and their radioactive products. Provides technical details and historical documents in several appendices.

Letcher, Trevor M., ed. *Future Energy: Improved, Sustainable and Clean*. Amsterdam: Elsevier, 2008. Considers the future of fossil fuels and nuclear power. Discusses renewable energy supplies—solar, wind, hydroelectric, and geothermal. Focuses on potential energy sources and currently underutilized energy sources for the future. Covers new concepts in energy consumption and technology.

Marion, J. B., and M. L. Roush. *Energy in Perspective*. 2d ed. New York: Academic Press, 1982. Written for a college survey course. Discusses energy consumption, energy sources, nuclear power, and the effects of nuclear radiation. Contains many good diagrams, tables, and photographs.

O'Very, David P., Christopher E. Paine, and Dan W. Reicher, eds. *Controlling the Atom in the 21st Century*. Boulder, Colo.: Westview Press, 1994. Focuses on the debates surrounding the legislation and licensing procedures involved in the operation of nuclear power plants.

Priest, Joseph. *Energy: Principles, Problems, Alternatives*. 6th ed. Dubuque, Iowa: Kendal Hunt Publishing, 2006. Designed for a college survey course on energy. Discusses nuclear fission power, breeder reactors, and fusion reactors. Contains interesting illustrations and tables, an appendix giving energy and consumption comparisons, and a glossary of terms.

Young, Warren. *Atomic Energy Costing*. Boston: Kluwer Academic Publishers, 1998. An economic analysis of the nuclear industry, focusing on cost-effectiveness, costs of operation, and other economic aspects associated with the production of nuclear energy and the operation of nuclear power plants.

See also: Earth Resources; Geothermal Power; Hydroelectric Power; Nuclear Waste Disposal; Ocean Power; Solar Power; Unconventional Energy Resources; Uranium Deposits; Wind Power.

NUCLEAR WASTE DISPOSAL

Nuclear waste disposal in a geologic repository is considered the safest and surest way to achieve isolation of the wastes from the surface environment. No one can guarantee that a repository will last forever, but its combined geologic characteristics will minimize damage in the event of failure.

PRINCIPAL TERMS

- **half-life:** the time required for half of the radioactive isotopes in a sample to decay
- **high-level wastes:** wastes containing large amounts of dangerous radioactivity
- **isotopes:** atoms of the same element that differ in the number of uncharged neutrons in their nuclei
- **leachate:** water that has come into contact with waste and, as a result, is transporting some of the water-soluble parts of the waste
- **low-level wastes:** wastes that are much less radioactive than high-level wastes and thus less likely to cause harm
- **permeability:** a measure of the ease of flow of a fluid through a porous rock or sediment
- **porosity:** a measure of the amount of open spaces capable of holding water or air in a rock or sediment
- **radioactivity:** the spontaneous release of energy accompanying the decay of a nucleus
- **solubility:** the tendency for a solid to dissolve
- **sorption:** the process of removing a chemical from a fluid by either physical or chemical means
- **transuranic:** an isotope of an element that is heavier than uranium and formed in the processing and use of nuclear fuel and plutonium

HIGH- VERSUS LOW-LEVEL WASTES

Nuclear waste disposal is a necessary evil in a world where radioactive isotopes and the energy produced by their decay are used, among other things, for generating electricity, diagnosing and treating diseases, and making nuclear weapons. Nuclear wastes are not all the same. Some wastes, because of their composition, are more dangerous than others, and some are more dangerous for longer periods of time. When evaluating a geologic site for disposal of nuclear wastes, the characteristics of the site must be identified so that the waste will be isolated from the earth surface environment for a minimally acceptable period of time. Ensuring the minimal length of isolation is most important for those wastes that

are dangerous for a long time and those that contain large concentrations of dangerous isotopes. Such wastes, usually called high-level wastes, consist of isotopes of uranium and plutonium produced in the use and processing of nuclear fuel. The length of isolation is not as important for those wastes that are dangerous for shorter times and less concentrated. These wastes are called low-level wastes and may include some long-lasting wastes but in lower concentrations. An example of a low-level waste is slightly contaminated garbage, such as disposable laboratory equipment, disposable medical equipment, and disposable gloves and overalls (used in the handling of radioactive materials). The length of time that a particular waste is dangerous depends on which radioactive isotopes are in the waste and on the half-life of each isotope. The half-life is the time required for half of the radioactive isotopes in a sample to decay.

During the natural process of radioactive decay, the nucleus of a radioactive isotope (the "parent") is changed into that of another isotope (the "daughter"), and energy is released. This radioactive decay from parent to daughter is usually one step in a series of many, as the original parent isotope changes into a nonradioactive, stable isotope. At each step, energy is released in the form of energetic particles or energetic rays. Thus, in a sample of uranium, the parent uranium isotopes are constantly decaying in a series of steps through various daughter isotopes until the sample consists of pure lead, the ultimate daughter isotope in the uranium decay series. It takes a long time for all the uranium to decay completely to lead. The half-life of one uranium isotope, uranium-238, is 4.5 billion years. The long half-life of uranium suggests that the half-lives for some of the daughter isotopes in the decay series are also long.

ISOLATION IN GEOLOGIC REPOSITORY

Exposure to radiation is never without risk. The problem with a long half-life is that the isotope is giving off energetic particles or rays that are dangerous to plants and animals for an extended period.

Thus, high-level wastes, where the isotopes are in concentrated form, must be isolated from the surface environment for a long time. Low-level wastes are less dangerous only because the concentrations of radioactive isotopes are lower. While they need to be isolated from the surface environment, they are not as destructive to life as high-level wastes.

Plants and animals are exposed to radiation through air, water, and food. Naturally occurring radioactive isotopes may be in the air (in the form of radon, for example) and taken into the lungs. Some radioactive isotopes are dissolved in the water we drink (potassium-40) and in the food we eat (carbon-14). To minimize excessive exposure to radiation, a geologic repository must isolate radioactive waste from the surface environment. A geologic repository is any structure in either rock or soil that uses, in part, the natural abilities of these materials to isolate the wastes from the surface environment. A geologic repository may be a shallow trench dug in the soil, partially filled with low-level waste and covered with the excavated soil, or it may be a cavern constructed deep underground in hard rock. Regardless, the purpose of the repository is to isolate the waste in such a way that there is no excess exposure of organisms to radiation as a result of the presence of the repository. In this sense, isolation means that the waste must be kept from contaminating the surrounding air, water, and food.

CHARACTERISTICS OF REPOSITORY

The rocks or soil of a geologic repository must have certain characteristics to isolate wastes properly. Some of the more important characteristics are porosity, permeability, mineral solubility, and sorption capacity. Porosity and permeability are related but are not the same. Porosity is a measure of the ability of a rock or soil to hold a fluid, either liquid or gas. Expressed as a percentage of the volume of the sample, porosity is the volume of open, or pore, space in the rock or soil. If the soil or rocks of a geologic repository have high porosity, then large amounts of water or air may contact the waste; however, contact with the waste does not necessarily mean that the contaminated water or air will reach the surface environment. If the rock or soil is also permeable—that is, if the pores are interconnected enough so that the water or air can flow from the repository to the surface—the repository is not likely to provide adequate isolation. Porosity can exist without permeability when many small pores contain a large volume of fluid, but the connections between pores are too small to allow fluid flow. Permeability can exist with small porosity when a few penetrating cracks in an otherwise solid rock allow easy fluid flow.

Mineral solubility is an important characteristic in cases where water may contact the waste. Soluble minerals—those that tend to dissolve in water—may be removed from the surrounding rock or soil by moving water. When minerals dissolve, a space is left behind that adds to the porosity, and may increase the permeability, of the repository.

The sorption capacity of a mineral is the ability of that mineral to remove a dissolved molecule or ion from a fluid. Sorption may occur by either chemical or physical means. A type of chemical sorption is the ion exchange that takes place in a home water-softening unit. Water containing problem ions (in the case of a repository, radioactive ions) flows over the exchanging minerals. The problem ions attach to the mineral and, in the process, force the ion that was attached originally into the solution. In this way, the water is cleansed of problem ions. In physical sorption, the water and the contaminant are attached to the mineral by the force of friction, which results in a thin layer of water attached to the mineral surface. The attached, or physically sorbed, water and any material dissolved in it are not moving.

All these characteristics of rock and soil are related in some way to the possibility that the waste will escape from the repository and be transported by a fluid to the surface. It is unlikely that any repository will provide isolation perfectly when all the characteristics are taken into consideration. Some characteristics, however, may be more important than others, and a weakness in one may be offset by strength in another.

DESCRIPTION OF REPOSITORY MATERIAL

Shallow burial of low-level wastes and deep burial of high-level wastes are preceded by careful description of the strengths and weaknesses of the geologic material of the repository. In the case of shallow burial, the repository material is usually soil or saprolite (near-surface rock that has been partially turned to soil by the actions of water). To determine the strengths and weaknesses of a particular soil for containing low-level waste, scientists must study the structure, hydrology, mineralogy, and chemistry of that soil.

When water, in the form of rain or snowmelt, enters a soil, in most cases it does not flow through the soil uniformly. Water tends to flow more easily through the soil along certain permeable paths. In nearly every soil there are preferred flow paths and isolated areas. Sometimes the structure is inherited from the original rock that broke down to soil. Fractures and other cracks contained in the original rock may be preserved as cracks in the soil. Structures inherited from the parent rock are especially evident in saprolite, which exists in a stage between rock and true soil.

Other soil structures are the products of soil processes. Such structures include the cracks developed when a soil dries out, the tunnels formed by burrowing organisms such as worms and ants, and the openings left when the roots of dead plants decay. All these structures are important because they determine the manner and speed at which water will flow through a soil.

The means and speed of water flow through soil are important aspects of its hydrology. The flow of water into a shallow burial trench is usually restricted by engineered, low-permeability barriers such as compacted clay or plastic liners. If the liners fail and water flows through and out of the wastes, the surrounding soil should slow the flow of contaminants in two ways: physically and chemically. Physically impeding, or retarding, the flow of contaminated water from the waste trench means that the soil structure does not provide permeable flow paths; the water is forced to flow through many small, interconnected pores. As a result, the waste-transporting water contacts more of the mineral material in the soil. Chemical retardation of wastes occurs when the contaminants in the slowly moving fluid interact with the minerals of the soil. Some of those minerals have the capacity to sorb contaminants, further slowing their migration. Other reactions between the minerals and the wastewater result in the chemistry of the water changing in such a way that the contaminants may become insoluble and form new solids in the soil. The effect is the same: The flow of contaminants from the burial trench is slowed or stopped.

Stability of Repository

Disposal of high-level wastes in a geologic repository requires that many of the same characteristics of the host rock be determined. In addition to having the appropriate hydrological, mineralogical, and chemical characteristics, the host rock is required to be reasonably stable for ten thousand years. For example, the modern hydrological characteristics of the proposed Yucca Mountain repository site in Nevada are ideal. The proposed repository will be located below the desert surface, out of the reach of downward-percolating rainwater and well above the nearest fresh groundwater. In the desert environment, rainfall is so infrequent that it cannot reach the repository from above, nor can the infrequent rainfalls raise the groundwater level to flood the repository. Construction of 8 kilometers of test tunnels in Yucca Mountain began in 1994, and evaluation of the site was to be completed by 2001, with construction to begin in 2005 and emplacement of high-level waste to start in 2010.

The problem is that climates have changed in the past—the desert Southwest of the United States used to be much wetter—and climates will undoubtedly change in the future. In addition, the repository may be disturbed, whether inadvertently or purposely, by human activities. The most likely inadvertent disturbance is of future generations, who, in a search for mineral deposits, will puncture the repository with drilling equipment. To avoid this scenario, the repository should be located in an area unlikely to yield mineral wealth. Deeply buried layers of rock salt, thought to be ideal repositories because of their stability and their resistance to water flow, are no longer being considered. Many of the rock salt deposits are already being mined for salt, and future generations may mine the remaining ones. In addition, some of these deposits are associated with oil and natural gas. Although these considerations make salt beds unsuitable for storing high-level waste, some are suitable for storing low-level waste.

The federal Low-Level Radioactive Waste Policy Act of 1980 delegates the responsibility for the disposal of low-level waste to the state in which that waste is produced. The Waste Isolation Pilot Plant (WIPP) was constructed 42 kilometers east of Carlsbad in the southeastern corner of New Mexico in order to develop and prove the required technology. By law,

WIPP can only accept low-level waste from the nation's nuclear weapons program; the first shipment arrived and was placed in storage during 1999. The waste is stored 650 meters below the surface in rooms excavated in an ancient, stable salt formation.

As part of its search for a means to dispose of nuclear waste, as prescribed by the Nuclear Waste Policy Act of 1982, the Department of Energy explored the possibility of storing waste in crystalline rocks—generally granite intrusions—in the eastern United States. Storage in such sites would have the advantage of shorter transport distance and geological stability. A provision in the act that allowed states and American Indian tribes to veto the construction of a site doomed the project from the outset, although a veto could be overridden by Congress. In the face of intense public opposition, the project was abandoned.

POLITICAL CONSIDERATIONS

Nuclear waste disposal, whether of low-level or high-level wastes, is a problem that will remain. More waste is being created every day. A geologic repository and its host rock or soil must safely isolate the wastes from the surface environment for a specified period of time. Should the repository and host material fail, plants and animals will be exposed to radioactive elements. The risk of exposure to radiation can never be eliminated, but careful design of the repository and careful determination of the host material's strengths and weaknesses can result in a disposal site that minimizes the risk.

Ultimately, the siting of a nuclear waste disposal facility will be a political decision. That political decision must be based on solid technical information and judgments. As nuclear wastes accumulate in temporary holding facilities (for example, spent fuel elements from nuclear reactor power generators are stored in large water-filled pools near the reactor), political pressures build to solve the disposal problem. An informed public, while pressing for a solution, will understand delays necessitated by the need to gather and interpret the data. It must also be understood that risk can be minimized but never eliminated.

Richard W. Arnseth

FURTHER READING

Bartlett, Donald L., and James B. Steele. *Forevermore: Nuclear Waste in America.* New York: W. W. Norton, 1985. Examines the history and future of nuclear waste disposal in the United States. Traces the political aspects of decision making in a technical field. Accessible to any reader with an interest in the subject, regardless of technical background.

Burns, Michael E., ed. *Low-Level Radioactive Waste Regulation: Science, Politics, and Fear.* Chelsea, Mich.: Lewis, 1987. Deals exclusively with the problems of low-level wastes. Covers a range of issues, from technical aspects of exposure risks to the politics among states when deciding where to locate a repository. Includes technical discussions and a wealth of data tables throughout. Includes extensive references for further reading and an adequate glossary.

Byrne, John, and Steven M. Hoffman, eds. *Governing the Atom: The Politics of Risk.* New Brunswick, N.J.: Transaction, 1996. Researches the environmental and social aspects of the nuclear industry. Careful attention is paid to government policies that have been implemented to monitor safety and accident prevention measures.

Carter, Luther J. *Nuclear Imperatives and Public Trust: Dealing with Radioactive Waste.* Washington, D.C.: Resources for the Future, 1989. Provides a detailed but not overly technical history of the radioactive waste repository story. Details the political and bureaucratic decision-making processes, and shows how ill-informed decisions led to early disasters. Contains a glossary of terms and extensive footnotes.

Gerber, Michele Stenehjem. *On the Home Front: The Cold War Legacy of the Hanford Nuclear Site.* 2d ed. Lincoln: University of Nebraska Press, 2002. Examines the measures that have been used at Hanford Nuclear Weapons Plant to dispose of hazardous waste, and the effects the disposal has on the environment. Includes illustrations, maps, index, and sixty-three pages of bibliographical references.

Grossman, Dan, and Seth Shulman. "A Nuclear Dump: The Experiment Begins (Beneath Yucca Mountain, Nevada)." *Discover* 10 (March, 1989): 48. A newsy, nontechnical article on the proposed Yucca Mountain, Nevada, high-level waste repository. Discusses some of the tests needed to characterize the rocks and their history. Highlights areas of disagreement on technical questions and indicates the complexity of such problems. Includes a few good illustrations of what the repository will look like. In general, a good introductory article for anyone with no background in the subject.

Hamblin, Jacob Darwin. *Poison in the Well*. Piscataway, N.J.: Rutgers University Press, 2008. Discusses the disposal of nuclear waste into the oceans during the 1970's. Discusses political, scientific, and popular knowledge. Relates how views and understanding of nuclear waste disposal have changed. Appropriate for general readers.

Lauf, Robert. *Introduction to Radioactive Minerals*. Atglen, Pa.: Schiffer Publishing, 2007. Discusses uranium and thorium as a source of nuclear energy. Provides the minerals' characteristics with nearly two hundred photos and micrographs. Organized by chemical groups; a useful resource for collectors and researchers.

League of Women Voters Educational Fund Staff. *The Nuclear Waste Primer: A Handbook for Citizens*. New York: Lyons, 1987. Introduces some of the technical jargon used in radiation chemistry and describes the magnitude of the problems of dealing with low- and high-level wastes, concluding with suggestions for civic action. Includes a brief but comprehensive glossary.

Letcher, Trevor M., ed. *Future Energy: Improved, Sustainable, and Clean*. Amsterdam: Elsevier, 2008. Discusses the future of fossil fuels and nuclear power; renewable energy supplies, including solar, wind, hydroelectric and geothermal power; and potential and underutilized energy sources. Covers new concepts in energy consumption and technology.

Noyes, Robert. *Nuclear Waste Cleanup Technology and Opportunities*. Park Ridge, N.J.: Noyes Publications, 1995. A complete description of the legislation enacted to handle the clean-up and disposal of radioactive and hazardous waste, as well as the environmental and social aspects associated with the disposal. Includes illustrations, index, and bibliography.

Rosenfeld, Paul E., and Lydia G. H. Feng. *Risk of Hazardous Wastes*. Burlington, Mass.: Elsevier, 2011. Provides case studies and other examples of waste management issues. Examines various sources concerning hazardous waste, industrial waste, and nuclear waste, among others. Chapters are separated by industry with references following each. Discusses current treatment, disposal, and transport practices. Highlights toxicology, pesticides, environmental toxins, and human exposure issues, along with risk assessment, regulations, and accountability. Includes numerous appendices with tables for a number of OSHA, NIOSH, ACGIH, ATSDR, WHO, and EPA limits and regulatory values.

Sutton, Gerard K.; and Joseph A. Cassalli, eds. *Catastrophe in Japan: The Earthquake and Tsunami of 2011*. Nova Science Publishers Inc., 2011. A number of reports on the effects of the 2011 Japanese earthquake and tsunami. Focuses on the agricultural and economic impacts, and the resulting nuclear crisis.

See also: Coal; Earth Resources; Earth Science and the Environment; Hazardous Wastes; Landfills; Land Management; Land-Use Planning; Mining Processes; Mining Wastes; Sedimentary Mineral Deposits; Strategic Resources.

OCEAN POWER

Ocean power encompasses several distinctly different approaches to power generation, which, if developed properly, promise potentially large amounts of clean, renewable energy. The importance of developing such alternative energy sources for a world largely dependent on fossil fuels—subject to escalating costs and exhaustibility, and creating environmental pollution—cannot be overstressed.

PRINCIPAL TERMS

- **marine biomass energy conversion:** the cultivation of marine plants, such as algae, for conversion of the harvest into synthetic natural gas and other end products
- **marine current energy conversion:** power from the transfer of kinetic energy in major ocean currents into usable forms, such as electricity
- **ocean thermal energy conversion:** power derived from taking advantage of the significant temperature differences found in some tropical seas between the surface and deeper waters
- **ocean wave power:** the use of wind-generated ocean surface waves to propel various mechanical devices incorporated as an electrical generating system
- **salinity gradient energy conversion:** power generated by the passage of water masses with different salinities through a special, semipermeable membrane, taking advantage of osmotic pressure to operate turbines
- **tidal flow power:** power from turbines that are sited in coastal areas to take advantage of the tidal flow's rise and ebb

TIDAL FLOW POWER

Attempts at harnessing ocean power involve the use of specialized technologies developed to exploit natural flows of energy within the marine environment. These energy flows are generated by the interaction of the ocean's waters with the effects of the sun's energy; the gravitational pull of other celestial bodies, such as the moon; and, to a much lesser extent, such influences as geothermal activity occurring on the sea bottom. Many engineering schemes have been devised to try to tap into these natural energy flows. These schemes recognize the fact that Earth's surface is mostly ocean—71 percent of it is covered by the sea—and that, due to an interplay of natural processes, this immense fluid environment is always in motion.

The most ambitious—and, as of 2012, most productive—ocean power schemes have been tidal power projects. The efficacy of these projects is dependent on how well engineered they are to take advantage of the key factors involved. Such factors include the character of a tide at a particular coastal locale as determined by local bottom topography, surface coastline geography, and the orientation of the coast to the open sea. Submarine topographic influences can accentuate the rise of an incoming tide, acting like a wedge to lift the oncoming bulge of tidal water. Thus, tides can reach up to 15 meters on coasts having the right tide-enhancing topography and orientation. The maximum rise and fall of a tide experienced at a particular location is important in tidal power, as it represents the amount of usable head; "head" is a term used in hydraulic engineering to describe the difference in elevation between the level at which water can flow by gravity down from an upper to a lower level, thus making itself available to do work. Unfortunately, only one hundred coastal sites worldwide are classified as having significant head and qualify as optimal candidates for tidal power installations. Scientists conservatively estimate that the global, dissipated tidal power amounts to 3 terawatts per year, or 3 trillion watts. Of this amount, perhaps only 0.04 terawatt would ever be exploitable from feasible tidal power sites.

The French government has been the world pioneer in realizing tidal flow power engineering schemes, constructing a large, functioning tidal energy station at the estuary of the La Rance River in

Brittany in northern France. At La Rance, a combination of factors produces a useful hydraulic head and has proved itself economically profitable for several decades. Twenty-four 10,000-kilowatt turbine generators operate within conduits inside the tidal dam, turning at the low speed of 94 revolutions per minute. The turbine blades are designed to operate bidirectionally, in response to either an incoming or an outgoing tide. Thus, the French plant exploits the free tidal water movement almost continuously.

OCEAN THERMAL ENERGY CONVERSION

Ranking alongside tidal flow power, in terms of economic feasibility and available technology, is the ocean thermal energy conversion (OTEC) approach. This method could ultimately be a large-scale global operation, as there are many more sites that can be used for OTEC than are available for tidal power. Projected, theoretical limits to this energy source are in the neighborhood of 1 terawatt per year. OTEC involves either the construction of floating, open-ocean plants or coastal, land-based plants that exploit the temperature differences existing between water masses at varying depths in the tropical seas. Only small pilot plants have been built thus far, and they have never been run for more than short periods.

The optimal conditions for the most efficient OTEC sites have been calculated to be those where an 18-degree Celsius temperature difference exists between the surface and depths in the range of 600 to 1,000 meters. Among the regions of the ocean that meet or approach these thermal conditions are Puerto Rico and the West Indies, the Gulf of Guinea, the Coral Sea, many of the Polynesian island groups, and the northwest African island groups. Only in these warm seas is there a large enough temperature difference between surface and deep sea. The actual process of converting the thermal difference of such areas involves a system in which a turbine is turned by heat from the warm, surface water layer. The heat is transferred through devices termed heat exchangers, which introduce the thermal energy into a closed system. A working fluid such as ammonia, contained in sealed pipes, propels the system through the process of controlled convection. The warmed ammonia is heated to a vapor in an evaporator unit that drives the turbine. Then the used vapor is conducted through a condenser unit, where cold water drawn from below cools it down to a liquid state

for another usage cycle. The cool water is brought into contact with the system by the deployment of a very long pipe, hundreds of meters in length, which projects through the thermocline, or boundary, between the upper and lower water masses. Although the efficiency of the system is typically low, only 2 to 3 percent, the thermal reservoir is immense and constantly replenished by the sun. To make such projects attractive economically, they need only be built on a sufficiently large scale.

MARINE CURRENT ENERGY CONVERSION

Similar to OTEC in its use of the major oceanic flows of thermal energy are plans to exploit large-scale currents. One such current, the Gulf Stream—often called the Florida Current—constantly conveys many millions of gallons past a given point in the ocean. One plan is a direct approach, involving large turbines placed within the main flow of the current. Ideally, the generating site would be close to major electrical consumers, such as coastal cities.

A good case can be made for implementing marine current energy conversion along the eastern coast of Florida. The city of Miami, a very large consumer of electrical power, is within ten miles of the Florida Current. This current is estimated to carry approximately 30 million cubic meters of water per second past the city at a rate sometimes reaching 2.5 meters per second. One scheme would involve anchoring a large cluster of special, slow-speed turbines, or water windmills, to the bottom; the cluster would ride midwater in controlled buoyancy. This complex would function at a depth ranging from 30 to 130 meters and stretch some 20 kilometers across the flow of the Florida Current directly adjacent to Miami. Estimates of the power output of this array are on the order of 1,000 megawatts, provided constantly on a twenty-four-hour, year-round basis. It would extract roughly 4 percent of the usable kinetic energy of the Florida Current at this point in its flow, which is calculated to be in the range of 25,000 megawatts. Other strong currents exist worldwide that also may be good potential sites for future turbine arrays.

OCEAN WAVE POWER

Another way to take advantage of oceanic energy is to utilize the kinetic energy of surface waves directly to power mechanical devices used to generate electricity. This approach capitalizes on the fact that waves raise and lower buoyant objects. If the floating object is also long and perpendicular to the ocean surface, with most of its mass below the waterline, it is inherently stable and less subject to damage. Such a device is embodied in several variations of a wave-powered pump that is beginning to see practical economic applications. In one form, a large, vertical cylinder floats in the waves. Inside, the lower end is open to the lifting and sinking of the water level in unison with wave motion. Because of its great length (tens of meters or more), it amplifies the wave motion in the column of air resting above the water. This motion propels air up and down through a double-flow electrical air turbine. As long as it remains in sufficiently deep water, it receives very little damage, no matter the magnitude of wave energy. A buoy-like device floats at the cylinder top, keeping out wave splash and snugly maintaining inside air pressure. The cylinder can be anchored by flexible moorings to the sea bottom, and electrical cables can feed the power output to shore.

Coastal-based variants also have been built solidly into rocky cliffs; they respond to wave-driven air pressure that enters from a conduit at its base. One example of the coastal-based type is the plant at Tostestallen in Norway. Numerous rocky coasts worldwide could be similarly utilized for wave power generation.

SALINITY GRADIENT ENERGY CONVERSION

One form of ocean power that has been envisioned but not implemented is the use of salinity gradients within the sea. This idea is still in the theoretical stage because of the lack of key materials necessary for the effective technologies to work. A major drawback at this point is the lack of appropriately tough and efficient synthetic membranes necessary for this energy to be economically practical. Salinity gradient energy works on the principle of extracting energy from osmotic pressure by the use of semipermeable membranes. Such membranes would most likely be fabricated from some type of plastic that possesses the correct, chemical properties and would take advantage of the osmotic pressure between water masses with different percentages of dissolved salts. Influenced by osmotic pressure, less salty water will naturally flow through a semipermeable membrane to the side of greater salinity. The membrane would therefore be designed to be permeable to the fresher water but impermeable to the saltier water.

Osmotic pressure would propel a controlled, one-way flow that could be employed to propel water-driven turbines for electricity.

A PROMISING ENERGY SOURCE

Global human population growth is still explosively on the rise. Going hand in hand with population growth is the trend of increasing worldwide urbanization and industrialization. Because of these trends, consumption of energy to run manufacturing processes, transportation, food production, climate control, and communication systems has escalated dramatically. Coupled with this ravenous energy use are the by-products of mounting energy consumption: widespread pollution and a general degradation of the world environment. The world industrial society already faces the specter of diminishing energy sources and the eventual, ultimate exhaustion of all fossil fuels. New energy sources, such as the nuclear option, seem to possess, so far, many drawbacks to widespread usage.

Renewable energy sources that are clean and have a low impact on the quality of the environment are ideal long-term solutions. Some alternative energy forms fitting this description are either technically feasible today or almost so. The various forms of ocean power that have been developed or are in theoretical stages are excellent candidates for helping to alleviate some of the world's more pressing energy-related problems. For at least some geographic areas, ocean power is not only an efficient and clean power source but is economically competitive with fossil fuels and nuclear power. Tidal flow power is an excellent example of the increasing viability of some forms of ocean power. By the end of the twentieth century, tidal power was producing less than 300 megawatts worldwide, although the largest site (producing 240 megawatts) has been in operation at La Rance River in France since 1967. As of 2011, fewer than a dozen small demonstration plants had been built. The most successful test OTEC system has been the one at Kailua-Kona, Hawaii, which has produced more than 50 kilowatts in net power. The ocean, however, remains Earth's largest solar collector, and supporters of ocean power remain optimistic. Forms of ocean power represent a means to live in harmony with the environment without compromising the use and demand for energy.

Frederick M. Surowiec

FURTHER READING

Anderson, Greg M., and David A. Crerar. *Thermodynamics in Geochemistry: The Equilibrium Model.* New York: Oxford University Press, 1993. An exploration of geochemistry and its relationship to thermodynamics and geothermometry. A thorough but somewhat technical resource. Recommended for persons with a background in chemistry and Earth sciences.

Andrews, John, and Nick Jelley. *Energy Science: Principles, Technologies, and Impact.* New York: Oxford University Press, 2007. Discusses various forms of energy, and environmental and socioeconomic impacts. Covers principles of energy consumption and includes information on the generation, storage, and transmission of energy. A strong mathematics or engineering background is required for some examples. Appropriate for undergraduates and professionals.

Bascom, Willard. *Waves and Beaches: The Dynamics of the Ocean Surface.* Rev. ed. Garden City, N.Y.: Anchor Press/Doubleday, 1980. An introduction to the subject of oceanography, emphasizing the role of wave and beach processes. Discusses energy derived from marine sources and presents the pros and cons of ocean power. Well illustrated throughout. Suitable for readers seeking a working knowledge of physical oceanographic processes.

Boyle, Godfrey, ed. *Renewable Energy.* 2d ed. New York: Oxford University Press, 2004. Provides a complete overview of renewable energy resources. Discusses various forms of energy, including hydroelectric energy and tidal power, and basic physics principles, technology, and environmental impact. Includes a further reading list.

Carr, Donald E. *Energy and the Earth Machine.* New York: W. W. Norton, 1976. A thorough survey of the primary energy sources that power the industrial world, including fossil fuels. Discusses ocean power sources within the scope of water-derived energy sources. Suitable for high school students or general readers seeking background on the subject.

Constans, Jacques A. *Marine Sources of Energy.* Elmsford, N.Y.: Pergamon Press, 1980. An accessible treatment of the subject of ocean power. Rich with explanatory diagrams, tables, drawings, and maps that help to explain the concepts involved. Appropriate for all readers, high school and above, especially those interested in the technical problems of ocean engineering.

Gage, Thomas E., and Richard Merrill, eds. *Energy Primer, Solar, Water, Wind, and Biofuels.* 2d ed. New York: Dell Publishing, 1978. Provides a wealth of information for those interested in the "nuts and bolts" of implementing alternative energy technologies. Suitable for those with a practical interest.

Gashus, O. K., and T. J. Gray, eds. *Tidal Power.* New York: Plenum Press, 1972. A classic book devoted exclusively to the generation of power through tidal ocean flow. Details the design and operating problems of the large-scale facility at La Rance, France, and the then-proposed facilities at the Bay of Fundy in Nova Scotia, Canada. Suitable for readers at the college level or above.

Hamblin, Kenneth W., and Eric H. Christiansen. *Earth's Dynamic Systems.* 10th ed. Upper Saddle River, N.J.: Prentice Hall, 2003. Introduces the reader to basic concepts such as Earth's gravity, rotation, and tides. Appropriate for readers without a background in physical geology.

Letcher, Trevor M., ed. *Future Energy: Improved, Sustainable, and Clean.* Amsterdam: Elsevier, 2008. Discusses the future of fossil fuels and nuclear power; renewable energy supplies, such as solar, wind, hydroelectric, and geothermal sources; and potential and underutilized energy sources. Covers new concepts in energy consumption and technology.

Meador, Roy. *Future Energy Alternatives.* Ann Arbor, Mich.: Ann Arbor Science Publishers, 1979. Surveys the major alternatives to fossil fuels, including ocean power options. Assumes no technical background and expertly introduces general readers to each energy alternative in turn. Appropriate for high school or college students, as well as general readers.

Ross, David. *Power from the Waves.* New York: Oxford University Press, 1995. Considers the ocean's potential power sources. Illustrations, maps, bibliography, and index.

Tarbuck, Edward J., Frederick K. Lutgens, and Dennis Tasa. *Earth: An Introduction to Physical Geology.* 10th ed. Upper Saddle River, N.J.: Prentice Hall, 2010. Provides a clear picture of Earth's systems and processes that is appropriate for the high school or college reader. Includes illustrations, graphics, and an accompanying computer disc. Bibliography and index.

Teller, Edward. *Energy from Heaven and Earth.* San Francisco: W. H. Freeman, 1979. Compares the major forms of energy use in industrialized countries, and the social and economic consequences of their use. Offers extensive discussions of energy-use policies and their bearing on the development of new energy sources. Introduces the overall energy picture for students or interested laypersons.

Wilhelm, Helmut, Walter Zuern, Hans-Georg Wenzel, et al., eds. *Tidal Phenomena.* Berlin: Springer, 1997. Collects lectures from leaders in the fields of Earth sciences and oceanography. Examines Earth's tides and atmospheric circulation. Complete with illustrations and bibliographical references.

Appropriate for readers without a strong knowledge of Earth sciences.

Wilson, Mitchell, et al., eds. *Life Science Library: Energy.* New York: Time, 1963. A profusely illustrated overview of the subject of energy. Outlines the history of energy-related physics and the growth of applied technology designed to exploit energy sources. Suitable for any reader at the high school level or beyond.

See also: Earth Resources; Geothermal Power; Hydroelectric Power; Mining Processes; Mining Wastes; Nuclear Power; Nuclear Waste Disposal; Radioactive Minerals; Solar Power; Strategic Resources; Tidal Forces; Unconventional Energy Resources; Uranium Deposits; Wind Power.

OFFSHORE WELLS

Offshore wells are drilled in favorable oil and gas areas that are inundated with oceanic (salt) or inland lake (fresh) waters. Most of the world's future hydrocarbon reserves may be discovered in offshore provinces.

PRINCIPAL TERMS

- **continental margin:** the offshore area immediately adjacent to the continent, extending from the shoreline to depths of approximately 4,000 meters
- **directional drilling:** the controlled drilling of a borehole at an angle to the vertical and at an established azimuth
- **drilling fluids:** a carefully formulated system of fluids and solids that is used to lubricate, clean, and protect the borehole
- **geophysics:** the quantitative evaluation of rocks by electrical, gravitational, magnetic, radioactive, seismological, and other techniques
- **hydrocarbons:** naturally occurring organic compounds that in the gaseous state are termed natural gas and in the liquid state are termed crude oil or petroleum
- **rotary drilling:** a fluid-circulating, rotating process that is the chief method of drilling oil and gas wells
- **sedimentary rock:** rock formed by the deposition and compaction of loose sediment created by the erosion of preexisting rock
- **seismology:** the application of the physics of elastic wave transmission and reflection to subsurface rock geometry

ESTABLISHING OFFSHORE DRILLING LOCATIONS

Offshore drilling is the extension of on-land oil and gas drilling techniques in waters that either cover or lie adjacent to landmasses of the earth, such as the Great Lakes of North America and the continental margins that surround each continent. The continental margins are the principal arena in which offshore drilling is conducted; they constitute approximately 21 percent of the surface area of the oceans and may contain a majority of the world's future reserves of oil and gas.

As of the last decade of the twentieth century, oil and gas production had been established off the coasts of more than forty countries in the continents of Africa, Australia, South America, North America, Asia, and Europe, and offshore drilling had been completed in the waters of more than half the nations on Earth. Drilling platforms capable of operating in water deeper than 1,800 meters probe the hydrocarbon potential of the outer continental margin. Even farther offshore, self-propelled drill ships capable of drilling more than 2,000 meters into the sea floor within waters as deep as 6,000 meters have been analyzing the deeper portions of the oceans since the early 1960's.

The specific offshore drilling location must be established at the outset. This is the responsibility of the geologist, who, by studying the various rock exposures and their structures, decides on the best surface site for the drilling equipment. Such exposures are submerged offshore, and geophysical methods of determining the geology of the sea floor are used. Offshore geophysics is an indirect technique of studying the rocks composing the sea floor by measuring their physical properties. A magnetometer towed behind an aircraft flying low over the water can be used to measure the magnetic properties of underlying rocks. With a gravimeter mounted in a slow-moving boat or aircraft, the gravity field associated with the sea floor can be analyzed. A combination of these methods will help the geologist locate a site underlain by rocks that may contain hydrocarbons. By far the most common offshore geophysical tool is the seismic reflection survey. Seismology depends on the artificial generation of an elastic sound wave and its transmission through the layered, sedimentary rocks underlying the sea floor within the continental margin zone. These waves reflect off sedimentary rocks and are transmitted back to the surface of the water, where a vessel will record the time difference between transmission and reflection. Millions of such combined reflection arrival times are interpreted as a cross-sectional view of the underlying sea floor. A series of intersecting seismic sectional views presents a simple three-dimensional portrayal of the best site for offshore drilling equipment.

DRILLING METHODOLOGY

The sole method used for drilling offshore wells is the rotary method. Rotary motion is supplied by diesel engines to a length of interconnected drill pipe (the drill string), to the bottom of which is attached the drill bit. Drilling bits come in a variety of styles designed to drill through differing types of sedimentary rock. Each type of drill bit contains an arrangement of high-strength alloy teeth that tear through the rock when rotated under pressure. As the hole deepens, new sections of drill pipe are added. Periodically, the inside of the borehole is lined with cemented casing, which prevents the hole from caving in. During drilling, drilling fluid is circulated through the pipe and hole. This circulating fluid, a mixture of water, special clays, and other minerals and chemicals, is necessary to maintain a safe temperature and pressure in the borehole and to clean the hole of newly created rock chips. The drilling fluid can be formulated to stabilize chemically active rock layers. Since it is dense, the drilling fluid supports (partially floats) 10 to 20 percent of the weight of the drill pipe—a significant consideration when drilling a deep well. All drilling activities take place within the derrick, an open steel structure that often is 60 meters tall. The derrick holds the draw works, whereby the drill string can be drawn out of the borehole and disassembled, one length at a time.

Should a drilling operation discover new reserves of oil or gas, production platforms must be constructed on-site after the movable drilling platform is deployed elsewhere. From these fixed production platforms, as many as fifty or more additional wells are drilled to determine the new hydrocarbon field's size and volume. Each of these boreholes is drilled from the same general location on the production platform; however, at a predetermined depth below the sea floor, individual boreholes veer away from one another, allowing different sectors of the field to be economically developed from one production platform. This process of deviation is termed directional drilling; it requires the assistance of a specialist, as the bottom positions of the boreholes must be very carefully controlled for effective hydrocarbon production. After the final directional hole has been drilled and the limits of the field fully defined, the production derrick is replaced with equipment used to gather the oil and gas from the many active wells flowing into the platform. An offshore pipeline is laid, connecting the producing platform to onshore pipeline and refining systems, and the new field begins production.

DRILLING PLATFORMS

The type of platform that will be used to contain the drilling equipment is critical, considering the inhospitable weather periodically encountered offshore. There are four types of platforms: the submersible, semisubmersible, and jack-up platforms, and the drill ship. A submersible platform is stabilized by flooding the hollow legs and pontoons of the platform and establishing seabed moorings. Since these platforms are in contact with the sea floor, they cannot be used in excessively deep waters. In such deep waters, a semisubmersible design is employed; the hollow pontoons are only partially flooded, permitting the platform to settle below the surface of the water but not to rest on the bottom. Inherent in this design is buoyancy that, along with anchor moorings, is sufficient to maintain the platform safely over the drilling site. The ratio of semisubmersible to submersible units in use is approximately 4:1.

By far, the most popular offshore unit is the jack-up platform. These movable structures are towed to the drill location and stationed by lowering massive steel legs to the seabed. In essence, the platform is "jacked up" on the legs to a level sufficient to protect the unit from storm waves. The fourth design is the drill ship, a free-floating, usually self-propelled vessel that contains an open-water drilling unit in the ship's center. The drill ship is kept on location by computer-activated propellers that maintain the horizontal and vertical motion of the standing ship within safe limits.

Regardless of the basic design, all offshore platforms contain a drilling derrick, storage and machinery housing, living and eating quarters, recreation and basic health facilities, and a helicopter landing pad. The entire structure must be of a size sufficient to house and maintain a normal working crew of forty to sixty individuals.

MULTIPLE BOREHOLE PLATFORMS

Because of the expense associated with the siting of an offshore drilling platform, more than one borehole is commonly drilled from the same surface location. More than sixty-five boreholes have been drilled from a single platform. Most are drilled into the seabed at a predetermined angle, allowing many wells to "bottom" in an oil or gas reservoir rock over a distance as much as 3 kilometers laterally from the platform site. In this manner, the hydrocarbon reservoir can be exploited economically with a minimum risk to the aquatic environment.

Multiple borehole platforms are designed to extend a safe height above sea level. Platforms in the Gulf of Mexico, the North Sea, and the Santa Barbara Channel of California have been designed to endure the most severe storms likely to occur within a one-hundred-year cycle, including hurricane-force storms and earthquake tremors. Special submersible drilling platforms are designed for operation in polar waters, where ice-free waters exist only for a short time period during summer. These structures, which must be able to withstand strong currents and floating pack ice, are protected below the waterline by a steel or cement caisson. Often the drilling deck is circular, allowing easy passage of floating ice.

Floating platforms, semisubmersibles, and drill ships move with wind and water currents. Unless compensated for, these motions affect drilling efficiency. When boreholes are drilled in very deep waters, compliant platforms that yield to weather and water currents are employed.

AN INTERNATIONAL PRIORITY

Offshore drilling for oil and gas, even with the very high costs dictated by its architectural, meteorological, engineering, and safety requirements, is an international priority; it is driven by need, economic return, and national security. Except for difficult-to-explore and environmentally sensitive regions, such as the Arctic, the South American interior, and central Africa, many onshore regions of the world have entered a mature stage of hydrocarbon exploration: Most of the accessible large-volume oil and gas fields have already been discovered. The United States is an excellent example, for it imports almost 50 percent of its daily petroleum requirements, leaving it vulnerable to foreign military and political disturbances. Studies have consistently shown that the best opportunities for increasing the reserves of American oil and gas lie in the continued exploration of the continental margins. Although not all offshore areas contain oil and gas, analyses conducted by the U.S. Geological Survey indicate that within the continental margin of the United States, as much as 40 billion barrels (a barrel equals 42 U.S. gallons, or about 159 liters) of oil and more than 5.7 trillion cubic meters of natural gas are yet to be discovered.

The degree and specific locations of offshore exploration and drilling activities depend on political, environmental, and economic factors, all of which affect petroleum products' availability and price. Certain offshore drilling regulations have already been established. In the United States, states' rights generally prevail to approximately 5,500 meters, or 3 nautical miles, offshore. From there out to designated international waters, federal government policies must be followed. The outer limits of American waters and those of other signatory nations are controlled by the 1982 Third United Nations Convention on the Law of the Sea, which allows nations to define Exclusive Economic Zones extending 200 nautical miles (370 kilometers) offshore. Because of economic and engineering constraints, these outer limits have not yet been tested widely with actual drilling. A common cause for the delay in exploration drilling is public concern for the marine environment. An example of a highly publicized and heavily studied American offshore oil spill is one that occurred in the Santa Barbara Channel, off California, in 1969; another is the 1989 tanker spill in Prince William Sound, in the Gulf of Alaska. In 2010, the Deepwater Horizon drill rig, operated by BP, sank in the aftermath of an explosion. Attempts to activate shutoff devices on the sea floor failed, and it took around six months to fully seal the wells. About 4.9 million barrels (210,000,000 U.S. gallons; 780,000 cubic meters) of oil were spilled—about 20 times the amount of the 1982 Alaska oil spill. The long-term effects of these spills are unknown. One study, however, has shown that between 1950 and 1980, the commercial fish-catch weight in the Gulf of Mexico increased fourfold, suggesting that offshore compatibility between the fishing and hydrocarbon industries is possible. Yet, as environmentally fragile continental margin areas, such as the shores of Alaska and Canada, continue to be evaluated, intensified concerns for the environment may delay exploration agendas. Finally, the costs of finding and producing oil and gas offshore will continue to climb as deeper waters are probed.

Albert B. Dickas

FURTHER READING

Baker, Ron. *A Primer of Offshore Operations.* 3d ed. Austin: University of Texas Press, 1998. Addresses international offshore operations. Covers the chemistry and geology of oil and gas, as well as the exploration, drilling, production, and transportation aspects of the business. Illustrated with color photographs and easy-to-understand diagrams.

Bar-Cohen, Yoseph, and Kris Zacny, eds. *Drilling in Extreme Environments: Penetration and Sampling on Earth and Other Planets*. Weinheim, Wiley-VCH. 2009. A comprehensive presentation of current drilling techniques and technology. Covers everything from ocean to extraterrestrial drilling and coring. Written by academics, research scientists, industrial engineers, geologists, and astronauts in a manner accessible to the layperson but useful for scientists and engineers.

Ellers, F. S. "Advanced Offshore Oil Platforms." *Scientific American* 246 (April, 1982): 39-49. Presents the methods of construction and emplacement of four different offshore oil platforms, all taller than the World Trade Center and designed to withstand 35-meter waves in water 200 meters deep. Intended for general readers.

Engel, Leonard. *The Sea*. Boston: Time-Life Books, 1967. An easy-to-read introduction to the general physical and chemical composition of the typical ocean. Well illustrated.

Gorman, D. G., and June Neilson, eds. *Decommissioning Offshore Structures*. New York: Springer, 1997. Focuses on the engineering and environmental aspects of the abandonment of offshore oil-drilling platforms and other structures. Bibliography and index.

Graham, Bob, et al. *Deep Water: The Gulf Oil Disaster and the Future of Offshore Drilling*. National Commission on the BP Deepwater Horizon Oil Spill and Offshore Drilling, 2011. Addresses the disaster in the Gulf of Mexico, events leading up to it, and repercussions to follow. The third part of this report focuses on future prevention of such disasters. Provides great detail from key players, scientists, and local residents.

Hall, R. Stewart, ed. *Drilling and Producing Offshore*. Tulsa, Okla.: PennWell Books, 1983. A semitechnical text that introduces the reader to every aspect of the offshore drilling business. Covers drilling, platforms, production, and maintenance, including diving and underwater construction.

Hyne, Norman J. *Nontechnical Guide to Petroleum Geology, Exploration, Drilling, and Production*. 2d ed. Tulsa, Okla: PennWell Books, 2001. Provides a well-rounded overview of the processes and principles of gas and oil drilling. Covers foundational material for the later practice and application material to build on. Covers some topics in detail. Appropriate for professionals in the oil drilling field, geologists, or students of similar fields.

Maclachlan, Malcolm. *An Introduction to Marine Drilling*. Herefordshire: Dayton's Oilfield Publications Ltd., 1987. Introduces the new offshore worker to the principal operations of the business. Provides insight into a way of life witnessed only by a small percentage of the general population. Contains a glossary of marine drilling terms.

Myers, A., D. Edmonds, and K. Donegani. *Offshore Information Guide*. Tulsa, Okla.: PennWell Books, 1988. Comprehensive guide for offshore operations. Provides some 1,600 references to offshore journals, directories, maps, legislation, and conference proceedings.

Segar, Douglas. *An Introduction to Ocean Sciences*. 2d ed. New York: W.W. Norton & Company, 2007. Comprehensive coverage of all aspects of the oceans and the oceanic crust. Readable and well illustrated. Suitable for high school students and above.

Sverdrup, Keith A., Alyn C. Duxbury, and Alison B. Duxbury. *An Introduction to the World's Oceans*. 8th ed. Columbus, Ohio: McGraw-Hill Science, 2004. A freshman-level review of the marine environment. Covers the physical, chemical, biological, and meteorological structure of the continental margins. Color plates review satellite and research submarine technology.

See also: Coal; Earth Resources; Geothermal Power; Hydroelectric Power; Land-Use Planning; Nuclear Power; Nuclear Waste Disposal; Ocean Power; Oil and Gas Distribution; Oil and Gas Exploration; Oil and Gas Origins; Oil Chemistry; Oil Shale and Tar Sands; Onshore Wells; Petroleum Industry Hazards; Petroleum Reservoirs; Solar Power; Strategic Resources; Well Logging; Wind Power.

OIL AND GAS DISTRIBUTION

Petroleum products, such as oil, and natural gas are essential for many everyday activities, but oil and gas are nonrenewable resources that are difficult to access and transport. A full understanding of the petroleum industry requires knowledge of the geologic processes behind petroleum reservoir formation and the technology used to find, extract, and process this critical resource.

PRINCIPAL TERMS

- **basin:** area containing thick sedimentary rock; usually rich in hydrocarbons
- **casing:** the metal pipe that provides the structure for a well
- **derrick:** apparatus from which a drill string and casing are suspended
- **drilling mud:** substance used to lubricate a drill and to prevent oil from flowing into the well during drilling
- **drill string:** the main drilling apparatus, made up of a drill bit, drill pipe, and collars to weigh down the bit
- **enhanced oil recovery:** the use of heat, chemicals, or gases to loosen oil from reservoir rocks
- **fracking:** hydraulic fracturing, which uses water and chemicals to break up shale and release natural gas
- **hydrocarbons:** compounds of hydrogen and carbon; the main components of petroleum and natural gas
- **improved oil recovery:** the process of pumping oil out or injecting gas and water to increase pressure in a reservoir, thereby forcing oil out
- **reservoir rock:** sedimentary rock that holds oil or gas in the spaces between particles
- **source rock:** rock containing organic materials that will be made into oil and gas
- **trap:** a type of rock formation that causes hydrocarbons to accumulate (trapping them), forming a reservoir
- **tree:** a wellhead; a fixture to control the flow of oil and gas; sometimes called a Christmas tree

WHERE ARE OIL AND GAS LOCATED?

Crude oil production is centered in Russia, Saudi Arabia, the United States, Iran, Iraq, and Venezuela, but more than 80 percent of the world's reserves are located in member countries of the Organization of Petroleum Exporting Countries, primarily Venezuela, Saudi Arabia, Iran, and the United Arab Emirates. Natural gas production, in contrast, takes place mainly in the United States, Canada, Iran, Algeria, and Russia. Reserves are primarily found in Russia, Qatar, the United States, and the United Arab Emirates.

Petroleum reservoirs are found in sedimentary rocks (sandstone, limestone, and dolomite). Igneous and metamorphic rocks form what is called basement rock, which underlies the sedimentary rock. Areas where the basement rock is at the earth's surface are called shields; these areas lack petroleum reservoirs. Natural gas is found in the same reservoirs with oil, but natural gas can withstand extremes of pressure and temperature that oil cannot. Therefore, it can often be found at depths without concomitant oil.

Oil and gas form when organic matter decays at a rate that is slower than the rate at which sediment accumulates on top of it. In a reservoir, the oil and gas exist in the spaces between particles of sedimentary rock. Rock that holds oil and gas in its pores is called reservoir rock. Rock containing organic material that will later form oil and gas is called source rock.

One type of reservoir is called a trap, a rock formation that blocks the movement of oil and gas so that they accumulate beneath the rock. The gas rises to the top and the oil remains beneath it. Water often exists in the trap, too. A stratigraphic trap is made of reservoir rock surrounded by shale. The shale below is source rock, and the shale above is a sealing cap for the reservoir.

FINDING PETROLEUM

In the early days of oil exploration, reservoirs could sometimes be located because they were leaky, meaning that the reservoirs' oil was visible at the earth's surface. These easily found reservoirs have long been tapped, so one of the most reliable ways to find petroleum is to drill test wells near or below current reservoirs.

To find a new oil field, one can use remote sensing methods to identify areas (namely, sedimentary basins) that might contain petroleum. Sedimentary rock has fewer magnetic minerals and lower density than magmatic rock and can easily be identified with magnetometry, gravimetry, reflection seismology, and basic knowledge of an area's geologic processes and formation. Thicker sedimentary rock is associated with larger amounts of petroleum. Seismology is helpful in offshore exploration because it measures the thickness of sedimentary rock on the ocean floor. Petroleum reservoirs also are associated with tectonic activity. Some of the largest oil fields in the world exist where the Arabian plate is colliding with the Eurasian plate.

After geologic information is used to locate potential reservoirs, an exploratory well is drilled to verify the presence of hydrocarbons—that is, hydrogen-carbon compounds, which make up the primary component of oil. The next step is the development of appraisal wells, which are used to gather more detailed data and to determine the shape and extent of the formation. During the drilling of an appraisal well, a geologist will keep a mud log, a record of the types of rock being penetrated; the log also will note if hydrocarbons are present.

Because hydrocarbons have a different electrical resistivity than water, instruments may be sent down the well to measure resistivity. These tools allow a reasonably accurate mapping of the reservoir, but nothing is certain until the well produces. A drill stem test is a highly controlled way of testing for actual production potential by using a valve mechanism to access the reservoir without fully opening it. After the test, the valve is sealed.

THE DRILLING PROCESS

There are two methods of drilling: cable-tool drilling, which involves using force to ram a hole into the earth to access the reservoir, and rotary drilling, which works in a manner similar to a household drill, with a bit that turns and cuts through the rock. Cable-tool drilling is expensive and has fallen out of use in most areas.

Drilling is used to make four types of wells: exploration, appraisal, production, and injection. After assessing geologic data, a company will drill an exploration well. If hydrocarbons are found, appraisal wells are drilled in the surrounding area to determine the extent and shape of the reservoir. Exploration or appraisal wells can, in some cases, become production wells. An injection well is meant for adding material

to a reservoir to increase the pressure and drive the hydrocarbons out of the production well.

Occasionally, a reservoir is located beneath something that will be difficult to drill through vertically (for example, a mountain). When this happens, a vertical well may be drilled some distance away, and the reservoir may be accessed by drilling a horizontal well that extends from a point along the vertical well.

Once the well is in place and producing, the initial flow of oil and gas can be extracted with comparative ease because the contents of the reservoir are under pressure. As the levels in the reservoir are drawn down, drive mechanisms are needed to force more reservoir contents out through the well. Some of these processes occur naturally as the gases expand. Sometimes the water below the oil expands, forcing the oil upward. If none of these processes occurs, or if they are no longer sufficient to extract remaining hydrocarbons, artificial lift can be employed through pumping or by injecting gas or water, or both. These processes are collectively known as improved oil recovery. In contrast, enhanced oil recovery employs chemicals, heat, and gases to loosen oil from the reservoir rocks.

DRILLING EQUIPMENT

The main drilling apparatus, called the drill string, comprises the drill bit, the drill pipe (through which drilling mud and fluids can be pumped), and collars to weigh down the bit. The initial drilling makes a hole large enough to contain the casing, the metal pipe that will provide the structure for the well. The space between the drill string and the casing is called the annulus.

After fluids are pumped down the drill pipe, they come back up, with the drill cuttings, through the annulus. During completion of the well, cement is pumped down the annular space to seal the spaces between sections of casing. The well is perforated at levels where hydrocarbons are present to allow the hydrocarbons to flow into the well. A wellhead (sometimes called a tree or a Christmas tree) is installed to control flow.

The drill string is suspended from a structure called a derrick. Derricks can be assembled and disassembled at different locations on land, or they can be mounted on drilling platforms offshore. Regardless of location, a derrick must be strong and tall enough to support all of the equipment needed for the well casing and drill string. Thus, for deep-water drilling, the derrick must support 1 million pounds or more.

A derrick is part of either a land rig, a jack-up rig (used in shallow water), or an offshore rig (a drill ship, submersible rig, or semisubmersible rig). The type of rig used depends upon location. In shallow waters, submersibles or jack-ups are used. In deep water, mobile rigs are more economical than fixed rigs. A drill ship lays down a template at the drill site on the ocean floor (along with a blowout preventer) and lines up the drill string.

A blowout preventer is a massive piece of equipment that will cut off the flow of oil from the reservoir in an emergency. Uncontrolled flow from the reservoir can cause explosions, fires, and other disasters, such as the rig disaster on the Deepwater Horizon in the Gulf of Mexico in 2010. A blowout preventer is like a giant clamp or shutoff valve placed either at the bottom of the ocean or on a rig platform. On land, it is placed beneath the rig.

A substance called drilling mud is used to lubricate the drill and to ease the way as the rock cuttings are extracted from the hole. The other important function of drilling mud is to prevent inflow of oil, gas, and water into the well before drilling is completed. To this end, the drilling mud must be heavier than whatever substance it is designed to keep out. Various chemicals and other additives are used to adjust the composition of drilling mud. A mud log is kept to record what the drill bit is penetrating (such as the type of rock) and whether hydrocarbons are present.

TECHNOLOGY USED IN STORAGE AND TRANSIT

Many oil fields have processing facilities. Oil is transported directly from the well to separation, treatment, and storage equipment through steel, plastic, or fiberglass flow-lines (pipes). One of the initial processing steps is to run the oil through a separator, which divides gas from liquid. A gas scrubber, in a process called stripping, draws off any trace liquids. This is an important step because liquid and gas hydrocarbons have different requirements for transit.

Gas may be fully processed on-site or may be processed so that it is transportable. Transportable gas excludes liquids to prevent condensation in the pipeline. Corrosive compounds also are removed, and then the gas is compressed. Gas pipelines are placed underground, unlike oil pipelines, which are usually above ground. Because of the high costs of transporting gas, the most economical and efficient use of natural gas is to use the gas close to the place of extraction.

Liquid petroleum is put into welded-steel stock tanks for storage. It is subsequently loaded onto trucks to be taken to a refinery, or it is transported through a pipeline. In the case of offshore wells with no nearby pipeline, oil may be put into floating production, storage, and off-loading vessel. A vessel used only for transit and not for processing is simply referred to as a floating storage and off-loading vessel.

After the *Exxon Valdez* oil spill in 1989 in Alaska, the U.S. Congress passed the Oil Pollution Act of 1990, which changed specifications for tankers transporting oil in the United States. The law required the phasing out of single-hulled tankers in favor of double-hulled tankers. In a double-hulled tanker, the containers of oil are protected by an inner hull and an outer hull, which are separated by an interstitial space for ballast water. Countries other than the United States have adopted similar oil transport legislation, but not all of this legislation requires the use of double-hulled tankers.

THE FUTURE OF OIL AND GAS

Because gas is more difficult and more costly to transport, and because the energy yield from gas tends to be lower than that from oil, gas resources are exploited primarily when they occur with oil. Some reservoirs contain only gas, but the distribution of gas is limited in comparison with oil. The types of gas resources that have been tapped in significant amounts are called conventional gases, and they can be extracted with existing technology.

As technology improves, humans may come to rely more upon what are called unconventional gases, including coal-bed methane, tight gas, shale gas, and methane hydrates. These gases, however, are not economical to access and extract. Coal-bed methane reserves are coal deposits with methane. The barrier to widespread extraction is the cost of processing and disposing of the water by-products.

Tight gas and shale gas are found in rock that does not easily release the gases. Extraction involves the use of more invasive methods than standard wells. Hydraulic fracturing (fracking), a process by which water and chemicals are injected into shale to break it up, is a controversial method used to extract shale gas because it is associated with groundwater contamination and other hazards. Methane hydrates, which are found in glacial deposits, have not been extensively documented or explored because of their inaccessibility.

One of the most high-profile changes in oil extraction is the drilling of ever-deeper wells. The well that was being drilled at the time of the Deepwater Horizon blowout reached 5,596 meters (3.48 miles) below sea level, and the rig was capable of drilling to nearly twice that depth. Even with the best technology, obtaining and transporting petroleum products is a risky enterprise, but one that is increasingly necessary to maintain industry at current levels.

J. D. Ho

FURTHER READING

Freudenburg, William, and Robert Gramling. *Blowout In the Gulf: The BP Oil Spill Disaster and the Future of Energy in America.* Cambridge, Mass.: MIT Press, 2011. Uses the 2010 Deepwater Horizon disaster as a starting point for a comprehensive discussion of the oil industry, particularly with regard to policy, resource accessibility, disaster prevention, and the future of petroleum exploration and extraction.

Hyne, Norman. *Nontechnical Guide to Petroleum Geology, Exploration, Drilling, and Production.* Tulsa, Okla.: PennWell, 2001. An excellent resource for study of the specific geology of hydrocarbon reservoirs, including types of rock and the geologic processes behind the formations in which reservoirs are most often found.

Maugeri, Leonardo. *Beyond the Age of Oil: The Myths, Realities, and Future of Fossil Fuels and Their Alternatives.* Santa Barbara, Calif.: Praeger, 2010. Describes in clear, concise language the different types of fossil fuels and issues with their use. Particularly useful is the chapter on natural gas, which is often given scant coverage in other publications.

Raymond, Martin, and William L. Leffler. *Oil and Gas Production in Nontechnical Language.* Tulsa, Okla.: PennWell, 2006. Describes in detail the different types of drilling, parts of drilling rigs, methods of drilling both onshore and offshore, and the process of drilling, step by step, including logging, testing, well completion, and hydrocarbon recovery.

Shepherd, Mike. *Oil Field Production Geology.* Tulsa, Okla.: American Association of Petroleum Geologists, 2009. Supplies photographs of drill bits and rigs, diagrams of drill strings and casings, and drawings of different types of reservoir formations, elucidating the mysteries of drilling equipment and the location of petroleum reservoirs.

Speight, James G. *An Introduction to Petroleum Technology, Economics, and Politics.* Hoboken, N.J.: John Wiley & Sons, 2011. Concisely provides background on the location, extraction, processing, and transportation of petroleum, but is valuable for giving economic and market context to the technical information.

See also: Offshore Wells; Oil and Gas Exploration; Oil and Gas Origins; Oil Chemistry; Oil Shale and Tar Sands; Onshore Wells; Petroleum Reservoirs; Well Logging.

OIL AND GAS EXPLORATION

Geologists seek to understand the geologic features with which oil and gas are associated in order to make exploration for these minerals less risky and more economical.

PRINCIPAL TERMS

- **dry hole:** a well drilled for oil or gas that had no production
- **lease:** a permit to explore for oil and gas on specified land
- **permeability:** a property of rocks where porosity is interconnected, permitting fluid flow through the rocks
- **porosity:** a property of rocks where empty or void spaces are contained within the rock between grains or crystals or within fractures
- **prospect:** a limited geographic area identified as having all the characteristics of an oil or gas field but without a history of production
- **regional geology:** a study of the geologic characteristics of a geographic area
- **reservoir:** a specific rock unit or bed that has porosity and permeability
- **seal:** a rock unit or bed that is impermeable and inhibits upward movement of oil or gas from the reservoir
- **source rock:** a rock unit or bed that contains sufficient organic carbon and has the proper thermal history to generate oil or gas
- **trap:** a structure in the rocks that will allow petroleum or gas to accumulate rather than flow through the area

EXISTENCE OF TRAPS

As the search for oil and gas begins, the exploration geologist identifies promising areas called prospects within a geologic region. A prospect is a potential oil or gas field. A prospect is not merely the extension of an existing field, as such work is classified as development geology. Rather, a prospect would be a potential new field some distance away from preexisting production. In order to locate prospects, a model of what such potential fields will look like needs to be developed. This model will be based on already discovered fields with well-known characteristics, although it is recognized that there will be differences between the two, and this knowledge will serve as a starting point for exploration. Despite differences

between prospects and models, all prospects have certain criteria that must be met to demonstrate that a prospect is viable or, in other words, drillable. If one of the criteria cannot be met, the prospect may be regarded as too risky to drill.

The first criterion that an exploration geologist must satisfy is proving the existence of a trap. Proving that a trap is present involves the use of several tools. First, the exploration geologist must refer to the model. Once the geologist knows which type of trap to look for, the specific tool is selected. The tool used to locate a trap is a map constructed on the basis of subsurface data. The type of map constructed will depend on the type of trap present in the model. If the trap is a structural trap—that is, created by the deformation of rocks by earth movements—a subsurface structure map showing folds and faults will be constructed, similar to a topographic map. The contours on this map connect points of equal elevation relative to sea level, except that they are in the subsurface rather than on the surface. When studied in the subsurface, hills on a structural contour map represent anticlines or domes, valleys represent synclines or depressions, and sharp cliffs usually indicate faults. If the trap in the model is of a stratigraphic type—that is, formed by features intrinsic to the rock layers themselves—a stratigraphic map showing thickness of a rock unit or bed, differing rock types (facies), or ancient depositional environments is developed. The data to construct such maps are generally derived from subsurface information about the rock units in question from previously drilled wells in the vicinity. In addition to well information, data from seismic, gravity, and magnetic surveys may add support to a prospect. Surface geologic data are considered but may not accurately reflect the geologic conditions of the subsurface.

QUALITIES OF A RESERVOIR

Once all the subsurface data have been analyzed and a potential trap located, a detailed examination of the rocks present in the prospect area is needed. Specifically, it must be demonstrated that a rock unit or bed within the trap can function as a reservoir for the oil or gas. The qualities that enable a rock to be a reservoir are porosity; pore spaces between grains, crystals, or open fractures; and permeability—the interconnection of pore spaces or fractures that will allow the flow of fluid (in this case, oil or gas). Demonstrating that these properties exist is usually done through a detailed study of the rocks by using special cylinder-shaped rock samples called cores collected from previously drilled wells. This core analysis is a very important aspect of petroleum geology and is the best way to describe and define reservoirs. Geophysical logs that measure certain physical properties of the rocks are also very valuable in defining a reservoir. These are collected by lowering instruments into the well and recording the rock properties along the way.

PRESENCE OF SOURCE BED AND SEAL

The next issue is demonstrating that a source bed for oil or gas exists in the prospect area. This feature is perhaps the most critical, as oil and gas are generated only under very specific chemical and physical conditions. A rock unit or bed must be located in the rocks of the prospect that have the proper chemical consistency and have been subjected to the proper temperatures needed to generate oil or gas. There also must be a pathway or mechanism to allow the oil or gas to migrate from the source bed to the reservoir, as the two will only rarely be the same unit. What is generally needed is a total organic carbon content of a bed greater than 1 percent and a temperature history that allows the organic matter to mature into hydrocarbons such as oil or natural gas. The rock type generally involved will be shale, but some fine-grained limestones and dolomites, as well as some cherts, will function as source beds. A pathway of migration for the generated oil and gas must be available. This is not a critical problem if the source bed occurs beneath the reservoir bed because of the tendency toward upward migration of oil and gas; their low density allows them to float upward on groundwater. If the source bed occurs above the reservoir, it can still function for the prospect if it has been dropped down by a fault below the reservoir unit. Direct downward migration of oil and gas is known to occur but is uncommon, and this process would have to be demonstrated as functioning in the prospect area if it were to explain the source for a prospect.

If these criteria are met, the next factor to be considered is the presence of a seal. As the tendency is for oil and gas to migrate upward, an impermeable rock unit or bed must be present above the reservoir bed to prevent the oil and gas from flowing through the reservoir unit rather than accumulating in the trap. If oil and gas cannot migrate upward, they will migrate laterally to the highest point (trap) in the last permeable bed (reservoir) they can enter. Seals can be of any impermeable rock type, but common examples are evaporites (rock salt and gypsum) and shales. Fine-grained limestones and dolomites, as well as some igneous and metamorphic rocks, can also function as seals.

FINAL CRITERION

The final geological criterion that must be satisfied is the time of oil and gas generation and migration, and its relationship to the time of trap formation. In order to determine when these events occurred, a detailed understanding of the geologic history must be attained. In general, the trap must have been formed before the generation and migration of oil and gas in the region. If this is not the case, the trap in question will not likely contain oil and gas. Many excellent traps exist that have been drilled and do not produce because of the timing difference.

If any dry holes (previously drilled wells without production) are near a prospect, the reason for their lack of production must be explained. Generally, this will be done by showing that the wells were drilled away from the prospect in question or that technical or economic problems were encountered. Geologists do not want a prospect with a dry hole in the middle unless that well was dry for a reason that the present model can explain.

OBTAINING A LEASE

If all of the preceding criteria have been met, the geologist may have a prospect, provided that a lease can be obtained. An oil and gas lease gives the holder the right to explore and produce oil and gas on the land in question. These are the only rights granted to the lease holder. This lease is usually for a specified term (five or ten years) and pays the landowner an annual rental fee plus a royalty interest in any production. If production is achieved on the lease, the term will not expire until production stops. Production cannot be halted to wait for better economic conditions, such as higher prices, as this would endanger the lease. The availability of acreage is a nongeologic factor in the exploration for oil and gas but a critical one nonetheless. A prospect without the potential of acquiring a lease is of little use, as it cannot be drilled and developed.

A PROCESS OF IDEAS

The process of oil and gas exploration is a difficult one because of the large degree of uncertainty involved. Advances such as high-resolution seismic profiling, computer analysis and management of well data, and the use of satellite photographs of the land surface have all improved the way geologists explore oil and gas deposits. Despite the benefits of improved technology, the exploration for oil and gas remains a process of ideas rather than equipment. This does not mean that old, established ideas will dominate the search for oil and gas. In fact, quite the opposite is true. Ideas can and do develop at a much faster rate than technology. Because of this, the potential for change in an approach is always present. Thus, a region that might be considered to have no oil and gas potential may appear differently to geologists who have a different or new idea about the geology of the area. Some new ideas affect all regions of the earth and tend to form a revolution of approaches in the oil and gas industry. Ideas over the last three decades have tended to view geology from a global perspective, and this has changed the exploration for oil and gas. The most important idea of this type is the theory of plate tectonics. This proposed mechanism resulted in a widespread reevaluation of the petroleum potentials of regions, resulting in increased discoveries of new fields.

The effect that the exploration of oil and gas has on society is a very clear one. Oil and gas are needed by almost everyone as sources of relatively inexpensive, clean, portable, and efficient energy. The present structure of the oil industry, including oil exploration and production, is sometimes criticized as being wasteful and inefficient. Even with the best methods, a large proportion of exploratory holes are dry, a fact of life inherent in the petroleum industry. In some nations, oil and gas exploration has been nationalized and is handled by a team of geologists in order to promote efficiency by eliminating competition for the same prospective new fields. This has been suggested as a model for the oil industry in the United States as a way to increase the known petroleum reserves.

Because the competitive system, in which many geologists develop varying working models of an area, is eliminated by a nationalized approach, only one model is present, and the chances of ever finding oil are greatly reduced. While many nationalized companies are very successful, their success stories often are in areas that simply had not been explored before, and the oil accumulations were very obvious. As these regions mature and oil and gas become harder to find, multiple ideas are needed to locate new fields. A competitive system ensures that as long as a profit can be made, exploration for oil and gas will continue.

Richard H. Fluegeman, Jr.

FURTHER READING

American Petroleum Institute. *Primer of Oil and Gas Production.* 3d ed. Washington, D.C.: Author, 1976. A basic introduction to the procedures and techniques of oil and gas production. Includes chapters on the origin and accumulation of oil and gas, and the properties of reservoirs. Written as a beginner's text in nontechnical language with helpful illustrations.

Baker, Ron. *A Primer of Oil Well Drilling.* 6th ed. Austin: University of Texas at Austin, 2000. An introduction to the procedures involved in oil and gas well drilling. Includes chapters on exploration and on oil and gas accumulations. Well illustrated with many on-site photographs. Suitable for high school or college students.

Gluyas, Jon, and Richard Swarbrick. *Petroleum Geoscience.* Malden, Mass.: Blackwell Science, 2004. Describes the tools and methods used in petroleum exploration. Includes limited images, drawings, and tables, as well as extensive references and further reading lists and an index. Appropriate for graduate students, academics, and professionals.

Hyne, Norman J. *Dictionary of Petroleum Exploration, Drilling, and Production.* Tulsa, Okla.: PennWell, 1991. Covers all terms associated with petroleum and the petroleum industry. Appropriate for beginners in the field. Illustrations and maps.

_____. *Nontechnical Guide to Petroleum Geology, Exploration, Drilling, and Production.* 2d ed. Tulsa, Okla.: PennWell, Corporation, 2001. Provides a well-rounded overview of the processes and principles of gas and oil drilling. Covers foundational material. Discusses exploration, refining, and distribution. Appropriate for professionals in the oil-drilling field, geologists, or students of similar fields. Illustrations, bibliography, and index.

LeRoy, L. W., and D. O. LeRoy, eds. *Subsurface Geology.* 4th ed. Golden: Colorado School of Mines, 1977. An important reference for all who are interested in the geology of the subsurface. Includes chapters on petroleum geology, but exploration for oil and gas is not the only focus of this book. Chapters on oil and gas are written by experts in the field, very complete, and well illustrated. A basic reference for the geology of oil and gas by oil-business professionals.

Levorsen, A. I. *Geology of Petroleum.* 2d ed. Tulsa, Okla.: American Associations of Petroleum Geologists, 2001. A full textbook on petroleum geology designed for undergraduates who have taken some basic geology courses. A reference for people with an interest in some of the more detailed aspects of petroleum exploration. The training textbook for many present-day exploration geologists. Includes a bibliography.

Li, Guoyu. *World Atlas of Oil and Gas Basins.* Hoboken, N.J.: John Wiley & Sons, 2011. Includes an overview section followed by information on oil and gas distribution throughout the countries of the world organized by continent or region. Covers each area briefly. Appropriate for undergraduates and oil professionals.

Owen, E. W. *Trek of the Oil Finders: A History of Exploration for Petroleum.* Tulsa, Okla.: American Association of Petroleum Geologists, 1975. A detailed history of the development of the petroleum industry throughout the world. Written from a historical perspective, in technical language only when necessary. Well indexed by subject, geographic location, and proper names. Chapters are organized by region.

Raymond, Martin S., and William L Leffler. *Oil and Gas Production in Nontechnical Language.* Tulsa, Okla.: PennWell Corporation, 2005. Provides a good overview of the industry. Includes pictures, charts, graphs, and drawings. Describes oil exploration, and how oil and gas are found and extracted. Written in a manner accessible to the layperson.

Selley, Richard C. *Elements of Petroleum Geology.* 2d ed. San Diego: Academic Press, 1998. Covers the specifics of oil and gas and their relationship to geology. Designed as a college textbook for students near the end of their coursework in geology and for geologists beginning careers in the petroleum industry. Requires a basic understanding of geological concepts. Contains subject and name indexes, useful illustrations, and appendices that include a well classification table, a glossary of oil terms and abbreviations, and a table of conversion factors.

West Texas Geological Society. *Geological Examples in West Texas and South-eastern New Mexico (the Permian Basin) Basic to the Proposed National Energy Act.* Midland: West Texas Geological Society, 1979. Designed to inform state and federal government members of the procedures and costs of exploration for oil and gas. Based on case histories and maps; diagrams and actual costs are included. A valuable reference, written in clear, nontechnical language.

See also: Offshore Wells; Oil and Gas Distribution; Oil and Gas Origins; Oil Chemistry; Oil Shale and Tar Sands; Onshore Wells; Petroleum Industry Hazards; Petroleum Reservoirs; Well Logging.

OIL AND GAS ORIGINS

Oil and gas are two of the most important fossil fuels. The formation of oil and gas is dependent on the preservation of organic matter and its subsequent chemical transformation into kerogen and other organic molecules deep within the earth at high temperatures over long periods of time. As oil and gas are generated from these organic materials, they migrate upward, where they may accumulate in hydrocarbon traps.

PRINCIPAL TERMS

- **fossil fuel:** a general term used to refer to petroleum, natural gas, and coal
- **geopolymer:** a large molecule created by linking together many smaller molecules by geologic processes
- **hydrocarbons:** solid, liquid, or gaseous chemical compounds containing only carbon and hydrogen; oil and natural gas are complex mixtures of hydrocarbons
- **kerogen:** fossilized organic material in sedimentary rocks that is insoluble and generates oil and gas when heated; as a form of organic carbon, it is one thousand times more abundant than coal and petroleum in reservoirs, combined
- **methane:** a colorless, odorless gaseous hydrocarbon with the formula CH_4; also called marsh gas
- **natural gas:** a mixture of several gases used for fuel purposes and consisting primarily of methane, with additional light hydrocarbon gases such as butane, propane, and ethane, with associated carbon dioxide, hydrogen sulfide, and nitrogen
- **petroleum:** crude oil; a naturally occurring complex liquid hydrocarbon, which after distillation yields a range of combustible fuels, petrochemicals, and lubricants
- **reservoir:** a porous and permeable unit of rock below the surface of the earth that contains oil and gas; common reservoir rocks are sandstones and some carbonate rocks

PRESERVATION OF ORGANIC MATTER

To become a fossil fuel, the organic matter in an organism must be preserved after the organism dies. The preservation of organic matter is a rare event because most of the carbon in organic matter is oxidized and recycled to the atmosphere through the action of aerobic bacteria. (Oxidation is a process through which organic matter combines with oxygen to produce carbon dioxide gas and water.) Less than 1 percent of the organic matter that is produced by photosynthesis escapes from this cycle and is preserved. To be preserved, the organic matter must be protected from oxidation, which can occur in one of two ways: the organic matter in the dead organism is rapidly buried by sediment, shielding it from oxygen in the environment, or the dead organism is transported into an aquatic environment in which there is no oxygen (that is, an anoxic or anaerobic environment). Most aquatic environments have oxygen in the water, because it diffuses into the water from the atmosphere and is produced by photosynthetic organisms, such as plants and algae. Oxygen is removed from the water by the respiration of aerobic organisms and through the oxidation of decaying organic matter. Oxygen consumption is so high in some aquatic environments that anoxic water is present below the near-surface oxygenated zone. Environments that lack oxygen include such places as deep, isolated bodies of stagnant water (such as the bottom waters of some lakes), some swamps, and the oxygen-minimum zone in the ocean (below the maximum depth to which light penetrates, where no photosynthesis can occur). Large quantities of organic matter may be preserved in these environments.

The major types of organic matter preserved in sediments include plant fragments, algae, and microbial tissue formed by bacteria. Animals (and single-celled, animal-like organisms) contribute relatively little organic matter to sediments. The amount of organic matter contained in a sediment or in sedimentary rock is referred to as its total organic content (TOC), and it is typically expressed as a percentage of the weight of the rock. To be able to produce oil, a sediment typically must have a TOC of at least 1 percent by weight. Sediments that are capable of producing oil and gas are referred to as "hydrocarbon source rocks." In general, fine-grained rocks such as shales (which have clay-sized grains) tend to have higher TOC than do coarser-grained rocks.

TRANSFORMATION OF ORGANIC MATTER

The organic matter trapped in sediment must undergo a series of changes to form oil and gas. These changes take place as the sediment is buried to great depths as a result of the deposition of more and more sediment in the environment over long periods of time. Temperature and pressure increase as depth of burial increases, and the organic materials are altered by these high temperatures and pressures.

As sediment is gradually buried to depths reaching hundreds of meters, it undergoes a series of physical and chemical changes, called diagenesis. Diagenesis transforms sediment into sedimentary rock by compaction, cementation, and removal of water. Methane gas commonly forms during the early stages of diagenesis as a result of the activity of methanogenic bacteria. At depths of a few meters to tens of meters, such organic compounds as proteins and carbohydrates are partially or completely broken down, and the individual component parts are converted into carbon dioxide and water, or are used to construct geopolymers—that is, large, complex organic molecules of irregular structure (such as fulvic acid and humic acid, and larger geopolymers called humins). During diagenesis, the geopolymers become larger and more complex, and nitrogen and oxygen content decreases. With increasing depth of burial over long periods of time (burial to tens or hundreds of meters over a million or several million years), continued enlargement of the organic molecules alters the humin into kerogen, an insoluble form of organic matter that yields oil and gas when heated.

As sediment is buried to depths of several kilometers, it undergoes a process called catagenesis. At these depths, the temperature may range from 50 to 150 degrees Celsius, and the pressure may range from 300 to 1,500 bars. The organic matter in the sediment, while in a process called maturation, becomes stable under these conditions. During maturation, a number of small organic molecules are broken off the large kerogen molecules, a phenomenon known as thermal cracking. These small molecules are more mobile than are the kerogen molecules. Sometimes called bitumen, they are the direct precursors of oil and gas. As maturation proceeds, and oil and gas generation continues, the kerogen residue remaining in the source rock gradually becomes depleted in hydrogen and oxygen. In a later stage, wet gas (gas with a small amount of liquid) and condensate are

formed. ("Condensate" is a term given to hydrocarbons that exist as gas under the high pressures existing deep beneath the surface of the earth but condense to liquid at the earth's surface.) Oil is typically generated at temperatures between 60 and 120 degrees Celsius, and gas is generated at somewhat higher temperatures, between about 120 and 220 degrees Celsius. Large quantities of methane are formed during catagenesis and during the subsequent phase, which is called metagenesis.

When sediment is buried to depths of tens of kilometers, it undergoes the processes of metagenesis and metamorphism. Temperatures and pressures are extremely high. Under these conditions, all organic matter and oil are destroyed, being transformed into methane and a carbon residue, or graphite. Temperatures and pressures are so intense at these great depths that some of the minerals in the sedimentary rocks are altered and recrystallized, and metamorphic rocks are formed.

MIGRATION AND TRAPPING

Accumulations of oil and gas are typically found in relatively coarse-grained, porous, permeable rocks, such as sandstones and some carbonate rocks. These oil- and gas-bearing rocks are called reservoirs. Reservoir rocks, however, generally lack the kerogen from which the oil and gas are generated. Instead, kerogen is typically found in abundance only in fine-grained sedimentary rocks such as shales. From these observations, it can be concluded that the place where oil and gas originate is not usually the same as the place where oil and gas are found. Oil and gas migrate or move from the source rocks (their place of origin) into the reservoir rocks, where they accumulate.

Oil and gas that form in organic-rich rocks tend to migrate upward from their place of origin, toward the surface of the earth. This upward movement of oil occurs because pore spaces in the rocks are filled with water, and oil floats on water because of its lower density. Gas is even less dense than oil and also migrates upward through pore spaces in the rocks. The first phase of the migration process, called primary migration, involves expulsion of hydrocarbons from fine-grained source rocks into adjacent, more porous and permeable layers of sediment. Secondary migration is the movement of oil and gas within the more permeable rocks. Oil and gas may eventually reach the surface of the earth and be lost to the atmosphere through a seep. Under some circumstances, however, the rising oil and gas may become trapped in the sub-surface by an impermeable barrier, called a caprock. These hydrocarbon traps are extremely important because they provide a place for subsurface concentration and accumulation of oil and gas, which can be tapped for energy sources.

There are a variety of settings in which oil and gas may become trapped in the subsurface. Generally, each of these traps involves an upward projection of porous, permeable reservoir rock in combination with an overlying impermeable caprock that encloses the reservoir to form a sort of inverted container. Examples of hydrocarbon traps include anticline

traps, salt dome traps, fault traps, and stratigraphic traps. There are many types of stratigraphic traps, including porous reef rocks enclosed by dense lime-stones and shales, sandstone-filled channels, sand bars, or lenses surrounded by shale, or porous, per-meable rocks beneath an unconformity. The goal of the exploration geologist is to locate these subsurface hydrocarbon traps. Enormous amounts of geologic information must be obtained, and often many wells must be drilled before accumulations of oil and gas can be located.

PHYSICAL AND CHEMICAL ANALYSES

The origin of oil and gas can be determined using physical and chemical analyses. Petroleum contains

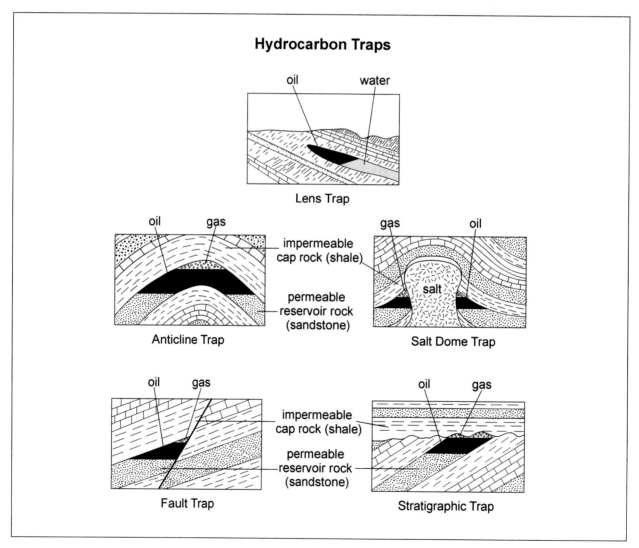

Hydrocarbon Traps

Lens Trap

Anticline Trap

Salt Dome Trap

Fault Trap

Stratigraphic Trap

compounds that serve as biological markers to demonstrate the origin of petroleum from organic matter. Oil can be analyzed chemically to determine its composition, which can be compared to that of hydrocarbons extracted from source rocks in the lab. Generally, oil is associated with natural gas, most of which probably originated from the alteration of organic material during diagenesis, catagenesis, or metagenesis. In some cases, gas may be of abiogenic (nonorganic) origin. Samples of natural gas can be analyzed using gas chromatography or mass spectrometry and isotope measurements.

Commonly, rocks are analyzed to determine their potential for producing hydrocarbons. It is important to distinguish between various types of kerogen in the rocks because different types of organic matter have different potentials for producing hydrocarbons. In addition, it is important to determine the thermal maturity or evolutionary state of the kerogen to confirm whether the rock has the capacity to generate hydrocarbons, or whether hydrocarbons have already been generated.

The quantity of organic matter in a rock, referred to as its TOC, can be measured with a combustion apparatus, such as a Leco carbon analyzer. To analyze for TOC, a rock must be crushed and ground to a powder and its carbonate minerals removed by dissolution in acid. During combustion, the organic carbon is converted into carbon dioxide by heating it to high temperatures in the presence of oxygen. The amount of carbon dioxide produced is proportional to the TOC of the rock. The minimum amount of TOC that is adequate for hydrocarbon production is generally considered to be between 0.5 and 1 percent TOC by weight.

INDIRECT METHODS OF ANALYSES

The type of organic matter in a rock can be determined indirectly, through study of the physical and chemical characteristics of the kerogen, or directly, by using pyrolysis (heating) techniques. The indirect methods of analysis include examination of kerogen with a microscope and chemical analysis of kerogen. Microscopic examination can identify different types of kerogen, such as spores, pollen, leaf cuticles, resin globules, and single-celled algae. Kerogen that has been highly altered and is amorphous can be examined using fluorescence techniques to determine whether it is oil-prone (fluorescent) or inert,

or gas-prone (nonfluorescent). Chemical analysis of kerogen provides data on the proportions of chemical elements, such as carbon, hydrogen, sulfur, oxygen, and nitrogen. A graph of the ratios of hydrogen/carbon (H/C) versus oxygen/carbon (O/C) is used to classify kerogen by origin and is called a van Krevelen diagram. There are three curves on a van Krevelen diagram, labeled I, II, and III, corresponding to three basic types of kerogen. Type I is rich in hydrogen, with high H/C and low O/C ratios, as in some algal deposits; this type of kerogen generally yields the most oil. Type II has relatively high H/C and low O/C ratios and is usually related to marine sediments containing a mixture of phytoplankton, zooplankton, and bacteria; this type of kerogen yields less oil than Type I, but it is the source material for a great number of commercial oil and gas fields. Type III is rich in oxygen, with low H/C and high O/C ratios (aromatic hydrocarbons), as in terrestrial or land plants; this type of kerogen is comparatively less favorable for oil generation but tends to generate large amounts of gas when buried to great depths. As burial depth and temperature increase, the amount of oxygen and hydrogen in the kerogen decreases, and the kerogen approaches 100 percent carbon. Hence, a van Krevelen diagram can be used to determine both the origin of the organic matter and its relative thermal maturity.

The potential that a rock has for producing hydrocarbons can be evaluated through a pyrolysis, or heating, technique, commonly called Rock-Eval. Rock-Eval yields information on the quantity, type, and thermal maturity of organic matter in the rock. The procedure involves the gradual heating (to about 550 degrees Celsius) of a crushed rock sample in an inert atmosphere (nitrogen, helium) in the absence of oxygen. At temperatures approaching 300 degrees Celsius, heating releases free hydrocarbons already present in the rock; the quantity of free hydrocarbons is referred to as S_1. At higher temperatures (300 to 550 degrees Celsius), additional hydrocarbons and related compounds are generated from thermal cracking of kerogen in the rock; the quantity of these hydrocarbons is referred to as S_2. The temperature at which the maximum amount of S_2 hydrocarbons is generated is called T_{max} and can be used to evaluate the thermal maturity of the organic matter in the rock. In addition, carbon dioxide is generated as the kerogen in the rock is heated; the quantity of CO_2 generated as the rock is heated to 390 degrees Celsius is referred to as S_3. (The temperature is limited to 390 degrees Celsius because at higher temperatures, CO_2 is also formed from the breakdown of inorganic materials, such as carbonate minerals.) These data can be used to determine the hydrocarbon-generating potential of the rock, the quantity and type of organic matter, and the thermal maturity. For example, $S_1 + S_2$, called the genetic potential, is a measure of the total amount of hydrocarbons that can be generated from the rock, expressed in kilograms per ton. If $S_1 + S_2$ is less than 2 kilograms per ton, the rock has little or no potential for oil production, although it has some potential for gas production. If $S_1 + S_2$ is between 2 and 6 kilograms per ton, the rock has moderate potential for oil production. If $S_1 + S_2$ is greater than 6 kilograms per ton, the rock has good potential for oil production. The ratio $S_1/(S_1 + S_2)$, called the production index, indicates the maturation of the organic matter. Pyrolysis data can also be used to determine the type of organic matter present. The oxygen index is S_3/TOC, and the hydrogen index is S_2/TOC. These two indices can be plotted against each other on a graph, comparable to a van Krevelen diagram.

FINITE FOSSIL FUELS

Oil and gas are derived from the alteration of kerogen, an insoluble organic material, under conditions of high temperatures (50 to 150 degrees Celsius) and pressures (300 to 1,500 bars). After oil and gas are generated, they migrate upward out of organic-rich source rocks and come to be trapped and accumulate in specific types of geologic settings. The search for oil and gas deposits trapped in the subsurface can be expensive and time-consuming, requiring trained exploration geologists. Once a promising geologic setting has been located, the only way to determine whether oil and gas deposits are actually present in the subsurface is to drill a well.

Oil and gas are two of the earth's most important fossil fuels. It is important to understand that a finite amount of these hydrocarbons is present within the earth. They cannot be manufactured when known reserves are depleted.

Pamela J. W. Gore

FURTHER READING

Durand, Bernard, ed. *Kerogen: Insoluble Organic Matter from Sedimentary Rocks.* Paris: Éditions Technip, 1980. Discusses various aspects of kerogen, ranging from its origin and appearance under the microscope to its chemical composition and structure. Written by specialists, mostly in English, but with a few articles in French. Contains a number of beautiful color plates illustrating the appearance of kerogen-rich rocks and organic microfossils (pollen, spores, acritarchs, dinoflagellates) as seen through the microscope.

Gluyas, Jon, and Richard Swarbrick. *Petroleum Geoscience.* Malden, Mass.: Blackwell Science, 2004. Describes the tools and methods used in petroleum exploration. Includes limited images, drawings, and tables. Provides extensive references, further reading lists, and an index. Appropriate for graduate students, academics, and professionals.

Hunt, John Meacham. *Petroleum Geochemistry and Geology.* 2d ed. New York: W. H. Freeman, 1996. Covers petroleum composition, origin, migration, accumulation, and analysis, and the application of petroleum geochemistry in petroleum exploration, seep and subsurface prospects, crude oil correlation, and prospect evaluation. Requires a background in chemistry and algebra.

Hyne, Norman J. *Nontechnical Guide to Petroleum Geology, Exploration, Drilling, and Production.* 2d ed. Tulsa, Okla.: PennWell, Corporation, 2001. Provides a well-rounded overview of the processes and principles of gas and oil drilling. Covers foundational material. Appropriate for professionals in the oil-drilling field, geologists, and students of similar fields.

North, F. K. *Petroleum Geology.* Boston: Allen & Unwin, 1985. Covers a wide variety of topics related to petroleum geology, including the nature and origin of petroleum; where and how oil and gas accumulate; exploration, exploitation, and forecasting; and the distribution of oil and gas. Designed to introduce students to many topics in exploration, drilling, and the basics of the origin of oil and gas with practical application. Well illustrated with maps and geologic cross-sections representing oil-producing areas around the world. Suitable for geologists and college students.

Peters, K. E. "Guidelines for Evaluating Petroleum Source Rock Using Programmed Pyrolysis." *The American Association of Petroleum Geologists Bulletin* 70 (March, 1986): 318-329. Provides information on Rock-Eval pyrolysis, one of the major analytical techniques for analyzing rocks to determine their hydrocarbon potential. Provides a brief summary of the technique and goes into detail using numerous examples, discussing some of the problems encountered in interpreting samples. Suitable for geologists and advanced college students.

Raymond, Martin S., and William L Leffler. *Oil and Gas Production in Nontechnical Language.* Tulsa, Okla.: PennWell Corporation, 2005. Provides a good overview of the industry. Includes pictures, charts, graphs, and drawings. Describes oil exploration, and how oil and gas are found and extracted from the earth. Appropriate for general readers.

Selley, Richard C. *Elements of Petroleum Geology.* 2d ed. San Diego: Academic Press, 1998. Covers the specifics of oil and gas and their relationship to geology. Designed for students near the end of their coursework in geology or for geologists beginning careers in the petroleum industry. Requires basic understanding of geological concepts. Contains subject and proper name indexes, useful illustrations, and appendices that include a well classification table, a glossary of oil terms and abbreviations, and a table of conversion factors.

Tissot, Bernard P., Bernard Durand, J. Espitalié, and A. Combaz. "Influence of Nature and Diagenesis of Organic Matter in Formation of Petroleum." *American Association of Petroleum Geologists Bulletin* 58 (March, 1974): 499-506. Discusses the generation of hydrocarbons and changes in kerogen that occur during burial. Provides a concise summary of the types of kerogen and depths at which oil and gas are generated. Well illustrated with graphs.

Tissot, Bernard P., and D. H. Welte. *Petroleum Formation and Occurrence.* 2d ed. Berlin: Springer-Verlag, 1984. One of the most comprehensive guides to the origin of petroleum and natural gas. Organized according to the production and accumulation of geologic matter (a geological perspective); the fate of organic matter in sedimentary basins (generation of oil and gas); the migration and accumulation of oil and gas; the composition and classification of crude oils and the influence of geological factors; and oil and gas exploration (application of the principles of petroleum generation and migration). Well illustrated with line drawings and graphs. Appropriate for college-level students and indispensable for geologists.

Waples, Douglas W. *Geochemistry in Petroleum Exploration.* Boston: International Human Resources Development Corporation, 1985. A leading reference in the field. Provides an overview of the origin of oil and gas. Concise and well illustrated with line drawings and graphs. Appropriate for college-level students.

Welte, Dietrich H., Brian Horsfield, and Donald R. Baker, eds. *Petroleum and Basin Evolution: Insights from Petroleum Geochemistry, Geology, and Basin Modeling.* New York: Springer, 1997. Explores the origins of oil and gas from the perspective of mathematical modeling of sedimentary basins. Somewhat technical, but illustrations and maps help clarify many of the difficult concepts. Bibliography.

See also: Diagenesis; Offshore Wells; Oil and Gas Distribution; Oil and Gas Exploration; Oil Chemistry; Oil Shale and Tar Sands; Onshore Wells; Petroleum Industry Hazards; Petroleum Reservoirs; Well Logging.

OIL CHEMISTRY

Crude oils are fossil organic chemicals that have been transformed by geologic processes into a complex mixture of many different chemical compounds called hydrocarbons. Although the composition of oils varies widely, the most abundant hydrocarbons in most oils are the paraffins, naphthenes, aromatics, and compounds with nitrogen, sulfur, and oxygen attached (NSOs). Less abundant compounds, called biomarkers, are true "geochemical fossils" that retain the original molecular structure and probable biological identity of the organisms from which the oil is derived.

PRINCIPAL TERMS

- **aromatic hydrocarbons:** ring-shaped molecules composed of six carbon atoms per ring; the carbon atoms are bonded to one another with alternating single and double bonds
- **biomarkers:** chemicals found in oil with a chemical structure that definitely links their origin with specific organisms; also called geochemical fossils
- **hydrocarbons:** natural chemical compounds composed of carbon and hydrogen, usually of organic origin; they make up the bulk of both petroleum and the tissues of organisms (plants and animals)
- **molecular weight:** a measure of the mass of the molecule of a chemical compound, as determined by both the total number and the size of atoms in the molecule
- **naphthenic hydrocarbons:** hydrocarbon molecules with a ring-shaped structure, in which any number of carbon atoms are all bonded to one another with single bonds
- **oxidation:** a very common chemical reaction in which elements are combined with oxygen—for example, the burning of petroleum, wood, and coal; the rusting of metallic iron; and the metabolic respiration of organisms
- **paraffin hydrocarbons:** hydrocarbon compounds composed of carbon atoms connected with single bonds into straight chains; also known as n-alkanes
- **saturated hydrocarbons:** hydrocarbon compounds whose molecules are chemically stable, with carbon atoms fully bonded to other atoms

Paraffin Hydrocarbons

The various hydrocarbon (organic matter) compounds found in petroleum differ from one another in two fundamental ways: the number of carbon and hydrogen atoms in the hydrocarbon molecule, and the shape of the molecule. Hydrocarbon molecules are classified according to the number of carbon atoms they contain. A simple numbering system is used in which C_2, for example, refers to a molecule with 2 carbon atoms, and C_3 is a molecule with 3 carbons. In most oils, compounds of very low molecular weight (less than C_5) are dissolved in the oil as natural gas. When the crude oil is pumped out of the ground, these molecules evaporate from the liquid petroleum and are either burned off or collected to be used as fuel.

Carbon atoms can be bonded to one another in straight chains, rings, or combinations of these basic forms. In many hydrocarbon molecules, small molecular fragments called side chains are attached like branches of a tree to the main chain or ring of the molecule. The simplest and most abundant petroleum hydrocarbons in crude oils are straight carbon chain molecules, referred to as the paraffin series. The smallest and simplest paraffin hydrocarbon molecule is methane, the most common component of natural gas. The methane molecule is composed of a central carbon atom bonded to four hydrogen atoms in a three-sided pyramid arrangement, or tetrahedron. The structure of paraffin molecules of higher molecular weight (C_2 to C_{30}) is made when additional carbon atoms are attached to the basic methane tetrahedron, making a carbon chain. As these carbon atoms are successively added, a chainlike arrangement of carbon atoms develops, with hydrogen atoms bonded to all the carbon atoms. This series of chainlike molecules is called the n-alkane series, with "n" denoting the number of carbon atoms in the chain. For example, the compound pentane, an abundant component of natural gas, is a five-carbon alkane chain (C_5H_{12}). Octane, the hydrocarbon compound by which gasoline is graded, is an eight-carbon alkane (C_8H_{18}). The highest molecular weight n-alkane found as a liquid in oil is heptadecane ($C_{17}H_{36}$), a hydrocarbon with a boiling point of 303 degrees Celsius.

All carbon atoms have four positions, called bonding sites, at which other atoms can attach themselves to form a molecule. In petroleum hydrocarbons, the carbon atoms usually are bonded to either hydrogen atoms or other carbon atoms, although some are bonded to nitrogen, sulfur, oxygen, or other atoms. Hydrocarbon molecules with all bonds joined to four other hydrogen or carbon atoms are called saturated hydrocarbons. For any saturated n-alkane in the paraffin series, the number of hydrogen atoms in the molecule (#H) can be predicted from the number of carbon atoms (n) in the chain by the simple equation $\#H = 2n + 2$. Note that for all n-alkane hydrocarbons the carbon-to-hydrogen ratio is always less than 1:2.

NAPHTHENIC HYDROCARBONS

If the ends of a paraffin hydrocarbon chain are linked together to form a ring, the result is the shape of the other abundant group of hydrocarbon compounds in crude oils, the naphthene series. Naphthenic molecules are composed of carbon atoms bonded together in rings; molecules with this molecular geometry are usually referred to as cyclic hydrocarbons or ring compounds. The simplest naphthenic hydrocarbon is cyclopropane (C_3H_6), a three-carbon ring molecule that occurs as a gas dissolved in crude oil. Cyclopropane, like other hydrocarbons of low molecular weight, bubbles out of the oil solution when it reaches the low-pressure conditions of the earth's surface in the same way that carbon dioxide gas bubbles out of a soft drink when the bottle is opened. In this way, it becomes part of the natural gas that is associated with crude oil production. The liquid naphthene with the lowest molecular weight is cyclopentane (C_5H_{10}); this compound and cyclohexane (C_6H_{12}) are the dominant cyclic hydrocarbons in most oils. Saturated naphthenic hydrocarbon molecules have hydrogen atoms bonded to all the carbon atom bonding sites in a manner similar to that of n-alkanes. In this case, two hydrogen atoms are bonded to each carbon, and the ratio of carbon to hydrogen for these compounds is 1:2.

AROMATIC HYDROCARBONS

Less abundant in oils than the paraffins and naphthenes are the aromatic hydrocarbons. This group of petroleum hydrocarbons, which constitutes from 1 to 10 percent of most crude oils, is so named because many of the compounds have pleasant, sometimes fruity, odors. The aromatics also have a carbon ring structure, but this structure has a different geometry from that of the naphthenes. Aromatic hydrocarbon molecules are formed of one or more six-carbon rings in which the carbon atoms are bonded to one another with alternating single and double bonds; that is, carbon atoms share two electrons with a neighbor. For this reason, the aromatic hydrocarbons have a carbon-to-hydrogen ratio of 1:1. The petroleum aromatic with the lowest molecular weight is benzene (C_6H_6), a chemical commonly used as an industrial solvent. Toluene, which consists of two rings joined along an edge, is the solvent that produces the familiar odor of correction fluid and many plastic cements.

Many of the aromatic hydrocarbons are cancer-causing (carcinogenic) substances; most potent are the high molecular weight, ringed aromatic molecules referred to as PAHs (poly-aromatic hydrocarbons). The first carcinogen ever discovered was benzopyrene, an aromatic molecule composed of five carbon rings. In the late 1800's, Sir Percival Pott linked this substance with cancer of the scrotum in London chimneysweeps. Their daily exposure to the aromatic hydrocarbons in coal tar and soot, coupled with their poor personal hygiene, was responsible for epidemic proportions of this disease.

NSOs

A small percentage of the hydrocarbons found in oils have distinctive molecular fragments bonded onto basic hydrocarbon structures. These nonhydrocarbon fragments most commonly contain nitrogen, sulfur, and oxygen; for this reason, hydrocarbon compounds with these attached fragments are called NSOs. NSO molecules tend to have much higher molecular weights than the hydrocarbon molecules described earlier. One of the most interesting NSO molecular fragments is the amino group; it contains nitrogen and has the formula NH_2. The amino group is the essential component of the amino acids, the building blocks for the many different proteins of which animal organs, muscles, and other tissues are composed. Amino acids are simple molecules that can form by organic or inorganic processes, but their presence in oil is best explained by inheritance from the organic matter source of the oil.

BIOMARKERS

The minor, or trace, hydrocarbon components in crude oil generally make up much less than 1 percent of the total oil; they are probably the most interesting of all petroleum compounds. Many of these chemicals have a chemical composition and molecular structure that definitely links their origin to specific organisms; for this reason, they are termed geochemical fossils or biological markers (biomarkers). Biomarker chemicals were synthesized by the organisms from which the oil originated and have been preserved through the long and complex history of sediment deposition and burial, oil formation (catagenesis), and migration of the oil from its source to the reservoir rock. Biomarkers are generally large molecules—molecules with much higher molecular weights than those of the more abundant oil hydrocarbons—that have a carbon-to-hydrogen ratio greater than 1:2.

Some of the long-chain paraffin hydrocarbons are among the best-known biomarkers. One relatively abundant group, the isoprenoid hydrocarbons, have chain-type molecules based on the isoprene group (C_5H_8); isoprene is the primary source of synthetic rubber. Isoprenoids are common in the waxes and chlorophyll of terrestrial green plants and are present in many crude oils and ancient sediments, which indicates that the petroleum isoprenoids are derivatives of chlorophyll, and that kerogen from terrestrial plants is a significant source of isoprenoid-rich oils. The most interesting plant-derived isoprenoid found in petroleum is pristane, a C_{15} carbon chain with four CH_3 molecular fragments, called methyl groups, attached to the main paraffin chain as side chains. Phytane is an isoprenoid similar to pristane but is composed of a C_{16} carbon chain with four methyl group side chains. The ratio of pristane to phytane is useful to geologists trying to determine the type of organic matter from which the oil was derived. Oil derived mostly from terrestrial plant tissue, for example, has a high pristane-to-phytane ratio (greater than 4:1), while oils from marine plankton have much lower ratios.

The study of the biomarker composition of oils has given geologists valuable insight into the origin

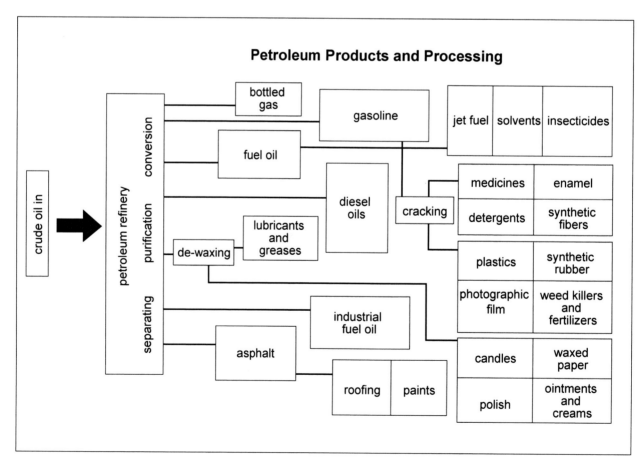

of petroleum, and is used as a valuable tool in oil exploration as well. Every oil has a unique biomarker composition that it inherited from the kerogens that generated the oil and the conditions of catagenesis. Certain biomarker compounds, even when present in exceedingly minute quantities, can be detected with the mass spectrometer. This "geochemical fingerprint" allows petroleum geologists to recognize distinctive chemical similarities between oils and their source sediments. The geochemical fingerprinting technique is routinely used to distinguish different oils from one another, to correlate similar oils from different areas, and to demonstrate a similarity between kerogen-bearing source sediments and the oil that was generated from them. It is also used to increase oil-exploration efficiency by serving as a critical clue to the presence of undiscovered petroleum. The technique also has been used at the sites of oil spills and polluted groundwater, and in other instances of oil pollution to identify the source pollutant; it has been successfully used as evidence in courts of law to determine the guilty party and to calculate damages for episodes of oil pollution.

CHROMATOGRAPHY

To determine the chemical composition of an oil, petroleum geochemists employ many different techniques, most of which are modifications of standard techniques of organic chemistry. Crude oils are composed of a vast number of individual chemical compounds that differ from one another in their molecular weight and shape, and in the distribution of electric charges on the outer portions of the molecules. These individual compounds are separated from one another by a technique called chromatography. Portions of the oil are slowly passed through a long glass or metal column packed with a chemical substance (usually an organic chemical) that attracts hydrocarbon molecules having certain size or charge characteristics. The attractive chemical inside this "chromatographic column" has a greater affinity for the heaviest or most highly charged molecules, so hydrocarbon molecules of different sizes and charges move through the column at different rates and emerge from the end of the column at different times. For example, the n-paraffins are separated from one another by their molecular weight, so the lightest of these molecules emerges first from the column, and the others come out in the order of their

carbon number. It should be noted that no column exists that separates all petroleum hydrocarbons, and several different chromatic columns are needed to separate an oil into its constituent chemicals. At the end of the column, a detector that is sensitive to molecular weight, charge, or other characteristics measures the amounts of each hydrocarbon compound as it emerges from the column and graphs the results. As with chromatographic columns, no one detector is able to separate all petroleum hydrocarbon compounds, and several different types are used. One of the most valuable and interesting of these detectors is the mass spectrometer. This detector breaks a hydrocarbon molecule into fragments as it leaves the chromatographic column and measures the molecular weight of these fragments. Laboratory studies have confirmed the way in which various known hydrocarbon compounds are fragmented in a mass spectrometer; this information is stored in the memory of a computer connected to the detector. During the analysis of an unknown hydrocarbon compound in an oil sample, its fragments are recognized and "reassembled" by the computer to determine the composition of the original hydrocarbon.

PETROLEUM PRODUCTS

Oil is a basic raw material, from which many thousands of products are made. The types and quantity of petroleum products obtained by refining crude oil are determined by its bulk chemical composition. For example, paraffin-rich oils with low aromatic content usually yield the highest quantity of hydrocarbon fuels, while highly aromatic oils refine into the best lubricating oils. Oils containing a large percentage of n-alkanes of low molecular weight (C_5 to C_{13}) yield high quantities of gasoline, kerosene, jet fuel, and expensive hydrocarbon solvents used in many manufacturing processes; oils rich in n-alkanes of higher molecular weight (C_{14} to C_{40}) yield high quantities of diesel fuel and lubricating oils. Oils dominated by hydrocarbons of high molecular weight (heavy oils) yield asphalt and the hydrocarbons used to make plastics and synthetics.

Oils from different areas of the world can be divided into various types based on their bulk chemistry. Oils composed primarily of paraffins, termed paraffin-based crudes, are the most sought-after of all oil types by the refining industry but represent only a small percentage of the oil being produced worldwide. Most paraffin-based crude oils in North America are of Paleozoic age (about 600 to 250 million years ago) and are produced from oil fields in the midcontinent region. One of these is the famous Pennsylvania crude, which has historically been the standard against which all oils are compared. Similar oils on other continents are usually much younger. Paraffin-based oils of Mesozoic age (about 250 to 65 million years ago) are produced in Chile, Brazil, and the Caucasus region. Paraffin-based oils of Cenozoic age (the last 65 million years) are found in Africa, Borneo, and China.

Crude oils dominated by naphthenic hydrocarbons, sometimes called asphalt-based oils, are relatively rare. Significant production of naphthenic oils occurs in the Los Angeles-Ventura area of Southern California and some oil provinces of the U.S. Gulf Coast, the North Sea, and South America. Highly aromatic oils are generally the heaviest and most viscous of all oil types; some are actually solids at surface temperatures, although they are generally liquid in the subsurface. Important deposits of this unusual oil type include the famous black oils of Venezuela, the very large Athabasca tar sand deposits of western Canada, and certain oils from West Africa. Oils of intermediate composition, called mixed-base oils, make up the bulk of worldwide petroleum production.

James L. Sadd

FURTHER READING

Barker, Colin. *Organic Geochemistry in Petroleum Exploration.* Tulsa, Okla.: American Association of Petroleum Geologists, 1979. Published for professional geologists as a short course in petroleum chemistry. General readers should be able to use the first half.

Chapman, R. E. *Petroleum Geology.* New York: Elsevier, 1983. Written to emphasize petroleum production and minimize the geological aspects. Useful resource for gaining an understanding of petroleum production and aspects of refining oil. Offers a good section on basic chemistry.

Hobson, G. D., and E. N. Tiratsoo. *Introduction to Petroleum Geology.* 2d ed. Houston: Gulf Publishing, 1979. An excellent text for the basics of petroleum geology. Accessible to most readers.

Jones, David S. J., and Peter P. Pujadó. *Handbook of Petroleum Processing.* Dordrecht: Springer, 2006. Covers processing of petroleum products from crude oil to gasoline, kerosene, lube oil, and other products. Discusses chemicals used in the refinement processes. Includes molecular diagrams and appendices with specific chapters. Well suited for advanced undergraduates and professionals in the oil industry.

Kelland, Malcolm A. *Production Chemicals for the Oil and Gas Industry.* Boca Raton, Fla.: CRC Press, 2009. Discusses environmental issues of the oil and gas industry, as well as the chemicals and reactions involved in extraction, refinement, and other treatments during production. Includes references, appendices, and an index.

Link, Peter K. *Basic Petroleum Geology.* 3d ed. Tulsa, Okla.: Oil and Gas Consultants International, 2007. A concise treatment of aspects of petroleum chemistry most important to the science and business of oil exploration.

North, F. K. *Petroleum Geology.* Boston: Allen & Unwin, 1985. One of the best general texts on all aspects of petroleum geology. Sections on petroleum chemistry are not as detailed as in some texts but explain basic information very well. Well illustrated.

Orszulik, Stefan T., ed. *Environmental Technology in the Oil Industry.* 2d ed. New York: Springer Science, 2010. Discusses various chemical processes and drilling methods. Discusses oilfield waste and pollutants-control methods. Covers new technology and chemistry of the oil industry.

Selley, Richard C. *Elements of Petroleum Geology.* 2d ed. San Diego: Academic Press, 1998. Covers the specifics of oil and gas and their relationship to geology. Appropriate for students near the end of their coursework in geology, or for geologists beginning careers in the petroleum industry. Requires basic understanding of geological concepts. Contains subject and name indexes, useful illustrations, and appendices that include a well classification table, a glossary of oil terms and abbreviations, and a table of conversion factors.

Taylor, G. H., et al., eds. *Organic Petrology.* Berlin: Gebreuder Borntraeger, 1998. Focuses on the geochemical analysis of coal, peat, and oil shales. Lengthy and well illustrated.

Tissot, B. P., and D. H. Welte. *Petroleum Formation and Occurrence.* 2d ed. Berlin: Springer-Verlag, 1984. A standard textbook for college-level courses in petroleum geology. Includes a section on petroleum biomarkers. Requires a basic organic chemistry background.

Zamel, Bernard. *Tracers in the Oil Field.* New York: Elsevier, 1995. Covers the use of radioactive tracers in oil and gas exploration and analysis. Illustrations and bibliography.

See also: Diagenesis; Earth Resources; Offshore Wells; Oil and Gas Distribution; Oil and Gas Exploration; Oil and Gas Origins; Oil Shale and Tar Sands; Onshore Wells; Petroleum Reservoirs; Well Logging.

OIL SHALE AND TAR SANDS

Although oil shale and tar sands are not generally thought of as potential sources of petroleum products, these resources have been developed to possibly replace dwindling oil fields. Tremendous reserves exist worldwide, but the current costs and technology preclude the use of oil shale and tar sands on a full scale.

PRINCIPAL TERMS

- **barrel:** the standard unit of measure for oil and petroleum products, equal to 42 U.S. gallons or approximately 159 liters
- **bitumen:** a generic term for a very thick, natural semisolid; asphalt and tar are classified as bitumens
- **fracking:** a controversial method of extracting natural gas from shale by fracturing it to liberate the gas
- **hydrocarbon:** an organic compound consisting of hydrogen and carbon atoms linked together
- **kerogen:** a waxy, insoluble organic hydrocarbon that has a very large molecular structure
- **oil shale:** a sedimentary rock containing sufficient amounts of hydrocarbons that can be extracted by slow distillation to yield oil
- **reservoir rock or sand:** the storage unit for various hydrocarbons; usually of sedimentary origin
- **retort:** a vessel used for the distillation or decomposition of substances using heat
- **tar sand:** a natural deposit that contains significant amounts of bitumen; also called oil sand

OIL SHALE RESERVES

Oil shales are typically fine-grained, stratified sedimentary rock. The term "oil shale" is actually a misnomer because the reservoir rock does not have to be shale. Organic matter is present in the pores of these rocks in the form of kerogen. Kerogen is produced over a long time as the original organic-rich sediments are transformed into complex hydrocarbons. Unlike oil and natural gas, which move relatively easily in the subsurface, hydrocarbons in oil shale migrate at an almost imperceptible rate or not at all, because the sedimentary parent rock often has a very low porosity and permeability, and the kerogen molecules form complex networks that adhere to the grains in the host rock. The richest known oil shales produce between 320 and 475 liters (2 to 3 barrels) of oil per ton of processed rock. Based on current technology, though, less than 7 percent of these reserves are recoverable.

When examined on a global basis, the United States has large oil shale reserves. Oil shales are located under more than 20 percent of the land area of the United States. Roughly 50 percent of the worldwide reserves are found in the Green River formation, a shale and sandstone unit that formed during the Eocene epoch about 50 million years ago. This formation, the only one in the western United States that has been extensively studied for oil shale potential, covers approximately 42,000 square kilometers of southwest and south central Wyoming, northeast Utah, and northwest Colorado. The thickest portions of the Green River formation are located in large structural basins that allowed large, shallow lakes to form in the topographically depressed areas. Subsequent deposition of rich organic sediments and later burial and thermal alteration led to the present deposits. Up to 540 billion barrels of oil exist in the rock units that have a thickness greater than 10 meters, a thickness that is sufficient to produce enough hydrocarbons to be cost-effective using present-day recovery methods. Estimates of the total oil shale reserves for the Green River formation range as high as 2,000 billion barrels of oil. Unfortunately, wide-scale development of these resources will probably not happen, because the extraction process requires large amounts of water, a commodity in small supply in the arid western United States. Approximately 3 liters of water is required to extract 1 liter of oil, so each barrel of oil requires that almost 480 liters of water be used to remove the oil from the reservoir rock. Major projects have been initiated in the Green River formation in western Colorado to set up entire cities to handle the processing of the reservoir rock and its eventual products. Almost all these projects, however, have been terminated or put on indefinite hold because of changes in the worldwide petroleum market.

Although the richest oil shale deposits are in the western United States, another 15 to 20 percent of worldwide reserves are found in Devonian and Ordovician rocks located from New York to Illinois and into southwestern Missouri. These deposits are not economically useful, however, because the cost to extract the small amount of oil present far exceeds the value of the oil produced. Several nations attempted using oil shale as a source for petroleum products during World War II, but postwar economic variations shut down most of them within a few years. A proposal to extract natural gas from these rocks by fracturing them underground—a process called "fracking"—generated widespread controversy in 2011. Advocates point to the availability of a new source of energy, while opponents are concerned about the potential contamination of groundwater and other environmental consequences.

TAR SAND RESERVES

Tar sands constitute another major potential source of unconventional oil reserves. These highly viscous deposits—sometimes referred to as tar, asphalt, and bitumen—probably formed as residues from petroleum reservoirs after the lighter, more hydrogen-rich crude oils migrated toward the surface. These porous sands contain asphaltic hydrocarbons, which are extremely viscous. Thus, the hydrocarbons are not bound up as tightly in the reservoir as they are in oil shales. G. Ronald Gray defines these heavy substances as one of the following: bitumen, an oil sand hydrocarbon that cannot be produced using conventional processes; extra-heavy oils; and heavy oils (see the bibliography). These three types are usually lumped together when discussing worldwide reserve estimates, which are set at about 5,000 billion barrels.

The seven largest tar sand fields have roughly the same amount of oil as do the three hundred largest conventional oil fields in the world. The largest tar sand deposits are located, in descending order, in northern Alberta, Canada; northeastern Siberia; and along the northern bank of the Orinoco River, Venezuela. The amount of heavy oil in place in the tar sands in the Athabasca deposit of Alberta is essentially equal to all the petroleum reserves found in the Middle East. Tar sand deposits are found in twenty-two states in the United States, the largest deposits being in Utah. Given the massive reserves of tar sands throughout the world, they could potentially supply modern society with as much oil as that obtained from conventional flowing wells. The major problem in the future is developing the technology to recover the usable hydrocarbon products in a cost-effective manner.

MINING TECHNIQUES

Once the hydrocarbons are detected in the reservoir rock, they must be extracted. One of two primary mining techniques is used to recover the oil. After the upper soil and rock layers overlying the oil shale are stripped away, the reservoir rock is removed, crushed, and transported to a large retort, where it is heated. This process involves raising the temperature of the rock and hydrocarbons to about 480 degrees Celsius, the temperature at which kerogen vaporizes into volatile hydrocarbons and leaves a carbonaceous residue. When oil shale is heated, the amount of organic matter that is converted to oil increases as the amount of hydrogen in the deposit increases. This vapor is then condensed to form a viscous oil. After the introduction of hydrogen, the mixture can be refined in a manner similar to that used to refine crude oil, which is drawn from the ground using conventional methods of drilling and pumping. One obvious problem with this method is that a considerable amount of heat (energy) must be expended in order to yield products that themselves are potential energy sources.

The second technique used to recover oil from oil shales involves in situ heating of the reservoir rock after it has been fractured with explosives or water under pressure. Researchers at the Sandia National Laboratories in New Mexico may have discovered a method to enhance the extraction of kerogen from the reservoir rocks. Underground fracturing of the rock by controlled explosions increases the number and size of passageways for air to pass through. Air is a necessary component to complete the chemically driven thermal reactions (essentially, combustion) that release the hydrocarbons from kerogen. Heating the rock by pumping in superheated water loosens the viscous hydrocarbons from the pores and cracks. Hydrocarbons are driven off by the heat, collected, and then pumped to the surface for further distillation and refining.

A key factor in the removal of hydrocarbon compounds from the ground is the carbon-to-hydrogen ratio. Carbon is twelve times as heavy as hydrogen, so carbon-rich heavy crude oils (those rich in aromatic hydrocarbons) are denser than conventional oil, which contains a higher percentage of hydrogen.

Carbon-rich crude oil yields smaller amounts of the more desirable lighter fuels, such as kerosene and gasoline. Heavy crudes, such as those derived from oil shale and tar sands, can be chemically upgraded by removing carbon or adding hydrogen in the refining process. These procedures, however, are rather complex and certainly add to the already high cost of extraction and refining.

Progress has been made in the recovery rate of hydrocarbons from reservoir rock. Extraction technology has increased the amount recovered from about 12 percent in 1960 to more than 50 percent in some instances by the late 1980's. The recovery rate is dependent on the percentage of complex hydrocarbons present, which varies from formation to formation. One main problem impeding the all-out effort to continue full-scale production is the economics of oil shale mining. When the price of oil, especially imported oil, is low and the amounts abundant, it is not feasible to consider oil shale as a source. Rising oil prices make oil shale more attractive, but investors will invest in oil shale only if they are confident prices will remain high enough to keep the recovery of oil shale profitable.

ENVIRONMENTAL CONCERNS

There are several major environmental problems with oil shale production. The refining process actually generates more waste than the amount of rock that is processed originally. The processed rock increases in volume by about 30 percent, because the hot water and steam used to extract the kerogen from the oil shale enter into the clay molecules present in the rock and cause them to expand. The problem then is what to do with the increased volume of material, as it will overfill the void produced by mining the rock. This expanded material also weathers very rapidly so that it will not remain in place and form a stable tailings pile. In situ retorting precludes the need to mine the shale and, thus, circumvents this disposal problem.

Another environmental concern is that of the air quality in the vicinity of the processing plant. Large amounts of dust being thrown into the atmosphere have an adverse effect on the air in the immediate area surrounding the plant. On a large scale, global air quality can be affected by increased oil shale production through increases in carbon dioxide. Research has shown that high-temperature retorting

methods (those using temperatures exceeding 600 degrees Celsius) may produce more carbon dioxide from the carbonate rocks containing the oil than the actual burning of the oil produced. As levels of carbon dioxide increase, the worldwide greenhouse effect increases. Additional carbon dioxide is generated through the combustion of free carbon, which is present in the kerogen.

Tar sands are formed in environments different from those in which oil shales form. While oil shales appear to form in lake environments, especially those characterized by sandstone and limestone deposits, tar sands are often found in conjunction with deltaic or nonmarine settings. Most of the major deposits of tar sands are in rocks of Cretaceous age or younger, whereas oil shales are often associated with older sedimentary formations. The lack of a cement other than asphalt to hold the sand grains together results in high porosities and permeabilities, thus affording the viscous tars an opportunity to flow. Another geologic setting that enhances the formation of large tar deposits includes areas with an impermeable layer overlying the deposits. The impermeable layer acts as a barrier to prevent the upward movement of the hydrocarbons.

The mining of tar sands must usually be done at or near the deposits, as large amounts of the sands are handled. The actual production of oil from tar sands in the Athabasca field involves large amounts of material being processed. For each 50,000 barrels of oil, almost 33,000 cubic meters of overburden must be removed and about 100,000 tons of tar sand mined and discarded. Extraction of bitumen and heavy oils from tar sands is relatively difficult because of the high viscosity of the hydrocarbons. Bitumen at room temperature is heavier than water and does not flow. The viscosity of these substances can be changed dramatically by applying heat. If tar sands are heated to about 175 degrees Celsius, the bitumens present flow readily and are capable of floating on water. In some cases, the injection of steam into the reservoir sands increases the flow, thus allowing the material to be pumped out of the ground. Hot water can also be used, but this method requires vast volumes of water, something not available in many regions where these sands are presently found. Underground combustion techniques require burning the tar sands underground and allowing the resulting heat to warm the bitumens to

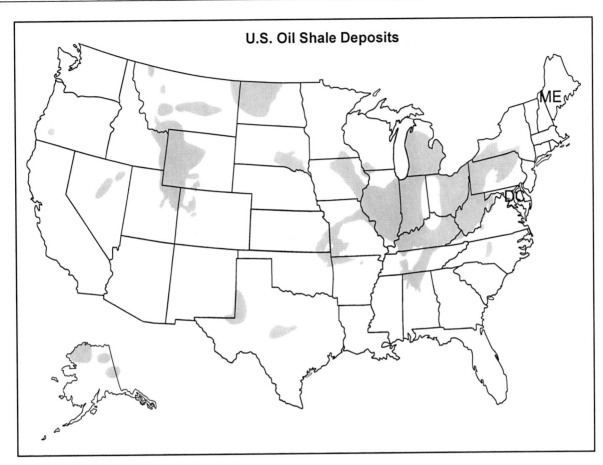

U.S. Oil Shale Deposits

the point at which they flow and can be pumped to the surface.

PROSPECTS FOR PRODUCTION

The existence of oil shales and tar sands has been known for several centuries. Deposits were used as a source of oil for lamps in Europe and colonial America. Native Americans used them to patch their canoes. When commercial production of petroleum expanded in the latter half of the nineteenth century, interest in developing and using oil shales and tar sands as a source of oil decreased. Interest increased again during World War II and in the mid-1970's as a result of shortages in petroleum imports. Once the short-term oil crises were over, though, both the federal government and private industry in the United States dropped most of the research and development associated with extracting oil from oil shale. The Canadian government has moved ahead with its development of extensive tar sand deposits and has been producing substantial amounts of oil

for many years. Oil can be produced from tar sands for roughly $30 per barrel, depending upon various factors.

Undoubtedly, interest in developing oil shale reserves in the United States will increase as the national reserves of petroleum diminish and imports become more scarce and expensive. Widespread production from oil shales will probably not occur until oil reaches a price of $50 to $60 per barrel, or until a technological breakthrough occurs. Such a breakthrough might include sonic drilling or genetic engineering. Sonic drilling uses a vibrating, rotating drill, and it can be faster and cheaper than other methods. At the end of the twentieth century, however, its use was limited to a depth of about 200 meters. Genetic engineering could involve the use of the many bacteria with the ability to break down hydrocarbons. It may be possible to genetically engineer them to efficiently convert kerogen into liquid or gaseous hydrocarbons. The mining method would then involve drilling holes into the oil shale and fracturing it with explosives below ground. Next, a nutrient solution containing the bacteria would be pumped into the formation; a sufficient time later, oil and gas could be extracted. If the bacteria were made dependent upon the nutrient solution, they would perish when the solution was exhausted.

It must be recognized that even the worldwide reserves of oil shale and tar sand are finite, and that other energy sources must be developed in the future. Even the most optimistic projections of Canadian oil sand production are far less than U.S. oil needs. As of 2011, Canadian oil sands produced about 1.1 million barrels per day, almost half of Canadian oil production but only about 5 percent of U.S. energy needs.

David M. Best

FURTHER READING

Carrigy, M. A. "New Production Techniques for Alberta Oil Sands." *Science* 234 (December, 1986): 1515-1518. A succinct, fact-filled discussion of tar sand production in Canada. Suitable for high school readers.

De Nevers, Noel. "Tar Sands and Oil Shales." *Scientific American* 214 (February, 1966): 21-29. A good review of the status of the development of these deposits in the earlier days of production. Compares the Green River formation oil shale fields and the tar sand sources in the Athabasca tar sand area of Alberta.

Duncan, Donald C., and Vernon E. Swanson. *Organic-Rich Shale of the United States and World Land Area.* U.S. Geological Survey Circular 523. Washington, D.C.: Government Printing Office, 1965. A classic work that provides much of the present-day knowledge of worldwide deposits of oil shale, in terms of both locations and reserves. Highly recommended as a key reference.

Gray, G. Ronald. "Oil Sand." In *McGraw-Hill Encyclopedia of the Geological Sciences,* edited by S. P. Parker. 2d ed. New York: McGraw-Hill, 1988. Discusses several geological factors that control the environments of formation of tar sands, providing specific information on worldwide reserves. Good source of data for those desiring to make comparisons of various areas.

Hughes, Richard V. "Oil Shale." In *The Encyclopedia of Sedimentology,* edited by R. W. Fairbridge and J. Bourgeois. New York: Springer, 1978. A brief discussion of the environments and characteristics of selected oil shale deposits. More than a dozen technical references are provided for those readers looking for the scientific basis and interpretation of oil shales.

Hyne, Norman J. *Dictionary of Petroleum Exploration, Drilling, and Production.* Tulsa, Okla.: PennWell, 1991. Covers all terms associated with petroleum and the petroleum industry. A great resource for the beginner in this field. Illustrations and maps.

_____. *Nontechnical Guide to Petroleum Geology, Exploration, Drilling, and Production.* 2d ed. Tulsa, Okla.: PennWell, Corporation, 2001. Provides a well-rounded overview of the processes and principles of gas and oil drilling. Discusses exploration, refining, and distribution. Covers foundational material. Appropriate for oil-drilling professionals, geologists, or students of similar fields. Illustrations, bibliography, and index.

Kelland, Malcolm A. *Production Chemicals for the Oil and Gas Industry.* Boca Raton, Fla.: CRC Press, 2009. Discusses environmental issues concerning the oil and gas industry. Focuses on the chemicals and reactions involved in extraction, refinement, and other treatments during production. Includes references, appendices, and index.

Lancaster, David E., ed. *Production from Fractured Shales.* Richardson, Tex.: Society of Petroleum Engineers, 1996. Covers oil reservoir engineering and oil shale reserves in the United States.

Loucks, Robert. *Shale Oil.* Bloomington, Ind.: Xlibris Corporation, 2002. A concise overview of the oil industry through the 1970's and 1980's. Includes both government and private industry data. May seem controversial to some readers. Provides support for the author's view of the shale oil industry and suggestions for the future of the industry and ecological conservation.

Nikiforuk, Andrew. *Tar Sands: Dirty Oil and the Future of a Continent.* Vancouver: Greystone Books, 2010. Discusses the use of tar sands in Canada, and political, social, and environmental issues related to the Alberta tar sands. While not scientifically or objectively written, this book does provide a perspective on an ongoing debate.

Smith, John W. "Synfuels: Oil Shale and Tar Sands." In *Perspectives on Energy, Issues, Ideas, and Environmental Dilemmas,* edited by L. C. Ruedisili and M. W. Firebaugh. 3d ed. New York: Oxford University Press, 1982. Discusses the definitions, characteristics, resources, and production of oil shales and tar sands in a well-outlined presentation. Reference list provides more sources to readers desiring either general or technical details.

Serves as an excellent starting point for someone researching the topic. Suitable for high school students.

Smith, John W., and Howard B. Jensen. "Oil Shale." In *McGraw-Hill Encyclopedia of the Geological Sciences*, edited by S. P. Parker. 2d ed. New York: McGraw-Hill, 1988. Includes several figures showing oil shale deposits in the United States. Explains the technical aspects of oil shale properties. An extensive bibliography lists more than a dozen technical references. Suitable for high school readers and beyond.

Snape, Colin. *Composition, Geochemistry, and Conversion of Oil Shales.* Boston: Kluwer Academic Publishers, 1995. Published in cooperation with the North Atlantic Treaty Organization (NATO) Scientific Affairs Division. Provides an overview of oil shale properties and production.

Yen, T. F., and G. V. Chilingarian, eds. *Oil Shale.* New York: Elsevier, 1976. Gathers twelve chapters authored by experts in the various facets of oil shale studies. Some articles are suitable for high school and college students; others are very specific and are suitable for researchers. Includes numerous references with each chapter.

See also: Biogenic Sedimentary Rocks; Offshore Wells; Oil and Gas Distribution; Oil and Gas Exploration; Oil and Gas Origins; Oil Chemistry; Onshore Wells; Petroleum Reservoirs; Well Logging.

ONSHORE WELLS

Several million wells have been drilled in the exploration for oil and gas within every environment of the earth. Testing, completion, and evaluation technologies of a completed well are similar whether the borehole is dug by the cable-tool or by the rotary method. Downhole, wire-line analyses have evolved from the simple electric log of the 1920's to a modern array of evaluative services.

PRINCIPAL TERMS

- **cable-tool drilling:** a repetitive, percussion process of secondary use in the boring of relatively shallow oil and gas wells
- **completion procedures:** all methods and activities necessary in the preparation of a well for oil and gas production
- **downhole tool:** a drill bit and motor mounted on the end of the drill string; fluid pumped into the drill string drives the downhole tool, and the drill string is not rotated
- **drilling fluids:** a carefully formulated system of fluids used to lubricate, clean, and protect the borehole during the rotary drilling process
- **drilling rig:** the collective assembly of equipment, including a derrick, power supply, and draw-works, necessary in cable-tool and rotary drilling
- **drill string:** the length of steel drill pipe and accessory equipment connecting the drill rig with the bottom of the borehole
- **hydrocarbons:** naturally occurring organic compounds that in the gaseous state are termed natural gas and in the liquid state are termed crude oil or petroleum
- **rotary drilling:** historically, the principal method of boring a well into the earth using a fluid-circulating, generally diesel-electric generated, rotating process
- **well log:** a graphic record of the physical and chemical characteristics of the rock units encountered in a drilled borehole

CABLE DRILLING

The location of the first drilling operation is lost to history, although it is known that the Chinese drilled for brine and water two thousand years ago using crude cable-tool methods. Similar methods of drilling were still being employed in the 1850's. By this process, a well is created by raising and lowering into the borehole a heavy metal bit suspended from a cable or rope. Gradually, the bit will pound its way through the rocks. With the addition of a jar, a mechanical device that imparts a sharp vertical stress to the bit, the process is greatly improved. Surface equipment, contained within a wooden derrick, or rig, was commonly steam driven and repeatedly withdrew the bit from the hole, allowing it to be again dropped to the bottom of the well. As the bottom of the hole fills with rock chips, a bailer is periodically used to remove this debris.

Cable drilling is a slow process. Its greatest advantage is easy identification of oil- and gas-producing rock units. Because minimal drilling fluids are used, the uncontrolled surface flows of encountered hydrocarbons (occurrences known as blowouts) are frequent. For this reason, cable drilling is most applicable within depths of 1,000 meters. As late as 1920, cable-tool rigs drilled as many as 85 percent of all wells completed in the United States.

ROTARY DRILLING

Introduced to the industry at Corsicana, Texas, in 1895, the rotary method was used to drill 90 percent of American wells in the 1950's. In rotary drilling, the drill bit is attached to connected sections of steel pipe, or drill string, and lowered into the borehole. Pressure is placed on the bit and the drill pipe is rotated, causing the bit to grind against the bottom of the borehole. In contrast to the cable method, new borehole depths are created by the rock being torn rather than pounded. When the drill bit becomes dull, the drill string is removed from the borehole, disassembled, and stacked within the tall mast, or derrick. A new bit is attached, and the drill string is reassembled.

The application of a drilling fluid system is a key element of the rotary method. Originally ordinary mud, drilling fluids have become a carefully formulated solution of water, clays, barite, and chemical additives. These fluids are circulated under pressure down the center of the drill string, extruded through the drill bit, and pumped back to the surface through the space between the drill string and the borehole.

These fluids serve several important functions: to lubricate and cool the drill bit, remove rock chips from the borehole, and protect the borehole from dangerous blowouts. Because of its mechanical advantages, the rotary method is approximately ten times faster than the cable method in drilling a borehole.

After the borehole is completed, and assuming commercial deposits of oil or gas are discovered, completion procedures are initiated. Because surface instruments cannot detect the presence of subsurface hydrocarbons, the rock units exposed in the wall of a borehole must be evaluated for the presence and quality of contained oil and gas. A preliminary analysis is conducted on the rock chips continuously brought to the surface by the drilling fluid system. A key component of this analysis is the identification of microscopic fossils, nicknamed "bugs," which indicate the age of the rock layers being drilled. The rock chips are far too small to preserve most fossils, but microscopic fossils survive the drilling process. Later, instruments lowered into the borehole determine the physical and chemical characteristics of the penetrated rocks and their contained fluids and gases. Should the presence of hydrocarbons be indicated, further testing is conducted to determine the economic value of the discovery. Finally, if economic payout is indicated, the borehole undergoes final completion procedures. Special production tubing systems are installed, and the oil or gas is pumped, or flows under its own pressure, from the rocks up the borehole and into a pipeline or surface storage system.

Rotary drilling procedures vary little with geographic location or climate. In urban areas, the derrick is covered with soundproof material and sometimes even disguised for aesthetic purposes. In sensitive areas, such as arctic regions and offshore operations, safety and environmental preservation precautions are mandated by state and federal law.

OTHER DRILLING METHODS

With the turbodrill method, or the downhole tool method, the drill string remains stationary while the drill bit rotates under the influence of circulating drilling fluid. Drilling very straight boreholes with minimum mechanical problems, this process excels at directional drilling, especially horizontal drilling. It also greatly reduces wear on the drill stem pipe. The use of a downhole tool is replacing the rotary drilling method. Although a rotary drill rig is still installed, its rotary capability is not used. The hammer drill, a combination of slow rotary motion coupled with percussion impact, produces a faster rate of rock penetration, but this method has not been widely accepted into practice. Experiments with vibration and sonic drills have proved unsuccessful or uneconomic thus far.

HISTORY OF A BOREHOLE

After the location for a borehole is determined, rotary drilling equipment is taken to the chosen site, an area of about 0.5 to 1 hectare. When the drill rig is assembled, sections of drill pipe, or drill string, are connected within the derrick. The drilling fluid hose is connected to the upper end of the drill string, while a drill bit is attached to the bottom. The rig is now ready to "make hole." The history of a borehole begins with its "spud-in" time, that moment when the ground is broken by the rotating bit.

The rotary table, located in the center of the rig floor and connected to powerful engines, rotates the drill string and attached bit. As the bit rotates, drilling fluid (termed mud) is pumped down the inside of the drill string and through openings in the bit. The density of this mud is carefully controlled so that as it exits the bit, it is capable of lifting rock fragments, or cuttings, from the bottom of the borehole, allowing the bit to rotate against a fresh rock surface. The drilling mud, with its contained cuttings, is circulated up the annulus (passage) between the wall of the borehole and the outside of the drill string. At the surface, cuttings are separated by flowing the drilling mud through a vibrating sieve. Periodically, a sample of cuttings is collected for geologic analysis. Finally, the cleansed mud circulates through the "mud pit," where, after cooling to surface temperature, it is pumped through the drilling fluid hose back into the drill string. While the borehole is being drilled, this mud system is in continuous circulation. Every 9 meters approximately, as the borehole becomes deeper, a new section of drill pipe is added to the drill string, increasing the depth capability of the rig.

At shallow depths, where the bit is penetrating loose soils and poorly consolidated rock formation, the drilling speed is measured in tens to hundreds of meters per day. With increased depths, penetration rates will diminish to as little as a meter per day, depending on rotation pressure and velocity and rock characteristics. At the surface, "conductor pipe" is driven 6 to 10 meters into the ground to protect the borehole against collapse. At depths below the conductor pipe, rock units containing fresh water are protected from drilling fluid contamination by lowering "surface casing" through the conductor pipe and into the borehole and injecting cement to hold it in place. At greater depths, progressively smaller radius "intermediate casing" may be cemented into the borehole, keeping the newly drilled hole open while sealing off unusually high-pressure or unusually low-pressure rock strata. Because each new series of casing must fit into the prior cemented casing, the borehole diameter becomes smaller with increased depth. When the borehole reaches programmed total depth (TD), the drilling process is complete. The next phase of activity involves testing for the presence and quantity of oil and gas.

The borehole is protected by cementing "production casing" through the depth of the production zone. Perforating guns, multibarrel firearms designed to fit into the borehole, are lowered to the target production depth and fired electrically. High-velocity bullets penetrate the casing cement and become embedded in the rock strata, creating pathways through the strata to the wall of the borehole. In some cases the rocks are fractured by explosives or high-pressure fluids. The holes are prevented from collapsing under the weight of the overlying rocks by injecting coarse sand into the holes. Oil or gas emitting from the rock through these pathways flows into installed production tubing and to the surface, where the hydrocarbons are either temporarily stored or directed to a nearby pipeline. At this point, the well is completed and "on line."

TESTING AND ANALYSES

Drill cuttings are periodically collected from the drilling fluid and analyzed in the field in converted mobile-home vehicles. These field tests determine rock type, contained minerals, density, pore space percentage, and association with either natural gas or crude oil (petroleum). Since drilling is an expensive operation, commonly costing millions of dollars, the majority of boreholes are subjected to additional analyses, termed well logging. Conducted by contracted specialists, well-logging operations involve lowering an elongated instrument called a sonde to the total depth of the borehole. As the sonde is slowly

pulled up the hole, it records various characteristics of the rocks exposed within the wall of the borehole and their contained fluids and gases. These characteristics, which include electrical resistivity, conductivity, radioactivity, acoustic properties, and temperature, are transmitted to the surface, where they are recorded and filed for future use. A basic property is the diameter of the borehole, which indicates the hardness or softness of the layers. Logging the rate of drilling also indicates the hardness or softness of the layers. It is common for four or five different logs to be recorded, while on a very important borehole, more than twice this number may be taken.

In the office, individuals trained in geology and engineering study the cuttings analyses and logging data and determine the presence and economic extent of oil or gas by calculating rock porosity, permeability, density, thickness, lateral extent, inclination, and pressure at various depths in the borehole. Should these analyses be pessimistic, the borehole is declared "dry and abandoned" and permanently sealed at several depths by cement plugs. Such is the fate of approximately six out of seven boreholes drilled in frontier (new) geographic regions or to unproven depths; such boreholes are termed wildcat wells. For the one in seven wildcat wells in which logging analyses indicate a chance of success, verification analyses in the form of drill-stem testing (DST) will be conducted.

Drill-stem testing equipment is attached to the base of the drill string and lowered to the rock depth to be tested. After this depth is physically isolated from the rest of the borehole, assuring a valid test, the DST tool is activated, allowing fluids or gases contained within the isolated rocks to flow into the drill string and to the surface. From DST, rock pressures and flow capacities are calculated. When DST verifies positive economic results determined by logging analyses, the commercial quantities of either oil or gas, or both, are declared, and the well is prepared for its final completion phase.

ECONOMIC AND POLITICAL CONSIDERATIONS

The fortunes of the American oil and gas drilling industry are closely tied to the market value of a barrel (42 U.S. gallons, or about 159 liters) of crude oil. As that value increases, so generally does the number of drilling rigs under contract. Adding confusion to this economy-to-rig-use relationship are such considerations as international politics, governmental policies, environmental concern, and marketplace competition for high-risk investment dollars. After a century and a half of drilling wells in the search for new reserves of oil and gas, terrestrial portions of the United States are considered a mature exploration province. The chances of discovering large new reserves of hydrocarbons on land are very small. The future lies in drilling within the offshore provinces (deep water of the Gulf of Mexico and the Atlantic eastern seaboard) and environmentally protected regions (northern Alaska and national parks and forestlands). In order to maintain a hydrocarbon-based energy economy while reducing dependence upon foreign hydrocarbons, oil and gas well-drilling and production programs may have to take place in these frontier exploration regions. Such programs must be governed by consensus regulatory, environmental, and economic policies until solar, nuclear, or some unforeseen resource assumes the dominant energy position and oil and gas wells no longer need be drilled into the earth.

Albert B. Dickas

FURTHER READING

Allaud, Louis A., and Maurice H. Martin. *Schlumberger: The History of a Technique.* New York: John Wiley & Sons, 1977. A historical account of the mineral-prospecting methodology invented in 1912 by Conrad Schlumberger and used in modified form in the evaluation of the majority of oil and gas wells drilled throughout the world.

Gray, Forest. *Petroleum Production for the Nontechnical Person.* Tulsa, Okla.: PennWell, 1986. Written for industry professionals working to master the technology and its terminology. Each chapter is accompanied by a series of exercises. Includes a detailed glossary.

A Guide to Practical Management of Produced Water from Onshore Oil and Gas Operations in the United States. Interstate Oil and Gas Compact Commission and ALL Consulting, 2006. Discusses regulations and industry trends of onshore wells. Examines water wells and oil wells, and covers oil and gas basins.

Hyne, Norman J. *Dictionary of Petroleum Exploration, Drilling, and Production.* Tulsa, Okla.: PennWell, 1991. Covers all terms associated with petroleum and the petroleum industry. A great resource for the beginner in this field. Illustrations and maps.

_____. *Nontechnical Guide to Petroleum Geology, Exploration, Drilling, and Production.* 2d ed. Tulsa, Okla.:

PennWell, Corporation, 2001. Provides a well-rounded overview of the processes and principles of gas and oil drilling. Suitable for professionals in the oil drilling field, geologists, or students of similar fields. Discusses the exploration, refining, and distribution. Illustrations, bibliography, and index.

Kennedy, John L. *Fundamentals of Drilling.* Tulsa, Okla.: PennWell, 1982. A basic presentation on oil and gas well drilling. Details the tools and methods used, and includes sections on economics and future trends, as well as an introduction to the industry.

Langenkamp, Robert D. *Oil Business Fundamentals.* Tulsa, Okla.: PennWell Books, 1982. Appropriate for those interested in learning business perspectives of oil and gas well drilling. Includes chapters on ownership rules, drilling, financing, and the marketing of hydrocarbons.

Nardone, Paul J. *Well Testing Project Management: Onshore and Offshore Operations.* Burlington, Mass.: Gulf Professional Publishing, 2009. Provides a useful overview of the rules, regulations, and common practices of the oil industry. Discusses well site operations and safety. Explains the well test planning process. An excellent guide for any persons involved in oil drilling.

Raymond, Martin S., and William L. Leffler. *Oil and Gas Production in Nontechnical Language.* Tulsa, Okla.: PennWell, 2005. Provides a good overview of the industry. Describes oil explorations, and how oil and gas are found and extracted from the earth. Appropriate for general readers. Includes pictures, charts, graphs and drawings.

Welker, Anthony J. *The Oil and Gas Book.* Tulsa, Okla.: PennWell, 1985. Written to bridge the communication gap between the oil and gas industry and members of the general public, such as bankers, investors, and newspersons. Covers such topics as partnerships, joint ventures, promotions, and working interests.

See also: Biogenic Sedimentary Rocks; Mining Processes; Mining Wastes; Offshore Wells; Oil and Gas Distribution; Oil and Gas Exploration; Oil and Gas Origins; Oil Chemistry; Oil Shale and Tar Sands; Petroleum Industry Hazards; Petroleum Reservoirs; Well Logging.

ORBICULAR ROCKS

Orbicular rocks are a type of intrusive igneous rock deposit, occurring primarily at the outer margin of plutonic dikes, which are cooled from magma. Orbicular rocks contain inclusions called orbicules that are composed of concentrated minerals layered in concentric spheres around a central core. Most orbicular deposits have been uncovered from geologic strata in Finland, though they have been found on every continent.

PRINCIPAL TERMS

- **biotite:** type of layered, silicate mineral composed of silica molecules bonded with magnesium and iron and appearing in dark brown and black varieties
- **carbonatite:** silicate mineral consisting of reduced levels of silica and rich in carbonate molecules formed from carbonic acid
- **dike:** type of plutonic rock formation consisting of cooled magma oriented perpendicular to the earth's surface and generally proceeding at a diagonal through the layers of surrounding parent rock
- **felsic:** mineral group characterized by high proportions of silicate minerals, including quartz, feldspar, plagioclase, and micas
- **horneblende:** type of silicate rock characterized by its dark color, which is the result of inclusions of iron and magnesium within the rock matrix
- **mafic:** igneous deposits of rocks consisting of silica enriched with high levels of iron, magnesium, sodium, and calcium and with low levels of potassium-rich feldspar
- **orbicule:** formation within orbicular igneous rocks consisting of similar materials to the parent rock; organized in concentric layers of crystals surrounding a core comprising smaller crystal clusters, larger individual crystals, or fragments of other rock types
- **petrology:** branch of geology that studies the formation and classification of rock types
- **plagioclase:** silicate mineral formed from molecules of silica bonded with sodium, aluminum, and calcium
- **plutonic:** rock formed from magma that cools in pockets beneath the earth's surface

OCCURRENCE OF ORBICULAR ROCKS

Orbicular rocks are igneous rocks that contain orb-like formations as inclusions within the rock body. Igneous rocks form when magma generated within the mantle of the earth rises toward the surface and cools to form solid structures. Orbicular rocks form within intrusive or plutonic igneous rock bodies, which are deposits that form when magma solidifies beneath the earth's surface rather than breaking through the crust to become lava.

Orbicular rocks can form within a variety of different types of intrusive rock deposits and most commonly form from magma that is high in silica, a molecule composed of silicon and oxygen that makes up the vast majority of most igneous rock. Orbicular rocks have been found in dikes, which are intrusive rock formations that form when magma rises toward the surface perpendicular to the existing rock, forming a horizontal or vertical wedge within the parent rock.

The vast majority of orbicular rocks come from small dikes, and orbicular sections of the deposit are generally found to occur at the margin of the igneous deposit. Orbicular rocks generally occur only within portions of an igneous deposit that are 3 to 5 meters (10 to 16.5 feet) in width and less than 5 m in depth. Many orbicular rocks have been discovered in isolated boulders that are the remains of ancient igneous dikes that have eroded and split into individual rocks.

In rare occasions, orbicular rocks have been found in sills, which are horizontal deposits of igneous rock that form within layers of the parent rock and are generally oriented parallel to the earth's crust. Alternatively, orbicular rocks may be part of larger intrusive deposits called batholiths, which result from the combination of subterranean magma pockets and that may extend for many kilometers beneath the earth's surface.

Orbicular rocks have been found in more than one hundred locations worldwide, but they are a relatively rare type of igneous formation; the exact geologic setting required for orbicular rock formation is still poorly understood. The vast majority of orbicular rocks have been uncovered in igneous deposits from Europe, specifically in Finland. This is not believed to represent a unique feature of Finland's geologic environment but rather is the result of dedicated efforts by Finnish geologists to complete detailed surveys of that country's mineralogical terrain. Significant quantities of orbicular rocks also have been uncovered from igneous deposits in the United States and in Indonesia.

CHEMICAL COMPOSITION OF ORBICULAR ROCKS

In general, the chemical petrology of orbicular rocks is similar to that of the igneous formations in which the rocks have been discovered. Morphological variations in the orbicular deposits are therefore related to the concentration and depositional orientation of the molecules within the formation, rather than to the presence of unique molecules not found in the surrounding parent rock.

Igneous rocks are classified according to both the chemical structure of the molecules found within the rock and the overall size of grains within the rock. Granitic rock is one of the most common types of igneous rock that contains orbicular inclusions. Granitic rock has visible grains and is composed largely of minerals derived from the molecule silica. Granitic rocks are classified as felsic, which are silicate rocks containing large quantities of feldspar, a common type of silicate mineral. Felsic rocks also contain other types of silica-based minerals, such as quartz and micas. Granitic orbicular rocks are the most common type of orbicular inclusion.

Orbicular deposits also can be found in mafic rocks, which are igneous rocks that contain a higher proportion of heavier elements, such as magnesium and iron. Unlike felsic rocks, mafic rocks contain lower levels of potassium but have higher levels of calcium and sodium along with iron and other metal deposits. Some igneous rocks are ultramafic, meaning they contain less than 45 percent silica, with the rest of the rock consisting largely of heavier elements like iron. In one rare occurrence, orbicular inclusions formed within a sample of carbonatite found in Finland. Carbonatite is a type of igneous rock containing high proportions of carbonate minerals, a molecular salt formed from carbonic acid.

Most orbicular rock deposits that are found are made of feldspar, one of the most common types of rock-forming minerals, accounting for more than 60 percent of the earth's crust. Feldspar is composed largely of silica molecules, with such elements as potassium, calcium, and sodium mixed into their matrix. Feldspar samples tend to be either rich in potassium, called K-feldspars, or rich in sodium or calcium.

Plagioclase is a type of feldspar rich in either calcium or sodium, with low potassium content. Plagioclase is one of the most common types of rock and is found in a variety of igneous deposits. Orbicular rocks also may contain quartz, which is a crystalline variety of silica that generally forms in superheated aqueous environments or within magma pockets.

Orbicular rocks contain inclusions of dark minerals within their matrices, which may represent different mineral types. Among the most common mineral inclusions in orbicular rocks is hornblende, which is a dark form of amphibole, a type of mineral consisting of silica molecules joined to either iron or magnesium atoms. Horneblende occurs in two basic varieties—one that is rich in iron, called ferrohornblende, and one that is rich in magnesium, called magnesiohorneblende.

Alternatively, orbicular rocks may contain biotite, which is a type of silicate rock from the group of minerals known as micas. The dark color of biotite is derived from iron atoms bonded to the silicate molecules within the rock. Biotite also may contain sheets of magnesium and aluminum atoms within the rock, further contributing to the mineral's dark color.

STRUCTURE OF ORBICULAR ROCKS

Orbicular plutonic rocks contain orb-like nodules of mineral called orbicules, which contain layers of minerals in a radial array surrounding an internal core, sometimes called the nucleus. The outer portion of an orbicule consists of concentric circles of minerals called shells.

The number of shells within an orbicule varies, and some orbicules may have only a single shell while others may have hundreds of shells surrounding the core. Between the shells are layers of minerals called subshells, which contain molecules similar to those of the surrounding rock but that may be organized differently.

The central core of an orbicule may consist of a cluster of small crystals, usually composed of quartz or a similar silica-rich mineral. Alternatively, some orbicules have cores consisting of one or a few large crystals, again generally derived from the quartz family of minerals. Some orbicular cores consist of fragments of parent rock, called xenoliths, which may differ from the type of rock contained within the remaining part of the orbicule, including the minerals of the shells and subshells.

Outer shells are usually composed of feldspar or plagioclase and therefore appear as lighter segments of rock. The portion between shells consists of plagioclase or granite with some darker minerals mixed into the matrix, making these areas somewhat darker than the outer shells. Subshells generally consist of plagioclase bonded to horneblende or biotite, making them darker than the outer shells or the area between shells. The typical orbicule therefore consists of alternating bands of light- and dark-colored rock surrounding the inner core, which tends to be light in color and to consist of quartz crystals or other silicate minerals. Many orbicule cores contain small xenoliths that generally appear as dark flecks or nodules within the crystals of the core.

The size of orbicules is related to the type of igneous rocks in the surrounding deposit. Large orbicules tend to develop within felsic rocks containing high proportions of quartz and feldspar. Orbicules in this type of sediment can range from 10 to 40 centimeters (4 to 16 inches) in diameter. Mafic igneous deposits generally contain smaller orbicules; similarly, ultramafic rocks contain the smallest and least developed orbicules.

FORMATION OF ORBICULES

Several hypotheses explain the origin of orbicules within igneous formations. Initially, the bands of various minerals were thought to be related to properties of liquid diffusion that affect the formation of mineral layers during the solidification process.

An alternative theory, developed in the 1970's, sees orbicules as the result of xenoliths of other rock types or crystals becoming part of (included in) a flow of magma. When this occurs, the magma surrounding the included segments cools such that "comb layering" occurs, in which crystals form like the teeth of a comb perpendicular to the surface of the inner core. Alternating layers within the orbicules are then seen as the result of successive periods of cooling, leading to comb layering of minerals around the forming nodule.

While comb layering appears to play a role in the formation of the shells and subshells contained within orbicules, chemical investigation has found that not all orbicules contain xenoliths within their core. Theories of orbicule genesis focus on the phenomenon of undercooling, which occurs when a sample of magma rapidly cools to temperatures significantly below those needed for solidification. In undercooling, crystals may form within the body of the magma and serve as attractors for layers of cooling material that form comb layers around the crystal structure. As this process continues, chemical properties within the mineral components cause them to layer into discrete segments of different types of minerals.

The orientation of orbicules within the matrix of an igneous deposit indicates that orbicules form while the surrounding rock is still cooling and that orbicules can shift within the rock matrix after formation. In some cases orbicules have been found where the outer shell has broken apart, perhaps after encountering disruptive contact with other orbicules as the inclusions shifted within the pliable rock matrix.

In general, the processes underlying the formation of orbicules is still an area of concerted research. The precise physical and chemical conditions required for the formation of orbicules appear to be relatively narrow in scope, as orbicular rock formations are rare among igneous deposits. There appears to be a narrow range of temperatures and pressures during which orbicule formation is possible, which explains why orbicules are relatively rare and why they tend to form only in certain portions of a developing dike, where temperatures fall within this range.

USES OF ORBICULAR ROCKS

Orbicular rocks contain attractive patterns of light and dark concentric circles, so they have been prized by rock collectors and are often used as decorative stones. Some orbicular rocks have been used as an alternative to marble for making counters and tabletops. A variety of collectible minerals can be made from polished and cut orbicular rocks.

Because of their rarity, orbicular rocks cannot be collected by amateurs in most locations in which they are found. In Finland, for instance, where the largest variety of orbicular rocks have been found, provisions prohibit removing orbicular specimens from rock quarries, thereby preserving the remaining orbicular deposits for geologic research.

Micah L. Issitt

FURTHER READING

Dahl, P. S., and D. F. Palmer. "The Petrology and Origin of Orbicular Tonalite from Western Taylor Valley, Southern Victoria Land, Antarctica." In *Antarctic Earth Science*, edited by R. L. Oliver, P. R. James, and J. B. Jago. New York: Cambridge University Press, 2011. Examines the petrology or orbicules found in deposits of tonalie uncovered in Antarctica. Discusses the theories used to explain the origin of orbicular rocks worldwide.

Fenton, Carol Lane, and Mildred Adams Fenton. *The Rock Book*. Mineola, N.Y.: Dover, 2003. A revised classic text from 1940 that discusses the variety of rocks and minerals accessible to amateurs. Provides a brief description of orbicular rock deposits found in the United States.

Gill, Robin. *Igneous Rocks and Processes: A Practical Guide*. Malden, Mass.: John Wiley & Sons, 2011. Detailed advanced text providing information on igneous rock formations and the structure of igneous rocks. Discusses comb layering and its relation to the structure of orbicular rocks.

Lahti, Seppo L., Paula Raivio, and Ilka Laitakari, eds. *Orbicular Rocks in Finland*. Espoo: Geological Survey of Finland, 2005. Collection of articles relating to the study of orbicular rocks in Finland. Contains information on deposition, petrology, occurrence, and measures to protect existing samples.

Vernon, Ron H. *A Practical Guide to Rock Microstructure*. New York: Cambridge University Press, 2004. Intermediate-level text detailing the analysis and classification of rock types based on microstructural components. Contains a discussion of orbicular rock deposits and the theories proposed to explain the formation of orbicules.

See also: Crystals; High-Pressure Minerals; Hydrothermal Mineralization; Igneous Rock Bodies; Plutonic Rocks; Quartz and Silica; Rocks: Physical Properties; Silicates.

OXIDES

Oxides represent one of the most important classes of minerals. They are the source of several key metals upon which the world is dependent; these metals include iron, aluminum, titanium, manganese, uranium, and chromium. Products manufactured from these metals touch virtually every aspect of modern living.

PRINCIPAL TERMS

- **crust:** the outer layer of the earth; it extends to depths of 5 kilometers to at least 70 kilometers, and is the only layer of the earth directly accessible to scientists
- **igneous rock:** a major group of rocks formed from the cooling of molten material on or beneath the earth's surface
- **metal:** an element with a metallic luster and high electrical and thermal conductivity; it is ductile, malleable, and of high density
- **metamorphic rock:** a major group of rocks that are formed from the modification of sedimentary or igneous rocks by elevated temperatures and/or pressures beneath the earth's surface
- **mineral:** a naturally occurring, solid chemical compound with a definite composition and an orderly internal atomic arrangement
- **ore:** mineral or minerals that, in a given deposit, contain large enough amounts of a valuable constituent to be worth mining for the metal(s) at a profit
- **quartz:** a very common silicate mineral
- **rock:** an aggregate of one or more minerals
- **sedimentary rock:** a major group of rocks formed from the breakdown of preexisting rock material, or from the precipitation of minerals by organic or inorganic processes

MINERAL CLASSIFICATION

Compositionally, the earth contains eighty-eight elements, but only eight of these constitute more than 99 percent (by weight) of the crust. These eight elements combine to form some two dozen minerals that make up more than 90 percent of the crust. Within the rocks and minerals of the crust, the element oxygen accounts for almost 94 percent of the total volume. On the atomic scale, that implies that most minerals are virtually all atoms of oxygen, with the other elements filling in the intervening spaces in orderly arrangements. Most minerals form because ions (atoms that have gained or lost one or more electrons) become mutually attractive. More precisely, ions with positive charges (cations), which have lost electrons, become attracted to ions with negative charges (anions), and if the charges are balanced and several rules of crystal chemistry are satisfied, a mineral will form. A simple example is the combination of the sodium cation and the chlorine anion in forming the mineral halite.

Ionic combinations include complex anions (radicals) that are strongly bound cation and anion groupings. These radicals take on a negative charge and will attract more weakly bound cations. Silica and carbonate are examples of these complex anions. These anion radicals and simple anions form the basis for one of the most widely used classification systems of minerals, with the following classes of minerals recognized: the native elements; sulfides and sulfosalts; oxides and hydroxides; carbonates; halides; nitrates; borates; phosphates, arsenates, and vanadates; sulfates and chromates; tungstates and molybdates; and silicates.

Although this classification is based entirely on chemical composition, subdivisions within these classes are based on both structural and additional chemical criteria. Of these eleven classes, the silicates dominate the crust, forming approximately 97 percent of this layer. Although all the other classes represent only 3 percent of the crust, these classes include the majority of the minerals that society has come to depend upon. In economic value alone, the oxides undoubtedly rank at or near the top of all the classes of minerals, including the silicates.

AO, A_2O, AND A_2O_3 OXIDES

Oxides are those minerals that have oxygen combined with one or more metals. Generally, the oxides are subdivided according to the ratio of the number of metals to the number of oxygens in the formula. Many of the oxides have relatively simple metal (A) to oxygen (O) ratios, so the following categories are recognized: AO, A_2O, AO_2, and A_2O_3. Some oxides have atomic structures in which different metals occupy different atomic (structural) sites. These minerals are commonly referred to as complex or multiple oxides, and most have the general formula AB_2O_4, where A and B are separate atomic sites.

In both the AO and A_2O oxides, there are no minerals that are considered common, although periclase (MgO) occurs in some metamorphic rocks. In the AO_2 oxides, however, several minerals are important, with two subdivisions recognized: the rutile group and the individual mineral uraninite. The three important minerals that occur in the rutile group are rutile (TiO_2), cassiterite (SnO_2), and pyrolusite (MnO_2). Rutile is a common minor mineral in a wide variety of quartz-rich igneous and metamorphic rocks. It is also found in black sands along with several other oxides. Pyrolusite is a very widespread mineral found in manganese-rich nodules on the floors of the oceans, seas, lakes, and bogs and is a major ore of manganese. Cassiterite is a common minor constituent in quartz-rich igneous rocks and is the principal ore of tin. Uraninite (UO_2) is a separate AO_2 oxide that, like cassiterite and rutile, is characteristically associated with quartz-rich igneous rocks. In several places it also occurs with gold in fossil stream deposits modified by metamorphism.

Even more common than the AO_2 oxides are the A_2O_3 oxides, which include the common minerals hematite (Fe_2O_3), corundum (Al_2O_3), and ilmenite ($FeTiO_3$). Hematite, the most widespread iron oxide mineral in the crust, occurs in a wide variety of conditions, such as metamorphic deposits, quartz-rich igneous rocks, and sedimentary rocks. Corundum is a widespread minor constituent in metamorphic rocks low in silicon and relatively rich in aluminum. It is also found in igneous rocks that have low silicon contents. Ilmenite is another mineral that typically occurs in small amounts in many types of igneous rocks and in black sands with several other oxides.

COMPLEX OXIDES AND HYDROXIDES

In the more complex oxides (AB_2O_4), the most common minerals occur in the spinel group, but the individual mineral columbite-tantalite (Fe,Mn)(Nb,Ta)$_2O_4$ is also commonly given consideration. The spinel group contains many minerals that have complex interrelationships. Spinel ($MgAl_2O_4$) is common in some metamorphic rocks formed at high temperatures and in igneous rocks rich in calcium, iron, and magnesium. Magnetite (Fe_3O_4) is a common minor mineral in many igneous rocks.

Relatively resistant to weathering, it also occurs in black sands and in very large sedimentary banded iron deposits. Chromite ($FeCr_2O_4$), another important mineral in the spinel group, is found only in calcium-poor and iron- and magnesium-rich igneous rocks. Unlike members of the spinel group, columbite-tantalite is a separate subdivision of the AB_2O_4 oxides because of its different atomic structure. Like many oxides, it is characteristically associated with quartz-rich igneous rocks.

The hydroxides are a group of minerals related to the oxides that include a hydroxyl group $(OH)^{-1}$ or the water (H_2O) molecule in their formulas. These minerals tend to have very weak bonding and, as a consequence, are relatively soft. Five hydroxides are briefly considered here: brucite ($Mg(OH)_2$), manganite ($MnO(OH)$), and three minerals in the goethite group (diaspore, $AlO(OH)$; goethite, $FeO(OH)$; and bauxite, a combination of several hydrous aluminum-rich oxides including diaspore). Brucite occurs as a product of the chemical modification of magnesium-rich igneous rocks and is associated with limestones. Manganite tends to occur in association with the oxide pyrolusite. Of the minerals in the goethite group, diaspore commonly is associated with corundum and occurs in aluminum-rich tropical soils. Goethite, an extremely common mineral, occurs in highly weathered tropical soils, in bogs, and in byproducts of the chemical weathering (breakdown) of other iron oxide minerals. Bauxite is also found in highly weathered tropical soils.

STUDY OF HAND SPECIMENS

There are several general approaches to the study and identification of the oxide minerals. They include the study of hand specimens, optical properties, the internal atomic arrangements and their external manifestations (crystal faces), chemical compositions, and the synthesis of minerals. In the study of hand specimens, minerals possess a variety of properties that are easily determined or measured. These properties both aid in identification and may make the minerals commercially useful. Several properties are important for the oxides and hydroxides. Luster describes the way the surface of the mineral reflects light. Many minerals have the appearance of bright smooth metals, and this sheen is referred to as a metallic luster. Some minerals, like hematite and goethite, do not readily transmit light but may exhibit this type of luster. Other minerals that are able to transmit light have nonmetallic lusters, regardless of how shiny the outer surface of the mineral may be. Some common nonmetallic lusters include glassy, vitreous, and resinous. Minerals that have metallic lusters also characteristically produce powders that have diagnostic colors. The color of the powdered mineral is called the streak. Hematite, for example, may be silvery gray in color, but its powder is red. The shapes of minerals can also be important. Corundum, for example, is typically hexagonal in outline; magnetite forms octahedra; and hematite often has thin plates that grow together in rosettes. Specific gravity—the ratio of the weight of a substance to the weight of an equal volume of water—is also an important diagnostic property of many of these minerals. Oxides tend to have much higher than average specific gravities. Specific gravity can be determined by weighing the mineral in water and out of water, or by floating it in a dense fluid whose density can then be measured. Hardness is another property that is used for identification purposes and makes some of the oxides useful. Hardness is simply the measure of a substance's resistance to abrasion. Some minerals, such as corundum, have particularly high hardnesses and therefore can be used commercially as abrasives.

MICROSCOPIC STUDY AND CRYSTALLOGRAPHY

The study of the optical properties of the oxides is conducted either in polarized light transmitted with a petrographic microscope or in reflected light. One important optical property of minerals in transmitted light is their refractive index, which is a measure of the velocity of light passing through them. The refractive index of minerals varies because light may travel at different velocities in different directions, and the differences can be used to identify the mineral. Minerals that are nontransparent, such as many of the oxides, are studied in reflected light with an ore microscope. Such properties as reflectivity, color, hardness, and reactivity to different chemicals are all considered.

The study of the orderly internal atomic arrangements within minerals and the associated external morphologies is called crystallography. The primary methods used to study these internal geometries are a variety of X-ray techniques that look at single crystals or powered samples of minerals. The most commonly used procedure is the X-ray diffraction powder method. This technique takes advantage of the principle that the internal atomic arrangement in every mineral is different from all others. X rays striking the powdered sample are reflected or diffracted at only specific angles (dictated by internal geometries). Thus, the measurement of the specific angles of diffraction provides enough information to identify the substance. Single crystals, however, are normally used in the more detailed studies with X rays. With respect to the study of crystal faces on mineral specimens, a goniometer, or a device for measuring the orientation of crystal faces, allows for the precise measurement of the angular relationships between these faces.

CHEMICAL AND SYNTHESIS STUDIES

Until the latter part of the twentieth century, most chemical analyses of minerals were conducted by methods generally referred to as wet-chemical analyses. In these analyses, the mineral is first dissolved in solution. The amounts of the individual elements in the solution are then determined either by the separation and weighing of precipitates, or by measurement via spectroscopic methods of elemental concentrations in solution. Over the years, the accuracy and speed of completing these analyses have been improved, but the method is laborious and requires the use of dangerous chemicals. The invention of the electron microprobe in the 1950's has revolutionized mineral analyses and has largely eliminated many of the problems and greatly decreased the time necessary to conduct most analyses. The beam of electrons that is used can be precisely focused on samples or areas of samples as small as 1 micron (10^{-3} millimeters) in diameter. Thus, not only small samples may be analyzed, but also individual crystals may be evaluated in several places to check for compositional variations.

Another important method used to study and understand the oxides and other minerals in the laboratory involves the synthesis of minerals. The primary purpose of these studies is to determine the temperature and pressure conditions at which individual minerals form. The roles of fluids and interactions with other minerals are also evaluated. High-temperature studies are typically conducted in furnaces containing platinum or tungsten heating elements, which can produce extreme temperatures. High-pressure studies are produced in large hydraulic apparatuses or presses. A device called the diamond anvil pressure cell uses two gem-quality diamonds to squeeze samples to pressures once requiring the use of huge presses. It has revolutionized high-pressure studies because it is very compact and allows materials to be examined under a microscope while being subjected to extreme pressures.

INDUSTRIAL USES

The oxides are one of the most important mineral groups because they have provided civilization with some very important metals. Iron, titanium, chromium, manganese, aluminum, and uranium are some of the key metals extracted from oxide ores. Iron is the second most common metal in the crust. Along with aluminum, manganese, magnesium, and titanium, iron is considered an abundant metal because it exceeds 0.1 percent of the average composition of the crust. The steelmaking industry uses virtually all the iron mined. Steel, an alloy in which iron is the main ingredient, also includes one or more of several other metals (manganese, chromium, cobalt, nickel, silicon, tungsten, and vanadium) that impart special properties to steel. Hematite, magnetite, and goethite are three of the most important ore minerals of iron.

Apart from its industrial uses, magnetite is the principal cause of magnetism in rocks and is the main reason that rocks can be mapped magnetically. Magnetic mapping is used to study buried or concealed rocks such as rocks buried under glacial deposits or on the ocean floor.

Unlike iron, which has been utilized for more than three thousand years, aluminum is a metal that did not gain prominence until the twentieth century. Since aluminum is light in weight and exhibits great strength, it is widely used in the automobile, aircraft, and shipbuilding industries. It is also utilized in cookware and food and beverage containers.

A third abundant metal that is primarily extracted from oxide and hydroxide minerals is manganese. Manganese is used in the production of steel. The other abundant metal in the crust that is produced from oxides is titanium. Titanium is important as a metal and an alloy because of its great strength and light weight. It is a principal metal in the engines and essential structural components of modern aircraft and space vehicles.

Representing less than 0.1 percent of the average composition of the crust, the group of scarce metals that occur as oxides and offer many important uses include chromium, uranium, tin, tantalum, and niobium. Of these metals, chromium and uranium are the most prominent. Chromium is utilized in the steel industry, where it is a principal component in stainless steel. In addition, because of its high melting temperature, chromite is used in bricks for metallurgical furnaces. Uranium is an important metal because it spontaneously undergoes nuclear fission and gives off large amounts of energy. Uranium is utilized in nuclear reactors to generate electricity. Tin is another scarce metal that for tens of centuries was utilized as an alloy in bronze. It has lost many of its older uses, but remains a prominent metal as new applications are developed. Tin is widely used in solders, type metal, and low melting point alloys. Tantalum is highly resistant to acids, so it is used in equipment in the chemical industry, in surgical inserts and sutures, and in specialized steels and electronic equipment. Niobium is used in the production of stainless steels and refractory alloys (alloys resistant to high temperatures) used in gas turbine blades in aircraft engines. Both have more recently been used in microelectronics. Col-tan, or niobium-tantalum ore, is mined in many of the same areas as conflict diamonds, and leads to many of the same problems, such as forced labor and conflict over mining areas.

Emery, a combination primarily of black corundum, magnetite, and hematite, is used as an abrasive. Several oxides also form gemstones. Rubies are the red gem variety of corundum, and sapphire is also a gemstone of corundum that can have any other color. Spinel, if transparent, is a gem of lesser importance. If red, the gem is called a ruby spinel.

Ronald D. Tyler

FURTHER READING

Berry, L. G., Brian Mason, and R. V. Dietrich. *Mineralogy*. 2d ed. San Francisco: W. H. Freeman, 1983. A college-level introduction to the study of minerals that focuses on the traditional themes necessary to understand minerals: how they are formed and what makes each chemically, crystallographically, and physically distinct from others. Descriptions and determinative tables include almost two hundred minerals (with twenty-eight oxides and hydroxides).

Bowles, J. F. W., et al. *Rock-Forming Minerals: Non-Silicates; Oxides, Hydroxides and Sulfides*. 2d ed.

London: Geological Society Publishing House, 2011. Organized by mineral into oxides, hydroxides, and sulfides. Includes physical and chemical characteristics of each mineral followed by experimental work with the mineral. Well indexed.

Craig, James R., David J. Vaughan, and Brian J. Skinner. *Earth Resources and the Environment*. 4th ed. Englewood Cliffs, N.J.: Prentice-Hall, 2010. A compact, well-illustrated introductory text that discusses the distribution and rates of use of a variety of mineral and energy resources. Covers a number of important metals that are derived from oxides and hydroxides. Indexed and includes a modest bibliography.

Deer, William A., R. A. Howie, and J. Zussman. *An Introduction to Rock-Forming Minerals*. 2d ed. London: Pearson Education Limited, 1992. A standard reference on mineralogy for advanced college students and above. Contains detailed descriptions of chemistry and crystal structure, usually with chemical analyses. Discussions of chemical variations in minerals are extensive.

Dietrich, R. V., and B. J. Skinner. *Rocks and Rock Minerals*. New York: John Wiley & Sons, 1979. Provides a relatively brief but excellent treatment of crystallography and the properties of minerals. Descriptions of minerals focus on the silicates. Considers several important oxides. Includes excellent illustrations, a subject index, and modest bibliography.

Frye, Keith. *Mineral Science: An Introductory Survey*. New York: Macmillan, 1993. Provides an easily understood overview of mineralogy, petrology, and geochemistry, including descriptions of specific minerals. Illustrations, bibliography, and index.

Hammond, Christopher. *The Basics of Crystallography and Diffraction*. 2d ed. New York: Oxford University Press, 2001. Covers crystal form, atomic structure, physical properties of minerals, and X-ray methods. Illustrations help clarify some of the more mathematically complex concepts. Includes bibliography and index.

Hurlbut, C. S., Jr., and Robert C. Kammerling. *Gemology*. 2d ed. New York: John Wiley & Sons, 1991. A well-illustrated introductory textbook for the reader with little scientific background. Covers the physical and chemical properties of gems,

their origins, and the instruments used to study them. Covers methods of synthesis, cutting and polishing, and descriptions of gemstones.

Klein, Cornelis, and C. S. Hurlbut, Jr. *Manual of Mineralogy*. 23d ed. New York: John Wiley & Sons, 2008. An excellent second-year college-level text that introduces the study of minerals. Topics include external and internal crystallography, crystal chemistry, properties of minerals, X-ray crystallography, and optical properties. Systematically describes twenty-two oxide minerals.

Lindsley, Donald H, ed. *Oxide Minerals: Petrologic and Magnetic Significance*. Washington, D.C.: Mineralogical Society of America, 1991. Part of the Mineralogical Society of America's Reviews in Mineralogy series. Covers topics relevant to oxide minerals and their magnetic properties, petrogenesis, and petrology. Extensive bibliography.

Perkins, Dexter. *Mineralogy*. 3d ed. Upper Saddle River, N.J.: Prentice Hall, 2010. Written without scientific jargon making it accessible to the layperson. Discusses the classification of minerals, mineral properties, optical mineralogy, and more.

Ransom, Jay E. *Gems and Minerals of America*. New York: Harper & Row, 1974. Intended for nonscientists interested in rock and mineral collecting. Introductory chapters identify basic mineral characteristics and their environments of formation. Later chapters focus on the locations and collection of gems and minerals throughout the United States. Considers a number of oxides.

Tennissen, A. C. *Nature of Earth Materials*. 2d ed. Englewood Cliffs, N.J.: Prentice-Hall, 1983. Written for the nonscience student and treats minerals from the perspective of both the internal relationships (atomic structure, size, and bonding) and external crystallography. Includes an excellent overview of the physical properties of minerals and the classification and description of 110 important minerals.

Wenk, Hans-Rudolf, and Andrei Bulakh. *Minerals: Their Constitution and Origin*. Cambridge, England: Cambridge University Press, 2004. Covers the structure of minerals, physical characteristics, processes, and mineral systematic. Multiple chapters devoted to nonsilicates, followed by chapters discussing silicates.

See also: Basaltic Rocks; Biopyriboles; Carbonates; Clays and Clay Minerals; Contact Metamorphism; Feldspars; Gem Minerals; Hydrothermal Mineralization; Kimberlites; Metamictization; Minerals: Physical Properties; Minerals: Structure; Non-Silicates; Pelitic Schists; Radioactive Minerals; Regional Metamorphism.

P

PEGMATITES

Pegmatites host the world's major supply of the rare metals lithium, beryllium, rubidium, cesium, niobium, and tantalum. They are also major sources of tin, uranium, thorium, boron, rare-earth elements, and certain types of gems.

PRINCIPAL TERMS

- **aplite:** a light-colored, sugary-textured granitic rock generally found as small, late-stage veins in granites of normal texture; in pegmatites, aplites usually form thin marginal selvages against the country rock but may also occur as major lenses in the pegmatite interior
- **crystal-liquid fractionation:** physical separation of crystals, precipitated from cooling magma, from the coexisting melt, enriching the melt in elements excluded from the crystals; this separation, or fractionation, leads to extreme concentration of incompatible elements in the case of pegmatite magma
- **exsolve:** the process whereby an originally homogeneous solid solution separates into two or more minerals (or substances) of distinct composition upon cooling
- **fluid inclusions:** microscopic drops of parental fluid trapped in a crystal during growth; inclusions persist indefinitely unless the host crystal is disturbed by deformation or recrystallization
- **incompatible elements:** chemical elements characterized by odd ionic properties (size, charge, electronegativity) that tend to exclude them from the structures of common minerals during magmatic crystallization
- **solidus/liquidus temperature:** the liquidus temperature marks the beginning of crystallization in magmas, and the solidus temperature marks the end; crystals and melt coexist only within the liquidus-solidus temperature interval
- **viscosity:** a property of fluids that measures their internal resistance to flowage; the inverse of fluidity or mobility

SIMPLE PEGMATITES

Pegmatites are relatively small rock bodies of igneous appearance that are easily distinguished from all other rock types by their enormous range of grain size and textural variations. Typically, fine-grained margins of aplite are abruptly succeeded by discontinuous interior layers of coarse, inward-projecting crystals, which, in turn, give way to zones of graphically intergrown quartz and alkali feldspar (that is, angular intergrowths, so called because they resemble a form of script). Most pegmatites contain numerous isolated pockets, rafts, and radial clusters of abnormally large, even giant, crystals and are cut by fractures filled with late-stage products. The spatial inhomogeneity of mineral distribution, rock texture, and chemical elements exhibited in pegmatites is unequaled in any other igneous product.

The majority of pegmatites are narrow lens-shaped, tabular, or pod-shaped masses measuring from a few meters to several tens of meters in length. Even large, commercially exploited pegmatites rarely exceed 1 kilometer in length and 50 meters in width. Nearly all pegmatites have a bulk composition approximating that of granite. They are composed predominantly of quartz, alkali feldspar, and muscovite, with minor quantities of tourmaline and garnet. Those containing only these minerals are termed simple pegmatites—"simple" referring to their mineralogy, chemistry, and internal structural features collectively.

RARE-MINERAL PEGMATITES

Complex pegmatites are composed of the same major mineral assemblage as simple pegmatites but, in addition, contain a great variety of exotic accessory minerals that host rare metals, such as lithium, rubidium, cesium, beryllium, niobium, tantalum, and tin. These pegmatites are also called rare-element or rare-mineral pegmatites, and they are of major economic and strategic importance. In contrast to simple types, they are complex in terms of mineralogy, chemistry, and internal structure. Beyond their economic value, rare-element pegmatites are of interest because they crystallize from the most evolved, or fractionated, granitic magmas in nature. They are the extreme product of extended crystal fractionation and therefore occupy a unique position between normal igneous rocks and hydrothermal vein deposits. The remainder of this article deals exclusively with this important class of pegmatites.

Major rare-mineral pegmatites may exhibit systematic internal variations in mineralogy and texture that are termed zoning. The usual zonal pattern begins with a thin selvage, or rim, of aplite, a few centimeters thick at most, which typically grades into the country rock. The aplite selvage is abruptly succeeded by a coarse-grained, muscovite-rich zone that, by decrease in muscovite abundance, passes into a zone dominated by quartz and feldspar that may also carry abundant spodumene, a lithium-bearing pyroxene. The innermost zone is composed mainly or exclusively of quartz. Ideally, the zones are crudely parallel structures symmetrically disposed with respect to the center line of the host pegmatite. In a real pegmatite, however, individual zones usually vary in terms of width, continuity, and position with respect to the center line of the body. The zonal pattern records the inward growth of a pegmatite from the enclosing walls of country rock. The bulk chemical composition of each individual zone is an approximate indication of the composition of the parental magma at the time the zone formed. It follows that the inward zone sequence is an approximate record of compositional changes that occurred in the evolving parent magma as the pegmatite formed. The existence of a common zonal pattern means that most pegmatite magmas evolve in a broadly similar fashion.

COMPOSITION OF RARE-MINERAL PEGMATITES

In spite of the general similarity in zone pattern shown by rare-mineral pegmatites as a class, no two bodies are ever exactly alike in terms of the size, shape, and spatial distribution of their zones. Yet the mineral assemblages and textures that compose individual zones are surprisingly consistent. Excluding the selvage, most zoned pegmatites are composed of five distinct mineral assemblages: (1) coarse microcline (potassium feldspar) with or without rare-earth minerals; (2) quartz with subordinate lithium minerals; (3) massive quartz (virtually monomineralic bodies); (4) albite (sodium feldspar) or fluorine-rich mica bodies that contain abundant tourmaline,

apatite, beryl and a wide variety of other minerals; and (5) mixed zones of sequential deposition (composed of very coarse-grained microcline and lithium minerals within a finer-grained matrix of albite).

Gross chemical inhomogeneity and remarkable concentrations of exotic minerals have earned zoned rare-mineral pegmatites a reputation for complex chemistry. In fact, the overall bulk composition of most such rock bodies is a rather simple one, dominated by the common elements oxygen, silicon, aluminum, sodium, and potassium, which compose the major pegmatite mineral phases of quartz, albite, microcline, and muscovite. Water and lithium oxide are next in importance, water forming generally 0.5 to 1.0 weight percent and lithium oxide being present in amounts up to roughly 1.5 weight percent. Oxides of phosphorus and iron usually constitute several tenths of a percent each. The elements calcium, magnesium, and manganese are present in trace amounts only. The total concentration of rare elements (beryllium, boron, cesium, rubidium, tantalum, niobium, tin, tungsten, uranium, thorium, fluorine, and rare-earths), for which pegmatites are famous, is somewhat less than 1 percent by weight in almost every case. This composition, except for the high lithium component, is essentially that of normal alkali granite. The lithium concentration in normal granites, however, is only 1.5 percent of that in typical lithium-bearing pegmatites; that is, pegmatites have about seventy times as much lithium as normal granite. The bulk composition of a pegmatite body is often assumed to be approximately that of its parent magma. That does not apply to the water content, because most pegmatite magmas exsolve water during crystallization, which then diffuses outward into the surrounding country rock.

FORMATION OF PEGMATITES

Experiments have firmly established that water, as a dissolved constituent in magma, lowers its melting temperature significantly. Water also lowers the viscosity (resistance to flow) of the residual melt, which is a major factor in promoting the growth of the large crystals that distinguish pegmatites from normal igneous rocks. Traditionally, pegmatites have been thought to be the products of complex interactions between water-saturated silicate melt of low viscosity, a separate coexisting aqueous fluid, and the enclosing rocks. Several important experiments, however, have shown that water has relatively little ability to dissolve certain major and trace elements concentrated in rare-mineral pegmatites. Also, fluid inclusion data from pegmatite minerals indicate that the trapped fluids are relatively dense and fall between silicate melt and simple aqueous solutions in terms of character. These data are important because they imply that pegmatite minerals crystallize from highly evolved silicate magma rather than whatever aqueous solution may be present. If that is true, it is likely that incompatible elements in the magma are responsible for the extreme concentration of rare metals in pegmatite magma.

Several lines of evidence suggest that boron, phosphorus, and fluorine play key roles in this regard. On account of their incompatibility with common silicate minerals, these elements are highly concentrated, along with water, in residual melt. High levels of these elements will have two major effects in silicate magmas. First, they will delay water saturation, or "boiling," by increasing the solubility of water in the residual melt, and second, they will promote concentration of such elements as lithium, sodium, potassium, rubidium, and cesium in that melt. Thus, the result will be a water-rich, sodium-aluminosilicate-rich, late-stage melt, from which albite, tourmaline, phosphate minerals, fluorine-rich micas, beryl, zircon, and niobium-tantalum-tin oxides can crystallize prior to boiling. Such a melt would possess the required low viscosity and low melting temperature (below 500 degrees Celsius) required to enable pegmatite magma to migrate significant distances from the parent pluton prior to crystallization.

High concentrations of boron, phosphorus, and fluorine in the late-stage melt can produce the additional important effect known as immiscibility. At some critical concentration, the single parent melt can split into two mutually insoluble "partner melts" with drastically contrasting chemical compositions. One partner melt will be very rich in silica and could form massive quartz zones enriched in lithium minerals. The remaining partner melt would be a strongly alkaline silicate melt capable of producing albite-dominated zones. The crystallization of tourmaline has been shown to be a very effective means of concentrating water in the residual melt and has been suggested to be a "triggering process" for rapid water saturation.

MAPPING PEGMATITES

In the past, studies of pegmatites have mainly concerned either the evaluation of their economic potential or their exotic mineralogy. Surprisingly few efforts have focused on pegmatites in a coherent and systematic fashion, and as a result, satisfying theories for their origin are poorly developed relative to other areas of igneous petrology.

A comprehensive study of pegmatites in a particular region would include not only the pegmatites but also the country rocks that enclose them and any bodies of exposed granitic rock. The study begins with preparation of a geologic map showing the distribution of pegmatite bodies, by type, throughout the area of interest. The map is compiled from firsthand, detailed observations of rock outcroppings obtained by numerous foot-traverses across the area. Additional useful information may be obtained from earlier geologic maps, aerial photographs, satellite imagery, and mine records. In most cases, the geologic map will show numerous, small, discontinuous pegmatite veins cutting older, high-grade metamorphic country rocks. These are known as external pegmatites. If any granite plutons are present in the area, they must be examined closely, as they are potential parent bodies of the pegmatites. Such granites may themselves host so-called internal pegmatites. External and internal pegmatites, even if derived from the same pegmatite magma (cogenetic pegmatites), generally will differ in mineralogy and zoning traits. It is a considerable achievement if a study can demonstrate that groups of external and internal pegmatites derive from the same parent body of granite. In such a case, the geologist has the rare opportunity to study the chemical and temporal relationships between parent rock, host rock, and an evolving, mobile pegmatite magma. This happy state of affairs is seldom realized because of the all-important "level of erosion." Since external pegmatites form by migration of pegmatite magma upward and away from the parent body along fracture pathways, they will normally be destroyed by erosion before their deeper parent and internal "relatives" are exposed.

By examining a large pegmatite-bearing area rather than a single pegmatite body (as is often done), it is possible to determine if systematic differences exist in the exposed pegmatites relative to host-rock type, local or regional fracture patterns in the host rock, or distance from igneous plutons.

Individual pegmatites can also be compared in terms of shape, size, orientation, mineralogy, and zone characteristics. Systematic variations of these parameters on the scale of a geological map constitute "regional zoning." Increased recognition of such effects is needed to provide greater insight into the operation of the pegmatite-forming process.

After regional geologic relationships are determined, individual pegmatites are mapped in the greatest detail possible (often at scales as large as 1 inch per 10 feet). The objective is to determine the size, shape, and zone sequence of each pegmatite body, which, in turn, will provide a basis for sampling and determining the composition of the pegmatite as a whole. Every effort is made to establish the correct sequence of zone crystallization, because the bulk compositions and fluid inclusion data from each member of the zone sequence can then be used to trace chemical changes in the pegmatite magma during the emplacement process.

COMMERCIAL VALUE

In the past, large and favorably located simple pegmatites were often worked for mica, quartz, and feldspar by the glass and ceramic industries, but production of this type has all but ceased. The post-World War II electronics revolution and other technological advances, such as lithium batteries, have fueled ever-increasing demands for the rare elements found in complex pegmatites.

Rare-element pegmatites with commercial grades and reserves that are sufficiently close to the surface for low-cost open-pit mining are distinctly uncommon. The few that do exist are the only sources for the elements lithium, beryllium, cesium, rubidium, tantalum, and niobium; they are important sources for boron, tin, tungsten, uranium, thorium, fluorine, and rare-earth elements (lanthanides) as well. Because many of the pegmatites in current commercial production are located in developing countries, some industrialized countries (the United States among them) stockpile rare-element commodities of strategic importance in case of national emergencies or supply disruptions.

Complex pegmatites are valued for reasons other than the commercial production of rare elements. From ancient times, they have been mined for precious and semiprecious gems. The intermediate zones of certain pegmatites are the major source of gem-quality topaz, tourmaline, and beryl (including

morganite, aquamarine, and emerald). Each year, hundreds of small, commercially subeconomic rare-element pegmatites are prospected by amateur and professional collectors of rare minerals and fine crystals. These pegmatites, in fact, are supporting a small but vigorously growing industry. Many (and perhaps even most) of the spectacular crystal specimens displayed in museums were obtained from subeconomic pegmatites found by amateur collectors.

Gary R. Lowell

FURTHER READING

Best, Myron G. *Igneous and Metamorphic Petrology.* 2d ed. Malden, Mass.: Blackwell Science Ltd., 2003. A popular university text for undergraduate majors in geology. A well-illustrated and fairly detailed treatment of the origin, distribution, and characteristics of igneous and metamorphic rocks. Discusses pegmatites and late-stage magmatic processes pertinent to pegmatite formation.

Cameron, Eugene N., R. H. Jahns, A. H. McNair, and L. R. Page. *Internal Structure of Granitic Pegmatites.* Urbana, Ill.: Economic Geology Publishing, 1949. This classic study of North American pegmatites during wartime illustrates the strategic nature of pegmatite deposits. Includes detailed maps and descriptions of American pegmatites and valuable descriptions of bodies subsequently removed by mining. Appropriate for professionals and college-level readers with some background in geology.

Desautels, Paul E. *The Mineral Kingdom.* Madison Square Press, 1972. One of many superbly illustrated books devoted to minerals and gems for hobbyists and serious collectors. A skillful blend of art, science, and history of minerals, with a useful index. Appropriate for high school readers and those interested in the aesthetic aspects of gems and crystals.

Grotzinger, John, et al. *Understanding Earth.* 5th ed. New York: W. H. Freeman, 2006. Covers the formation and development of Earth. Appropriate for high school students and general readers. Includes an index and a glossary of terms.

Guilbert, John M., and Charles F. Park, Jr. *The Geology of Ore Deposits.* 4th ed. Long Grove, Ill.: Waveland Press, Inc., 1986. A splendid new edition of a traditional college text for undergraduate geology majors. Includes a discussion of pegmatites. Includes excellent photographs of pegmatite textures and outcrops, and a comprehensive review of traditional American perspectives on pegmatites.

Hall, Anthony. *Igneous Petrology*. 2d ed. New York: John Wiley & Sons, 1996. Discusses the occurrence, composition, and evolution of igneous rocks and magmas. Focuses on a different rock group in each chapter. Emphasizes the role of water in magmas in the sections on granite. Appropriate for college-level readers, but includes illustrations that are useful to the general reader.

Hutchison, Charles S. *Economic Deposits and Their Tectonic Setting*. New York: John Wiley & Sons, 1983. An economic geology text for undergraduate majors in geology. Offers modern international perspectives on various classes of ore bodies, including pegmatites. Presents the Varlamoff spatial classification of pegmatites, which is widely known outside the United States, as well as an excellent cross-section of the famed Bikita pegmatite.

King, Vandall T. *A Collector's Guide to the Granite Pegmatite*. Atglen, Pa.: Schiffer Publishing, 2010. Covers the distribution of pegmatite crystals in North America and from sites around the world. Discusses the structure of the deposits, and provides information on pegmatite crystals, minerals, and gem pockets. Includes color photos and a further reading list.

Klein, Cornelis, and Cornelius S. Hurlbut, Jr. *Manual of Mineralogy*. 23d ed. New York: John Wiley & Sons, 2008. Provides an excellent index of mineral names and tabulates chemical formulas, crystal systems, and physical properties of all the common mineral species and many of the rare ones encountered in pegmatites. This basic mineralogy text is essential for anyone embarking on a study of rare-mineral pegmatites. A brief pegmatite summary is provided.

Laznicka, Peter. *Giant Metallic Deposits*. 2d ed. Berlin: Springer-Verlag, 2010. Discusses the location, extraction, and future use of many metals. Organized by topic rather than by metal. Discusses geodynamics that result in large metal ore deposits and the composition of metals within deposits, as well as hydrothermal deposits, metamorphic associations, and sedimentary associations.

Norton, James J. "Sequence of Minerals Assemblages in Differentiated Granitic Pegmatites." *Economic Geology* 78 (August, 1983): 854-874. Updates the classic zone sequence of Cameron (1949) by constructing one that takes into account the Bikita and Tanco pegmatites. Appropriate for college-level readers with some background in geology. Includes useful data tables and bibliography for anyone with a serious interest in pegmatites.

Simmons, William, et al. *Pegmatology*. Rubellite Press, 2003. A reference dedicated to the topic of pegmatites. Contains about 300 color images, charts, and graphs. Written for undergraduates as an introduction to pegmatites.

See also: Aluminum Deposits; Building Stone; Cement; Coal; Contact Metamorphism; Diamonds; Dolomite; Fertilizers; Gem Minerals; Gold and Silver; Igneous Rock Bodies; Industrial Metals; Industrial Nonmetals; Iron Deposits; Manganese Nodules; Metamorphic Rock Classification; Metamorphic Textures; Metasomatism; Platinum Group Metals; Salt Domes and Salt Diapirs; Sedimentary Mineral Deposits; Uranium Deposits.

PELITIC SCHISTS

Pelitic schists are formed from fine-grained sedimentary rocks. They are an important metamorphic rock type because they undergo distinct textural and mineralogical changes that are used by geologists to gauge the temperatures and pressures under which the rocks were progressively modified.

PRINCIPAL TERMS

- **equilibrium:** a situation in which a mineral is stable at a given set of temperature-pressure conditions
- **index mineral:** an individual mineral that has formed under a limited or very distinct range of temperature and pressure conditions
- **isograd:** a line on a geologic map that marks the first appearance of a single mineral or mineral assemblage in metamorphic rocks
- **metamorphic zone:** areas of rock affected by the same limited range of temperature and pressure conditions, commonly identified by the presence of a key individual mineral or group of minerals
- **mica:** a platy silicate mineral (one silicon atom surrounded by four oxygen atoms) that readily splits in one plane
- **mineral:** a naturally occurring chemical compound that has an orderly internal arrangement of atoms and a definite chemical formula
- **mudrock:** a collective term for sedimentary rocks composed of fine-grained products derived from the physical breakdown of preexisting rocks (shales, for example), which break along distinct planes, and mudstones, which do not
- **pelitic:** an adjective for mudrocks and the metamorphic rocks derived from them
- **progressive metamorphism:** mineralogical and textural changes that take place as temperature and pressure increase
- **sedimentary rock:** a rock formed from the physical breakdown of preexisting rock material or from the precipitation—chemically or biologically—of minerals
- **texture:** the size, shape, and arrangement of crystals or particles in a rock

AGENTS OF METAMORPHISM

Metamorphism literally means "the change in form" that a rock undergoes. More precisely, metamorphism is a process by which igneous and sedimentary rocks are mineralogically, texturally, and, occasionally, chemically modified by the effects of one or more of the following agents or variables: increased temperature, pressure, or chemically active fluids. Pelitic schists have long been recognized as one of the best rock types to gauge and preserve the wide range of mineralogical and textural changes that take place as metamorphic conditions progressively increase. It is perhaps easiest to understand the conditions under which metamorphism occurs by excluding those conditions that are not generally considered metamorphic. At or near the earth's surface, sedimentary rocks form from the physical or chemical breakdown of preexisting rocks at relatively low pressures and temperatures (less than 200 degrees Celsius). Igneous rocks form from molten material at high temperatures (650 to 1,100 degrees Celsius), and at low pressures for volcanic rocks or at high pressures for those formed at great depth. The conditions that exist between these two extremes are those that are considered metamorphic.

Although all the agents of metamorphism work together to produce the distinctive textures and minerals of regionally metamorphosed rocks, each has its own special or distinctive role. Temperature is considered the most important factor in metamorphism. It is the primary reason for recrystallization and new mineral growth. Two types of pressure affect metamorphic rocks. The pressure caused by the overlying material and uniformly affecting the rock on all sides is referred to as the confining pressure. An additional pressure, referred to as directed pressure, is normally caused by strong horizontal forces. While confining pressure has little apparent influence on metamorphic textures, directed pressures are considered the principal reason that several types of metamorphosed rocks develop distinctive textures. Chemically active fluids are important because they act both as a transporting medium for chemical constituents and as a facilitator of chemical reactions during metamorphism. As metamorphic conditions increase, fluids play an important role in aiding metamorphic reactions, even though they are concurrently driven out of the rocks. Most metamorphic rocks preserve mineral assemblages that represent the highest conditions achieved. As metamorphic conditions start to fall, reactions are very slow to occur because adequate fluids are normally not available. If the metamorphism involves significant additions and losses of constituents other than water, carbon dioxide, and other fluids, this process is referred to as metasomatism. For example, if the sodium content of a rock markedly increases during metamorphism, this is referred to as sodium metasomatism.

TYPES OF METAMORPHISM

Geologists recognize several types of metamorphism, but one, regional metamorphism, is predominant. Regional metamorphism takes place at depth within the vast areas where new mountains are forming. The products of regional metamorphism are exposed in a variety of places. One of the best sites is found on shields—broad, relatively flat regions within continents where the thin veneer of sedimentary rocks has been stripped away, exposing wide areas of deeply eroded ancient mountain belts and their regionally metamorphosed rocks, complexly contorted and intermingled with igneous rocks. The cores of older eroded mountain belts, such as the Appalachians and the Rockies, also commonly expose smaller areas of very old regionally metamorphosed rocks. Pelitic schists mainly occur in regionally metamorphosed rocks. They may also occur in contact metamorphic deposits, which are small, baked zones that formed immediately adjacent to igneous bodies that invaded pelitic rocks as hot molten material. In contrast to the thousands of square kilometers typically occupied by regionally metamorphosed rocks, individual contact metamorphic deposits are rarely more than several kilometers wide.

FOLIATION

Fine-grained sedimentary rocks (shales and mudstones) that formed from the physical breakdown of other rock types are commonly referred to as pelites. Because pelitic sediments can be hard to distinguish, the collective label "mudrock" has come into widespread use. Pelitic rocks undergo very marked textural and mineralogical changes that geologists use to gauge the conditions of formation in a particular area. "Schist," as used in the term "pelitic schist," has several definitions. The name may be applied broadly to any rock that is foliated or contains minerals that have a distinct preferred planar orientation. "Schist"

also has a much narrower definition, referring only to foliated, coarse-grained rocks in which most mineral grains are visible to the unaided eye. This foliation, or preferred orientation, is imparted upon the rock because of deformation and recrystallization in response to a pronounced directed pressure at elevated temperatures.

The development of foliation is the most diagnostic feature of rocks that have undergone regional metamorphism. Directed pressure produces a variety of foliations that change as a function of the conditions of formation. An unmetamorphosed pelitic rock is typically a very fine-grained, soft rock that may or may not be finely layered (containing closely spaced planes). At low temperatures and pressures (low grades of metamorphism), pelitic rocks remain fine-grained, but microscopic micaceous minerals form or recrystallize and align themselves perpendicular to the directed pressure. This produces a dense, hard rock called a slate, which readily splits parallel to the preferred orientation. Any layering may be partially or totally obliterated. The accompanying texture is called a flow or slaty cleavage. At slightly higher conditions, micaceous minerals become better developed but remain fine-grained. The foliated rock produced takes on a pronounced sheen and is referred to as a phyllite. As conditions continue to increase, grain size increases until it is visible. This change produces a texture referred to as schistosity, and the rock is called a schist. At very high conditions, the micas that were diagnostic of lower conditions start to become unstable. The rock, referred to as a gneiss, remains foliated but does not readily split as the slates, phyllites, and schists do. The foliation, or gneissosity, takes the form of an alternating light and dark banding. At higher conditions, gneissic rocks gradually grade into the realm where igneous rocks form.

MINERALOGIC ZONES

The first study to show that a single rock type can undergo progressive change on a regional scale was that of British geologist George Barrow toward the end of the nineteenth century in the Highlands of Scotland, southwest of Aberdeen. His work indicated that pelitic schists and gneisses contained three discrete mineralogic zones, each represented by one of three key minerals: staurolite, kyanite, and sillimanite. Barrow suggested that increasing temperature was the controlling agent for the zonation observed. Later work by Barrow and other geologists confirmed this hypothesis and broadened the mapping over the entire Highlands, revealing additional mineral zones that formed at lower metamorphic conditions. From lowest to highest temperatures, several mineralogical zones were recognized. In the chlorite zone, there are slates, phyllites, and mica schists generally containing quartz, chlorite, and muscovite. In the biotite zone, mica schists are marked by the appearance of biotite in association with chlorite, muscovite, quartz, and albite. Biotite occurs in each of the higher zones. A line marking the boundary of the chlorite and biotite zones is referred to as the biotite isograd. (An isograd is a line that marks the first appearance of the key index mineral that is distinctive of the metamorphic zone.) In the almandine (garnet) zone, mica schists are characterized by the presence of almandine associated with quartz, muscovite, biotite, and sodium-rich plagioclase. The garnet first appears along the almandine isograd and occurs through all the higher zones. In the staurolite zone, mica schists typically contain quartz, muscovite, biotite, almandine, staurolite, and plagioclase. In the kyanite zone, mica schists and gneisses contain quartz, muscovite, biotite, almandine, kyanite, and plagioclase. Staurolite is no longer stable in this zone. In the sillimanite zone, mica schists and gneisses are characterized by quartz, muscovite, biotite, almandine, sillimanite, and plagioclase. Sillimanite forms at the expense of kyanite. These zones—also called Barrovian zones—are recognized throughout the world.

STUDY OF PELITIC ROCKS

The study of pelitic schists can be conducted from a number of perspectives. The foundation of all studies, however, is the kind of fieldwork that George Barrow conducted in Scotland, which includes careful mapping, sample collection, rock description, and structural measurement.

Once the data and samples are returned to the laboratory, other methods of investigation may be employed. Rocks are commonly studied under the binocular microscope or powdered and made into thin sections to be analyzed in polarized light transmitted through the sample beneath the petrographic microscope. The preparation of the standard thin section involves several steps: cutting a small block of the sample so that it is 2.7 millimeters by 4.5 millimeters on one side; polishing one side and glueing it to a glass slide; and cutting and grinding the glued sample to a uniform thickness of 0.03 millimeter. Observations and descriptions of these thin sections are essential to the study of pelitic rocks because they commonly provide information about the interrelationships among the minerals present and clues to the metamorphic history of the sample.

Whole-rock chemical analyses of pelitic rocks are also important. Standard analytical procedures, atomic absorption, and X-ray fluorescence spectroscopy are all methods used to determine the amounts of the elements within these rocks. Chemical analyses of individual minerals also provide important information about pelitic rocks.

MINERAL ANALYSES

Although mineral analyses have traditionally been conducted by the same analytical techniques as whole-rock analyses, problems obtaining material pure enough for analysis have plagued scientists. Since the 1960's, almost all published mineral analyses have been produced by the electron microprobe, which greatly improves the accuracy of the analyses. These whole-rock analyses and individual mineral analyses are commonly plotted together in several ways on triangular graph paper in order to illustrate the kinds of mineral associations that can exist at different metamorphic conditions.

In experimental petrology, metamorphic minerals are grown under a variety of controlled equilibrium conditions to provide geologists with a better understanding of the actual conditions for their formation. Although such studies have greatly enhanced knowledge of naturally occurring metamorphic reactions, the field is not without controversy. The behavior of the reactions of kyanite to sillimanite, andalusite to sillimanite, and kyanite to sillimanite as temperatures increase shows the great complexity of this endeavor. The various studies conducted on these minerals that occur in pelitic schists are not in agreement, so many geologists remain deeply divided about the exact conditions under which these three minerals are stable.

Physical chemistry and thermodynamics provide geologists with tools to predict how minerals in idealized, simplified chemical reactions theoretically behave. Such comparisons are extremely important in making judgments about more complex natural environments.

ECONOMIC USES

The only important pelitic rock type that is useful as such is very low-grade slate that is quarried for flagstone, roofing material, countertops, blackboards, billiard tables, and switchboard panels. Yet pelitic rocks do contain a number of minerals that are extracted for their usefulness.

Staurolite, almandine-garnet, and kyanite are all minerals from pelitic rocks that have some use as gemstones. In addition, staurolite crystals grow together and may form crosses, which are sold as amulets called fairy stones, although most objects sold as "fairy stones" are not genuine staurolite. (The name "staurolite" is derived from the Greek word meaning "cross" because of the mineral's diagnostic crystal shape.) Garnet, from pelitic schists, also has some use as an abrasive for sandblasting and spark plug cleaning, and is a common ingredient in sandpaper, which is often called "garnet paper." In addition, sillimanite and kyanite are used as refractory (high-temperature) materials in porcelain and spark plugs.

Graphite is another mineral that may occur in or be associated with pelitic schists and forms from carbon-rich sediments. Deposits near Turin, Italy, occur in micaceous phyllites, schists, and gneisses. Graphite has a variety of uses, including the production of refractory crucibles for the making of bronze, brass, and steel. It is used with petroleum products as a lubricant and blended with fine clay in the "leads" of pencils. It is also used for electrotype and in steel, batteries, generator brushes, and electrodes.

Another mineral, pyrophyllite, is found in very low-grade aluminum-rich pelitic rocks and has properties and uses similar to those of talc. Pyrophyllite is used in paints, paper, ceramics, and insecticides, and as an absorbent powder. When heated, it can expand and create worm-like textures, leading to the term "vermiculite." Vermiculite is often used as a potting soil medium. A special fine-grained variety of pyrophyllite called agalmatolite is prized by the Chinese for the carving of small objects.

Ronald D. Tyler

FURTHER READING

Bates, Robert L. *Geology of the Industrial Rocks and Minerals.* Mineola, N.Y.: Dover, 1969. Deals with rocks and minerals that are extracted because of their economic importance. Considers slate, graphite, pyrophyllite, kyanite, and corundum, all products of metamorphosed pelitic rocks, in terms of their properties and uses, production, and occurrences. Includes an index and an excellent bibliography.

Blatt, Harvey, Robert J. Tracy, and Brent Owens. *Petrology: Igneous, Sedimentary, and Metamorphic.* 3d ed. New York: W. H. Freeman, 2005. Deals with the descriptions, origins, and distribution of igneous, sedimentary, and metamorphic rocks. Pelitic rocks figure prominently in the treatment of the occurrences, graphic representations, the processes of metamorphism, and the mineral changes during metamorphism. Includes an index and bibliographies for each chapter.

Dietrich, Richard V., and B. J. Skinner. *Rocks and Rock Minerals.* New York: John Wiley & Sons, 1979. Focuses on the description and identification of rocks and minerals through simple methods intended for use in the field. Provides a good overview of metamorphism, emphasizing the important role that pelitic rocks play in the study of regional metamorphism.

Ernst, W. G. *Earth Materials.* Englewood, N.J.: Prentice-Hall, 1969. A compact but excellent introduction to the study of rocks and minerals. Considers metamorphism and emphasizes the importance of textural and mineralogical changes that pelitic rocks undergo during regional metamorphism. Briefly addresses the mineral reactions among the index minerals kyanite, sillimanite, and andalusite.

Gillen, Cornelius. *Metamorphic Geology.* Winchester, Mass.: Allen & Unwin, 1982. A readable, well-illustrated introduction to metamorphism that focuses on textures and field relationships. Provides a good discussion of pelitic rocks and Barrovian zones. Emphasizes the strong tie between metamorphic processes and mountain-building events. Includes a glossary and a short bibliography.

Grotzinger, John, et al. *Understanding Earth.* 5th ed. New York: W. H. Freeman, 2006. Introduces the high school and college reader without a geology background to the subject of metamorphism and metasomatism. Includes index and illustrations.

Howie, Frank M., ed. *The Care and Conservation of Geological Material: Mineral Rocks, Meteorites, and Lunar Finds.* Oxford: Butterworth-Heinemann, 1992. Examines mineralogy and the processes associated with the field. Provides a good treatment of the fine minerals that make up schists and pelitic schists.

Lindsley, Donald H., ed. *Oxide Minerals: Petrologic and Magnetic Significance.* Washington, D.C.: Mineralogical Society of America, 1991. Offers an overview of petrology and petrogenesis studies. Focuses on oxide minerals and their magnetic properties. Offers a detailed section on the geochemical makeup of schists.

Mason, R. *Petrology of the Metamorphic Rocks.* 2d ed. Berlin: Springer, 2011. Written for the second-year geology student, this text deals with metamorphic rocks from the perspective of descriptions in the field and beneath the petrographic microscope. Discussion of pelitic rocks covers regional and contact metamorphism. Includes a glossary, bibliography, and index.

Miyashiro, Akiho. *Metamorphism and Metamorphic Belts.* New York: Springer, 1978. Covers several types of metamorphism but focuses on regional metamorphism. Stresses the mineralogical reactions that occur in pelitic rocks (and other rock types) that best illustrate the concepts of progressive metamorphism. Includes an index and a fine bibliography.

Oldershaw, Cally. *Rocks and Minerals.* New York: DK, 1999. Useful to new students who may be unfamiliar with the rock and mineral types discussed in classes or textbooks.

Passchier, Cees W., and Rudolph A. J. Trouw. *Microtectonics.* Berlin: Springer, 2005. Includes a chapter covering foliations and lineations in detail. Discusses deformation, shear zones, porphyroblasts, natural microgauges, and study techniques. Well-organized and logical explanations. Designed for advanced graduates and researchers with a good grasp of structural geology and microscopic petrology.

Tennisen, A. C. *Nature of Earth Materials.* 2d ed. Englewood Cliffs, N.J.: Prentice-Hall, 1983. Written for the nonscientist. Covers the nature of atoms and minerals and igneous, sedimentary, and metamorphic rocks. Includes a section on the uses of these materials. Well illustrated. Includes a bibliography and subject index.

Terabayashi, Masaru, et al. "Silicification of Politic Schist in the Ryoke Low-Pressure/Temperature Metamorphic Belt, Southwest Japan: Origin of Competent Layers in the Middle Crust." *Island Arc* 19 (2010): 17-29. Discusses silicified rock in contact with biotite schist and politic schist. Discusses greenschists as well.

Thompson, Alan Bruce, and Jo Laird. "Calibrations of Modal Space for Metamorphism of Mafic Schist." *American Mineralogist* 90 (2005): 843-856. Evaluates reactions of mafic rock metamorphism into greenschist, blueschist, and amphibolites-facies. Discusses the use of modal change data to determine pressure and temperature gradients in mafic schist.

Thompson, James B., Jr. "Modal Spaces for Pelitic Schists." *Reviews in Mineralogy and Geochemistry* 46 (2002): 449-462. Provides a historical context for modal space, and provides an example of its use in studying politic schists. Provides methodology useful for wide application of modal spaces.

Winkler, H. G. F. *Petrogenesis of Metamorphic Rocks.* 5th ed. New York: Springer-Verlag, 1979. Addresses the chemical and mineralogical aspects of metamorphism. Emphasizes the principle that mineral reactions in common rock types can be used to determine metamorphic conditions. Separate chapters deal with pelitic and other important rock types. Includes index and short bibliographies after each chapter.

See also: Blueschists; Contact Metamorphism; Hydrothermal Mineralization; Metamorphic Rock Classification; Metamorphic Textures; Metasomatism; Regional Metamorphism; Sub-Seafloor Metamorphism.

PETROLEUM INDUSTRY HAZARDS

High-profile disasters, such as oil spills or rig explosions, often come to mind in discussions of petroleum industry hazards. More common, yet just as hazardous, are the day-to-day workings of the industry: extracting pressurized, flammable, toxic resources from deep in the earth and then processing and transporting those resources to their final destination. A large part of petroleum extraction and distribution involves dealing with these safety concerns.

PRINCIPAL TERMS

- **aromatic hydrocarbons:** a toxic hydrocarbon group that includes benzene, toluene, and xylene
- **benzene:** a carcinogenic aromatic hydrocarbon found in crude oil; an important ingredient in gasoline
- **blowout preventer:** a massive device used at a rig wellhead to function as a shutoff valve if the pressure in a reservoir gets out of control
- **boom:** a metal or plastic barrier used to prevent the spread of oil during a spill
- **dispersant:** a chemical that breaks spilled oil into droplets that will sink instead of remain on the surface of the water
- **fracking:** hydraulic fracturing; uses water and chemicals to break up shale and release natural gas
- **hydrocarbons:** compounds of hydrogen and carbon; the main component of petroleum and natural gas
- **kick:** a "belch" during uncontrolled pressure in a reservoir; can lead to a blowout or a gusher
- **methane:** the main component of natural gas
- **skimmer:** a vessel used to skim oil from the surface of water after an oil spill

CHEMISTRY AND TOXICOLOGY

Although the word *petroleum* technically refers only to crude oil, the petroleum industry encompasses the extraction and processing of both crude oil and natural gas. Hydrocarbons—compounds of hydrogen and carbon—are the main components of crude oil and natural gas, and they occur in various states: solid (for example, paraffin), liquid (for example, hexane and benzene), and gas (for example, methane, ethane, and propane). Gaseous hydrocarbons are most familiar in the form of natural gas and liquid hydrocarbons in the form of gasoline.

The hydrocarbons in crude oil are either aromatic (carbon atoms in a ring) or aliphatic (carbon atoms in an open chain). Aromatic hydrocarbons (benzene, toluene, and xylene) have a lower molecular weight, are more difficult to break down in the environment, and are more toxic both in the immediate sense—causing visible sickness—and in the long-term sense because their presence in the air (such as near refineries) is associated with an elevated incidence of cancer and leukemia.

After events such as ocean oil spills hydrocarbons remain in the ocean and are metabolized by organisms at both low and high levels of the food chain; the hydrocarbons remain in their bodies to be consumed by other animals or to be passed to their young. Death, internal organ damage, and reduced ability to swim or maintain body temperature can result. Stationary organisms in intertidal zones are particularly vulnerable, as are juveniles of any species.

In low concentrations, hydrocarbons in the ocean can cause behavioral changes in fish, such as an inability to swim properly. Birds and mammals may lose buoyancy and insulation when oil coats their feathers or fur. By grooming and preening, these animals may ingest the oil, which is harmful to their systems and can affect reproduction and offspring. Toxic hydrocarbons also can sink to the bottom of the ocean and contaminate sediments, which are absorbed by bottom dwellers and detritus feeders.

Petroleum also contains heavy metals, which can pollute the air or water and become a health concern. The majority of environmental pollution associated with the petroleum industry comes not from dramatic, one-time events but from the steady emissions related to automobile use and everyday refinery operations and from runoff on urban roads.

DRILLING

Because petroleum reservoirs are below ground, sometimes at great depths, they must be accessed by drilling. Drilling for oil and natural gas on land or offshore is inherently risky because it involves accessing pressurized flammable materials on a large scale and often at distances and depths that are difficult to monitor. In addition, drilling for oil frequently takes place in remote locations, such as the Arctic, where disaster aid is not immediately available.

During drilling, a substance called drilling mud is used to counteract corrosion, lubricate the drill bit, and maintain sufficient pressure to prevent oil and gas from rushing uncontrollably into the well. Drilling mud is made of water, oil, solids like bentonite or barite, chemicals, and sometimes other additives that affect viscosity and weight. If the drilling mud cannot exert sufficient pressure, formation fluids are able to push up against it and enter the well. This is called a kick. When a kick is detected, the composition of the drilling mud is altered to try to counteract the pressure of the reservoir contents. If a kick is not controlled, it can become a blowout (also called a belch), an uncontrolled spewing of oil and gases from the reservoir into the well and out of the well to the surface.

Oil gushers, as blowouts are commonly known, are not a cause for celebration, as they are sometimes shown to be in popular media. Instead, gushers have the potential to cause worker deaths, explosions, fires, and environmental damage. To prepare for the possibility of a blowout, oil wells are equipped with blowout preventers, which can be massive in scale. Their purpose is to clamp down on the wellhead and contain any pressurized contents. When a blowout preventer fails, uncontrolled oil and gas erupt into the ocean or onto the land surrounding the well.

Historically, most drilling has occurred to access oil rather than natural gas, though some natural gas occurs alongside oil. It is cheaper and easier to transport oil, and most natural gas is distributed close to its source for economic reasons. Recently, however, there has been more interest in drilling for gas, particularly with a process called hydraulic fracturing, or fracking, which is used to split rock to access gas. Fracking has been combined with horizontal drilling to create a web of wells that maximize access to pockets of gas. Huge amounts of chemicals are necessary for fracking, and the integrity of the concrete well casings is paramount in preventing water contamination. Fracking is highly controversial because of risks that the extensive rock fracturing will allow toxic substances to reach groundwater supplies.

TRANSPORTATION OF OIL AND GAS

Once oil and gas have been extracted from a reservoir, they must be transported to a refinery to be made into usable products. One of the primary methods of transportation is the oil tanker. Tankers can be used to move oil from offshore rigs to onshore refineries or to transport oil from a remote onshore location to a refinery.

The *Exxon Valdez* was a single-hulled tanker whose 1989 accident and subsequent oil spill led to U.S. legislation to increase tanker safety. The Oil Pollution Act of 1990 required the phasing out of single-hulled tankers and the phasing in of double-hulled tankers for transit within the United States. Double-hulled tankers have an outer hull, an interstitial space for ballast water, then another hull surrounding the tanks of oil. Double-hulled tankers are generally considered safer in terms of their ability to contain oil if the tanker is grounded, but they are less stable because of their higher center of gravity. Countries other than the United States have different specifications for tankers, and some single-hull designs are considered to be as safe as double-hull designs.

Oil is also transported through pipelines. The Trans-Alaska pipeline, for instance, runs above ground, transecting Alaska from Prudhoe Bay to Valdez on Prince William Sound, a distance of roughly 1,300 kilometers (800 miles). Made of steel, the pipeline is vulnerable to corrosion and wear, and like any kind of pipe, it experiences buildup from the contents it transports. A device called a smart pig is periodically run through the pipeline to keep it clear, but even with routine maintenance, pipelines deteriorate and experience breakage.

Pipeline corrosion led to a March, 2006, spill of more than 1 million liters (270,000 gallons) of crude oil on Alaska's North Slope tundra. In the case of the Trans-Alaska pipeline, a winter spill is less harmful than a summer spill because it is more likely to be contained before extensive seepage into the ground or impact to flora and fauna.

OIL SPILLS

Oil spills can occur on land or offshore. On land, a priority is to prevent the spill from entering any water systems, above or below ground, and to keep the spill from reaching areas of human habitation. A localized spill may occur at an oil well or may occur during transit.

Pipeline spills result from corrosion of the pipe or from damage by vandals. An illegal tap of a Mexican pipeline in 2010 triggered an explosion that killed twenty-eight people and caused extensive spillage. An Exxon pipeline ruptured in 2011, pouring oil into the Yellowstone River in Montana.

The oil spills most often noted by popular media occur offshore from well blowouts or accidents during transit by tanker. After a spill in the ocean, a number of natural processes affect the oil. The more volatile compounds in the petroleum will evaporate into the air. The water will disperse the oil instead of mixing with it because oil is lighter than water. Seawater in particular is even less likely to dissolve oil than is freshwater. Sunlight, to some extent, will break down the oil, making it slightly more soluble. Microorganisms also will begin to oxidize the oil.

Cleanup efforts can be mechanical or chemical. Mechanical cleanup includes the use of booms and skimmers. Booms are generally made of plastic or metal, and they serve as simple barriers to the spread of the oil slick. Skimmers are boats that travel along the slick and physically skim the oil from the water. It also is possible to instigate a controlled in situ burn to remove some of the surface oil.

Chemical dispersants can be applied to break up the oil into droplets that will sink instead of remain on the surface of the water. This reduces the chances that the oil will reach sensitive coastal areas, but the toxicity of dispersants makes them controversial because large quantities are required for treatment. Because of their toxicity to marine life, oil spill cleanups must address cleaning both animals and their habitat, including beaches and intertidal rocks. This involves steam cleaning and other painstaking procedures.

THE *EXXON VALDEZ* AND DEEPWATER HORIZON DISASTERS

The *Exxon Valdez* was a giant oil tanker that ran aground at Bligh Reef in Prince William Sound, Alaska, in 1989, puncturing its single hull. A combination of factors, including the remote location and poorly organized response efforts, led to the spilling of approximately 42 million liters (11 million gallons) of oil into the water. It was the largest spill in U.S. history at the time.

A number of cleanup methods were employed to mitigate the effects of the spill. These included burning the oil, using booms and skimmers, and applying chemical dispersants. Both the burn and the dispersants were less effective than they might have been because of poor weather and water conditions. As a result, the oil coated much of the shoreline, affecting birds and seals. Whales, porpoises, and other species were affected in less immediately visible ways.

The Deepwater Horizon explosion in April, 2010, in the Gulf of Mexico made the *Exxon Valdez* spill seem mild in comparison. Deepwater Horizon was a semisubmersible oil rig capable of drilling a well deeper than 10,668 meters (35,000 feet), a depth greater than the height of Mount Everest. The bottom of the well was 5,596 m (18,360 ft, or 3.48 mi) below sea level, and it was under tremendous pressure.

The first sign that something was wrong came when the crew experienced lost returns: The amount of drilling fluid they were sending down was not coming back up again, a situation that indicated leaks in the well walls. The well then began to kick, a warning sign of a potential blowout.

Other signs of trouble existed, including high pressure readings and more return fluids. Hydrocarbons were making their way past the blowout preventer (BOP) and into the well. The crew finally tried to activate the BOP but the BOP was unable to do its job. Highly flammable and explosive gases and liquids surged from the well and the rig caught fire and exploded, killing eleven people and injuring sixteen more.

Five million barrels of oil spilled into the Gulf of Mexico. As with previous spills, dispersants and physical containment measures (booms) were implemented. The flow was finally halted, albeit temporarily, with a capping stack (a type of seal). The stack placement was followed with a static kill, the insertion of heavy drilling mud and subsequent cementing of the well. Later, a relief well was completed, and the reservoir was fully sealed. Investigators failed to find any one cause of the disaster. Though the failure of the BOP was the most immediate reason for the blowout, many other events led to that outcome.

THE LOCAL PETROLEUM INDUSTRY

Much oil and gas exploration and extraction takes place in remote locations. Occasionally, what is normally out of sight is brought to the public's attention.

Public concerns over extensive hydraulic fracturing arise because the process might affect drinking water. Oil spills make headlines because they affect coasts and beaches, bringing crude oil in contact with humans and animals. The Deepwater Horizon explosion had a huge impact on marine life and ecosystems in the Gulf of Mexico and, by extension, on the economy of the Gulf states, which rely heavily on tourism and seafood industries.

Refining is one petroleum industry process that often takes place near population centers. In refining, crude oil is heated and different products (called cuts) are drawn off based upon their boiling points. Gasoline is one of the first products to separate, while heavier products require higher temperatures for extraction. In addition to gasoline, kerosene, diesel, and heavier oils and paraffin are produced. A refinery unit called a cracker can take heavier cuts and break them down to extract gasoline, for which there is more demand.

Petroleum processing entails high temperatures and the use of dangerous chemicals. A number of things can go wrong during refining. The cracking units process thousands of barrels of oil each day, and shutting down the units severely curtails productivity. In 2010, an estimated 7,711 kilograms (17,000 pounds) of benzene spewed into the air of Texas City, Texas, sickening residents. (The hydrogen compressor of the refinery's ultracracker unit had failed, but the refinery continued to process.) Other refinery pollutants include hexane, butane, heptane, pentane, and propylene, which affect the respiratory system and the central nervous system in various ways when they are present in large quantities.

Refinery accidents can be more dramatic than clouds of pollution. The volatile chemicals employed during distillation require careful handling and processing. Small-scale explosions of pressurized pipes can injure or kill workers, but large-scale explosions also occur. For instance, at an isomerization unit (which alters the molecular structure of hydrocarbons to form products such as gasoline additives) at one refinery in the United States, incorrect pressure readings, alarm failure, and the inability of the blowdown drum to vent a sudden rush of gases contributed to an explosion that killed fifteen workers. Flaring equipment and a more effective venting system might have averted the disaster, but keeping refineries updated and in safe working condition is expensive.

J. D. Ho

FURTHER READING

Fischetti, Mark. "The Human Cost of Energy." *Scientific American* 305, no. 3 (2011): 96. Primarily through the use of graphics, this short piece compares the oil and natural gas industries with other energy industries with regard to worker deaths. Also addresses health issues brought on by particulate pollution related to fossil fuel power plants.

Lerner, Steve. *Sacrifice Zones: The Front Lines of Toxic Chemical Exposure in the United States.* Cambridge, Mass.: MIT Press, 2010. A collection of case studies of American cities and towns affected by various types of pollution, including chemicals from oil refineries. Of particular interest is the chapter on a Brooklyn, New York, oil spill.

Magner, Mike. *Poisoned Legacy: The Human Cost of BP's Rise to Power.* New York: St. Martin's Press, 2011. In addition to detailing the events leading to the Deepwater Horizon disaster, this work chronicles lower-profile tragedies, such as refinery explosions, pollution from broken refinery equipment, and the effect of the Deepwater Horizon blowout on the coastal communities of the Gulf of Mexico.

Mooney, Chris. "The Truth About Fracking." *Scientific American* 305, no. 5 (2011): 80-85. In clear language, explains the process of fracking, which is used to obtain natural gas. Elucidates how fracking differs from simple hydraulic fracturing and discusses why communities in which fracking occurs are concerned about their water supplies.

National Commission on the BP Deepwater Horizon Oil Spill and Offshore Drilling. *Deep Water: The Gulf Oil Disaster and the Future of Offshore Drilling.* Washington, D.C.: Author, 2011. This single resource covers almost every aspect of the Deepwater Horizon disaster, including the technical aspects of what went wrong and the impact of the spill on residents.

U.S. Forest Service. *Final Environmental Impact Statement: Changes Proposed to Existing Yellowstone Pipeline Between Thompson Falls and Kingston.* Missoula, Mont.: Author, 2000. Outlines the history of a major pipeline, describes the environmental and political issues involved in its placement and maintenance, and examines the decision-making process with regard to safety and other concerns.

See also: Offshore Wells; Oil and Gas Distribution; Oil and Gas Exploration; Oil and Gas Production; Onshore Wells; Well Logging.

PETROLEUM RESERVOIRS

A petroleum reservoir is a body of rock that contains crude oil, natural gas, or both, that can be extracted by a well. Two conditions must be met in tapping such resources. First, to contain and permit extraction of its fluids, the reservoir rock must be porous and permeable. Second, there must be a reservoir trap, a set of conditions that concentrates the petroleum and prevents its migration to the earth's surface.

PRINCIPAL TERMS

- **field:** one or more pools; where multiple pools are present, they are united by some common factor
- **natural gas:** a flammable vapor found in sedimentary rocks, commonly but not always associated with crude oil; it is also known simply as gas or methane
- **permeability:** a measure of the rate of flow of fluids through a porous medium
- **petroleum:** a dark green to black, flammable, organic liquid commonly found in sedimentary rocks; it looks like used crankcase oil and is also called crude oil or liquid hydrocarbon
- **pool:** a continuous body of petroleum-saturated rock within a petroleum reservoir; a pool may be coextensive with a reservoir
- **porosity:** the percentage of pore, or void, space in a reservoir
- **reservoir:** a body of porous and permeable rock; petroleum reservoirs contain pools of oil or gas
- **trap:** a seal to fluid migration caused by a permeability barrier; traps may be either stratigraphic or structural
- **well log:** a strip chart with depth along a well borehole plotted on the long axis and a variety of responses plotted along the short axis; there are many varieties, including borehole logs, geophysical logs, electric logs, and wireline logs; information may be obtained about lithology, formation fluids, sedimentary structures, and geologic structures

SEDIMENTARY RESERVOIR ROCK

Petroleum (from the Latin) literally means "rock oil." Petroleum reservoirs are volumes of rock that contain or have the potential to contain hydrocarbons (crude oil and natural gas) that can be extracted by wells. A reservoir may contain one or more pools—that is, continuous bodies of oil or gas. One or more related pools form a field. The definition of a pool depends on economic as well as geologic considerations—as the price of oil goes up, the size of a pool goes down. There are four components to a petroleum reservoir: the rock, pores, trap, and fluid.

In the vast majority of cases (more than 99 percent), the reservoir rock is a sedimentary rock. Sedimentary rocks form when unconsolidated sediment, deposited on the earth's surface by water or wind, becomes a solid rock. Two types of sedimentary rocks contain 95 percent of the world's petroleum: sandstones and carbonates. Sandstone reservoirs contain approximately half of the United States' petroleum. Freshly deposited sand is made up of mineral grains, mostly quartz, deposited from water in rivers, along shorelines, and in the ocean in the shallow water adjacent to the continent; less often, it is deposited by wind in sand dunes. Sand-sized grains range from 0.06 to 2.00 millimeters, or smaller than a sesame seed.

Carbonate rocks include limestones and dolostones. Limestones are made of the mineral calcite (calcium carbonate, $CaCO_3$). Sand-sized grains of calcite form when organisms, such as clams and corals, extract calcite (and aragonite, a related calcium carbonate mineral) from water to build their skeletons. When the animal dies, the skeletons are broken into fragments, which are then sorted by waves and currents and deposited along beaches or on the sea floor. Very fine-grained, mud-sized calcite and aragonite may be deposited with the sand-sized grains. This fine carbonate comes from calcareous algae (marine plants). In some cases, corals and algae may combine their activity to construct a wave-resistant mass, a coral reef. Reefs form the best potential reservoir rock. Dolostones, although they account for a relatively small portion of the world's total carbonate mass, contain almost 80 percent of the petroleum in carbonate reservoirs in the United States. Dolostones are dominated by the mineral dolomite. Dolomite forms by the reaction of preexisting calcium carbonate minerals with magnesium-rich solutions. How this replacement takes place is not well understood.

POROSITY AND PERMEABILITY

A petroleum reservoir has the same general attributes as a water aquifer; it is a porous and permeable body of rock that yields fluids when penetrated by a borehole. Porosity is a measure of the pore space in a reservoir, the space available for storage of petroleum. It is usually expressed as a percentage. A freshly deposited unconsolidated sand contains about 25 percent intergranular pore space, normally occupied by water. As the sand is buried deeper and deeper, the grains interpenetrate, and the pores are reduced in size and percentage. At the same time, cements will precipitate in the pores. Thus, the pore space decreases, and the sand becomes a solid—a sandstone. To be a petroleum reservoir, a rock needs 10 percent porosity or more. A good reservoir will have 15 to 20 percent porosity. Sandstone reservoirs seldom exceed 25 percent porosity, but carbonate reservoirs can have up to 50 percent porosity. This greater porosity is a result of very high porosity in the open structure of reef rocks and cave systems of some limestones.

Porosity can be divided into primary and secondary types. Primary porosity is the porosity present at the time the rock was deposited. It depends on several factors, including the roundness and size range of the grains (sorting). Secondary porosity forms after the rock is solidified, when grains and cements are dissolved. Secondary porosity is most common in limestones but is also important in sandstones.

Permeability is a measure of the rate of flow of fluids through a porous medium. In general, the more porous the rock, the more permeable it is likely to be; however, the relation is not simple. For example, Styrofoam (a type of polystyrene plastic) has very high porosity but almost no permeability, which is why Styrofoam cups can hold liquids.

RESERVOIR TRAPS

A petroleum trap is a geometric situation in which an impermeable layer of rock (a caprock) seals a permeable, petroleum-bearing reservoir rock from contact with the earth's surface. A common type of caprock is shale, or consolidated, lithified mud. There are two common types of traps: structural and stratigraphic. Structural traps form where the originally horizontal sedimentary rock layers are disrupted by warping (folding) and breaking (faulting). A much sought-after type of fold trap is an anticline. Anticlinal folds have a cross-section that is concave down (like the letter *A*). Because oil and gas are lighter than water, they migrate to the high point on the anticline

and are prevented from reaching the surface by an impermeable caprock. Fault traps occur where a break in a rock layer brings a reservoir rock into contact with an impermeable rock. Salt-dome traps form where bodies of salt flow upward and pierce overlying rock layers. Salt domes form both anticlinal and fault traps on their crest and margins. They are common on the Gulf Coast of the United States. Stratigraphic traps form where the permeability barrier is a result of lateral or vertical changes in the rock, changes that result from the conditions under which the original sediment was deposited. Stratigraphic traps include buried river channels, beaches, and coral reefs. Geologic methods dominate the search for stratigraphic traps.

Reservoir fluids consist of water, crude oil, and natural gas. When sediments are deposited, they are (or soon become) saturated with water, usually salt water. The oil forms from the alteration (called maturation) of organic material buried in sedimentary rocks. This process takes place in the absence of oxygen at temperatures of 60 to 150 degrees Celsius when the sediments are buried. Natural gas forms by further thermal alterations of the hydrocarbons (thermogenic gas) or by low temperature alteration of near-surface organic material (biogenic gas, the gas that can be seen bubbling up from the bottom in swamps). After formation, the oil and gas must be expelled from its source rock and migrate into a reservoir rock. As oil is lighter than water, it migrates up the water column to the highest location in the trap. If there is no trap, the oil will seep onto the earth's surface and be destroyed by oxidation. Although oil seeps are common, in most cases the oil will be prevented from reaching the surface by a trap. The trapped reservoir fluids then stratify themselves on the basis of density, the lightest at the top (natural gas) and the densest on the bottom (water).

AREAS OF CONCENTRATION

Petroleum reservoirs are not uniformly distributed in time or space; certain areas of the world and certain intervals of geologic time have a disproportionate amount of the world's petroleum reserves. One unifying characteristic of these major areas of concentration is the presence of a basin. A basin is an area where rocks thicken from the margin to basin center, rather like a mud-filled saucer. The geologic conditions in basins favor the formation of oil and gas in the basin center and its migration into traps along the basin margin.

Most of the world's oil is in the Middle East, in such countries as Saudi Arabia (264 billion barrels), Kuwait (104 billion barrels), Iraq (143 billion barrels), and Iran (151 billion barrels). Major Latin American reserves are in Venezuela (296 billion barrels) and Mexico (12 billion barrels), while Africa's largest reserves are in Nigeria (37 billion barrels) and Libya (47 billion barrels). The United States has 19 billion barrels in reserves, of which 85 percent is in four states: Alaska (largely in the North Slope fields around Prudhoe Bay), Texas (both along the Gulf Coast salt domes and in the Permian Basin of West Texas), California (the southwest region of the state), and Louisiana (Gulf Coast salt domes). Oil reserve figures are, at best, moderately good indications of actual oil reserves. Members of OPEC may raise their reserve figures to obtain higher production quotas, and petroleum companies may understate or fail to estimate their reserves for tax and regulatory reasons. While the world's major producing areas are widely separated, they were much closer together 100 million years ago (in the Cretaceous period). Many of them shared a common setting, a now vanished seaway between Eurasia and Gondwanaland that geologists call the Tethys Sea.

In terms of time, very little oil is found in rocks older than 500 million years, although there is record of oil up to a billion years old. Rocks of Jurassic and Cretaceous age (about 65 to 200 million years old) contain 54 percent of the world's oil and those about 35 to 55 million years old, from the Eocene epoch, contain 32 percent of the world's oil. It is not clear whether this young age (geologically speaking) is related to the origin of oil or results from the fact that deep wells cost more money so that older rocks are less thoroughly drilled. It may be that many older petroleum traps no longer exist because of erosion.

EXPLORATION FOR RESERVOIRS

The exploration for petroleum reservoirs has two intimately related aspects, one geophysical and the other geological. In geophysical exploration, the scientist, called an exploration geophysicist, uses the physics of the earth to locate petroleum reservoirs in structural traps. The principal technique is seismic reflection profiling. In this approach, the geophysicist sets off deliberate explosions and uses the energy reflected from subsurface rock layers to interpret the folds, faults, and salt domes that may be present. Seismic reflection is similar to sonar. Variations in the gravity, magnetic, and heat-flow characteristics of the earth may also point to the location of oil and gas pockets in the subsurface.

Geological exploration dominates in the search for stratigraphic traps. Petroleum geologists use facies models to predict the location and extent of petroleum reservoirs. Facies models are based on information about the size, internal characteristics, and large-scale association of modern sediment accumulations. Data based on well logs, obtained from previously drilled wells, are used to prepare structure maps, which show the "topography" of the reservoir surface, and thickness variation (isopachous) maps. Data from well logs, well cuttings, and cores are used to prepare lithofacies maps, which show the lateral variation in the rocks. These variations influence porosity and permeability trends in the reservoir. Such maps can then be compared to the facies models to make predictions about the size and location of the edges of stratigraphic traps in the subsurface. Commonly applied facies models for sandstones include those for rivers, beaches, deltas, and sand dunes. Those for carbonate rocks include beaches, reefs, and dolostones.

Seismic stratigraphy is a technique that combines geology and geophysics. In this approach, the data from artificial explosions are used to interpret the depositional system of the rocks in a way that is similar to radar and sonar. Echoes from the explosions are used to map the three-dimensional structure of the rocks. For example, a delta system has a structure that is distinct from that of sediments deposited in the shallow waters of the continental shelf. On the whole, the most favorable "prospect" is the portion of a trap closest to the earth's surface. Since oil and gas rise to the top of a trap, this location is the most likely volume to contain petroleum.

ECONOMIC CONSIDERATIONS

Petroleum supplies a major portion of the world's fossil energy and, consequently, is an important element in the complex international play of economic forces. In other words, what decides whether a body of rock is a petroleum reservoir is not simply geology but also the reservoir's economic and political setting. The basic problem is that petroleum is a nonrenewable resource, present in finite amounts, and not randomly distributed in the earth's subsurface. Millions of oil and gas wells have been drilled in the continental United States, making it the most mature country in the world from the point of view of petroleum exploration. The "easy" oil has been found, and it is becoming harder to find the fewer and fewer undiscovered and economically exploitable petroleum pools. On a more positive note, however, as drilling density has increased, so has knowledge. Better information, better understanding of how and when petroleum enters a reservoir, and better techniques for finding traps have improved the success ratio of oil drilling.

David N. Lumsden

FURTHER READING

Dandekar, Abhijit Y. *Petroleum Reservoir Rock and Fluid Properties.* Boca Raton, Fla.: CRC Press, 2006. Examines the geological characteristics of petroleum reservoirs. Discusses electrical, mechanical, and physical rock properties. An excellent resource for graduate students.

Gerritsen, Margot G., and Louis J. Durlofsky. "Modeling Fluid Flow in Oil Reservoirs." *Annual Review of Fluid Mechanics* 37 (2005): 211-238. Explains fluid flow equations. Discusses the challenges of crude oil properties and high pressure when accessing oil reservoirs. Covers gas injection processes and carbon dioxide sequestration.

Gluyas, Jon, and Richard Swarbrick. *Petroleum Geoscience.* Malden, Mass.: Blackwell Science, 2004. Provides a good description of tools and methods used in petroleum exploration. Includes images, drawings, and tables, as well as extensive reference and further reading lists and an index. Appropriate for graduate students, academics, or professionals.

Haun, John D., and L. W. LeRoy, eds. *Subsurface Geology in Petroleum Exploration.* Golden: Colorado School of Mines, 1958. Covers a number of specific techniques of use to petroleum geologists: the analysis of well cuttings, cores, and fluids; well-logging methods and interpretation; subsurface stratigraphic and structural interpretation; geochemical and geophysical methods; well drilling; formation testing; and well evaluation. Appropriate for general readers.

Hunt, John Meacham. *Petroleum Geochemistry and Geology.* 2d ed. New York: W. H. Freeman, 1996. Covers many topics, including composition, origin, migration, accumulation, and analysis of petroleum and the application of petroleum geochemistry in petroleum exploration; seep and subsurface prospects; crude oil correlation; and prospect evaluation. Requires some background in chemistry and algebra.

Hyne, Norman J. *Dictionary of Petroleum Exploration, Drilling, and Production.* Tulsa, Okla.: PennWell, 1991. Covers terms associated with petroleum and the petroleum industry. A great resource for the beginner in this field. Illustrations and maps.

_____. *Nontechnical Guide to Petroleum Geology, Exploration, Drilling, and Production.* 2d ed. Tulsa, Okla.: PennWell, 2001. Provides a well-rounded overview of the processes and principles of gas and oil drilling. Discusses exploration, refining, and distribution. Appropriate for oil-industry professionals and general readers. Illustrations, bibliography, and index.

King, Robert E., ed. *Stratigraphic Oil and Gas Fields.* Tulsa, Okla.: American Association of Petroleum Geologists, 1972. Discusses various exploration techniques and case histories of specific reservoirs, pools, and fields. Case histories include information on discovery, development, and trap mechanisms. Appropriate for professional geologists and nonspecialists.

Levorsen, A. I. *Geology of Petroleum.* 2d ed. Tulsa, Okla.: American Association of Petroleum Geologists, 2001. This classic textbook, which was very popular from the mid-1950's through the late 1970's, influenced and unified the way petroleum geologists classify traps and view their discipline. Appropriate for advanced undergraduate and graduate geology students as well as nonspecialists.

Li, Guoyu. *World Atlas of Oil and Gas Basins.* Hoboken, N.J.: John Wiley & Sons, 2011. Includes an overview section followed by information on oil and gas distribution throughout the countries of the world, organized by continent or region, which is covered briefly. Appropriate for undergraduates and oil-industry professionals.

Moore, Carl A. *Handbook of Subsurface Geology.* New York: Harper & Row, 1963. Aimed at undergraduate geology students who intend to go into the oil industry. Stresses the preparation of diagrams pertinent to exploration geologists (structure maps, thickness variation maps, lithofacies maps, and cross-sections) and the interpretation of well logs and evaluation of a formation test.

Selley, Richard C. *Elements of Petroleum Geology.* 2d ed. San Diego: Academic Press, 1998. Intended for undergraduate geology majors. Topics include the properties of oil and gas, exploration methods, generation and migration of petroleum, reservoir and trap characteristics, and basin classification. Requires a background in elementary geology, algebra, and chemistry.

Welte, Dietrich H., Brian Horsfield, and Donald R. Baker, eds. *Petroleum and Basin Evolution: Insights from Petroleum Geochemistry, Geology, and Basin Modeling.* New York: Springer, 1997. Explores the origins of oil and gas from the perspective of mathematical modeling of sedimentary basins. Somewhat technical, but illustrations and maps help to clarify many of the difficult concepts. Bibliography.

See also: Offshore Wells; Oil and Gas Distribution; Oil and Gas Exploration; Oil and Gas Origins; Oil Chemistry; Oil Shale and Tar Sands; Onshore Wells; Well Logging.

PLATINUM GROUP METALS

Although platinum group metals are among the rarest elements known, their chemical inertness, high melting points, and extraordinary catalytic properties make them an indispensable resource for modern industrial society.

PRINCIPAL TERMS

- **catalyst:** a substance that facilitates a chemical reaction but is not consumed in that reaction
- **immiscible liquids:** liquids not capable of being mixed or mingled
- **layered igneous complex:** a large and diverse body of igneous rock formed by intrusion of magma into the crust; it consists of layers of different mineral compositions
- **mafic/ultramafic:** compositional terms referring to igneous rocks rich (mafic) and very rich (ultramafic) in magnesium- and iron-bearing minerals
- **magma:** molten rock material that solidifies to produce igneous rocks
- **placer:** a mineral deposit formed by the concentration of heavy mineral grains such as gold or platinum during stream transport
- **reef:** a provincial ore deposit term referring to a metalliferous mineral deposit, commonly of gold or platinum, which is usually in the form of a layer
- **specific gravity:** the ratio of the weight of any volume of a substance to the weight of an equal volume of water
- **troy ounce:** a unit of weight equal to 31.1 grams; used in the United States for precious metals and gems

EARLY DISCOVERY OF PLATINUM

The average crustal abundance of the platinum group metals is not known exactly but is comparable to that of gold. Platinum, like gold, can be found as a pure metal in stream placer deposits, where its density and resistance to corrosion have resulted in concentration of the metal during stream transport. Platinum appears to have been recovered from placer deposits since earliest times. A hieroglyphic character forged from a grain of platinum has been dated from the seventh century B.C.E., and ancient South American metalsmiths used platinum as an alloy to improve the hardness of gold as early as the first millennium B.C.E.

The name "platinum" is derived from the Spanish word for silver, to which it was originally considered to be inferior. Spanish conquistadors called the metal *platina del Pinto* (little silver of the Pinto) after its discovery in placer gold deposits of the Rio Pinto. Although the metal looked like silver, it proved to be more difficult to shape. Also, with a density similar to that of gold, platinum in small quantities mixed with gold was difficult to detect. Fear of its being used to degrade gold and silver led to a temporary ban against the importation of platinum to Europe. This was an ironic reversal from today's perspective because platinum is frequently more costly than gold nowadays. The metal was eventually brought to Europe and was described by Sir William Watson in 1750.

LATER DISCOVERY OF OTHER GROUP METALS

The platinum group metals consist of six elements so named because they occur with and have similar properties to platinum. In addition to platinum, they include osmium, iridium, rhodium, palladium, and ruthenium. These five metals were discovered in the first half of the nineteenth century by scientists who examined the residue left when crude platinum was dissolved in aqua regia, a mixture of hydrochloric and nitric acids. Because of their resistance to corrosion, all six platinum group metals, along with gold, are referred to by chemists as the noble metals. Economic geologists classify these metals together with silver as the precious metals. The weight of precious metals, like that of gems, is given in troy ounces.

Four of the platinum group metals were discovered by British scientists in 1803. Smithson Tennant discovered osmium and iridium. The name "osmium" was taken from the Greek word *osme* (meaning "smell"), because the metal exudes a distinctive odor, actually toxic osmium tetroxide, when it is powdered. Osmium is bluish white and extremely hard. Its melting point of about 3,045 degrees Celsius is the highest of the platinum group. The specific gravity of osmium has been measured at 22.57, making it the heaviest known element.

Iridium was named by Tennant for the Latin word *iris*, a rainbow, because of the variety of colors produced when iridium is dissolved in hydrochloric acid. Iridium is white with a slight yellowish cast. Like osmium, it is very hard, brittle, and dense. It has a melting point of 2,410 degrees Celsius and is the most corrosion-resistant metal known, rivaling osmium as the densest metal.

Rhodium and palladium were first discovered by William Hyde Wollaston, who named rhodium for the Greek word *rhodon* (a rose) because of the rose color produced by dilute solutions of rhodium salts. The metal is actually silvery white, has a melting point of about 1,966 degrees Celsius, exhibits a low electrical resistance, and is highly resistant to corrosion. Wollaston named palladium for Pallas, a recently discovered asteroid named for the ancient Greek goddess of wisdom. Palladium is steel-white, does not tarnish in air, and has the lowest specific gravity (12.02) and melting point (1,554 degrees Celsius) of the platinum group metals. Like platinum, it has the unusual property of absorbing enormous volumes of hydrogen.

The existence of ruthenium was proposed in 1828 but not established until 1844 by Russian chemist Karl Karlovich Klaus. He retained the previously suggested name ruthenium in honor of Ruthenia, the Latinized name for his adopted country of Russia. Ruthenium is a hard, white, nonreactive metal with a specific gravity slightly greater than that of palladium (12.41) and a melting point of 2,310 degrees Celsius.

BUSHVELD COMPLEX

Important deposits of platinum group metals are found in the Bushveld complex of South Africa, the Stillwater complex of Montana, and at Sudbury, Ontario. A large deposit similar to that at Sudbury exists at Norilsk in Siberia.

The Bushveld complex holds a special place as history's greatest source of platinum, and it still contains the world's largest reserves of platinum group elements. A large layered igneous complex, Bushveld is located north of the town of Pretoria in the northeast corner of South Africa and covers an area roughly the size of the state of Maine. It formed 1.95 billion years ago when an enormous intrusion of mafic magma, the largest known mafic igneous intrusion, was injected and slowly cooled in the earth's crust. As cooling and solidification occurred, the denser, mafic minerals became concentrated downward in the magma chamber, and the igneous rock became stratified with ultramafic layers at greater depth and layers of increasingly less mafic rocks upward.

Placer platinum was discovered in South Africa in 1924 and was subsequently traced by Hans Merensky to its source, a distinctive igneous layer that became known as the Merensky Reef. The reef is located in the lower part of the Bushveld complex, about one-third of the distance from the base to the top. Although commonly less than 1 meter in thickness, it has been traced for 250 kilometers around the circumference of the complex, and nearly half of the world's historic production of platinum group metals has come from this remarkable layer.

The average metal content in the layer is about one-third of a troy ounce per ton of rock, or about 1 part platinum group metals in 100,000 parts rock, or 10 parts per million—thousands of times the normal crustal abundance of these metals. Platinum is the most abundant metal extracted from the reef. Other platinum group minerals are, in order of abundance, palladium (27 percent), ruthenium (5 percent), rhodium (2.7 percent), iridium (0.7 percent), and osmium (0.6 percent). Significant quantities of gold, nickel, and copper are present as well. The mining of such a narrow layer is so labor-intensive that each South African miner produces only about 30 ounces of platinum group metals per year.

STILLWATER COMPLEX

The Stillwater complex is a large-layered, mafic to ultramafic igneous complex remarkably similar to Bushveld. It is exposed for about 45 kilometers along the north side of the Beartooth Mountains in southwest Montana. The Stillwater area has long been famous for its large but low-grade chromium-rich layers, and platinum was discovered there in the 1920's. Serious exploration for economic concentrations of platinum, however, was initiated in 1967 by the Johns-Manville Corporation. This led, in the 1970's, to identification of the J-M Reef, a palladium- and platinum-rich horizon between 1 and 3 meters thick, which, like the Merensky Reef, can be traced through most of the complex.

The Stillwater complex formed 2.7 billion years ago. Like Bushveld, the complex is layered with ultramafic igneous rocks at the base and mafic rocks higher up. The J-M Reef lies slightly above the ultramafic zone. It has an average ore grade of 0.8 ounce of platinum group metals per ton of rock with a 3:1 ratio of palladium to platinum. Mining of the J-M Reef commenced in 1987. The ore is concentrated at the mine site and shipped to Antwerp, Belgium, for refining. The J-M Reef is the only significant source of platinum in the United States, and the Stillwater mine is projected to be in production until about 2020.

Sudbury Complex

The Sudbury complex, just north of Lake Huron in southeast Ontario, Canada, is similar in many ways to Bushveld and Stillwater, but it is not conspicuously layered. Nickel and copper are the main products, with platinum group metals produced as a by-product. Nickel was discovered in the Sudbury area in 1856. At that time, the region was largely wilderness, and government survey parties were engaged in running base, meridian, and range lines in preparation for a general survey and subdivision of northeastern Ontario. Considerable local magnetic attraction and the presence of iron were noted during the survey. An analysis of the rock showed that it contained copper and nickel as well. The Sudbury magma formed 1.85 billion years ago as the result of a large meteorite impact and now appears as an elliptical ring of mafic igneous rock 60 kilometers long by 27 kilometers wide. Some fifty ore deposits are found along and just outside its outer edge. Platinum group elements are extracted at Sudbury from the residues left over after smelting of nickel and copper.

Research into Origin of Deposits

Recent industry demands for platinum have stimulated research into the origin of platinum group metal deposits. Much of this research has been directed toward understanding the world's great mafic igneous complexes. It has long been recognized that the origin of the Merensky and J-M reefs is tied to the formation of the layering within these mafic igneous complexes. Geologists have firmly established that the mafic magmas originate in the earth's mantle and that they derive trace amounts of the platinum group metals from their mantle source rocks. It is also well known that as these magmas crystallize, various minerals are precipitated from the magma in a fairly well-established sequence. Geologists have long believed that the layered mafic igneous complexes represent the settling into layers of precipitated mineral grains according to density. Repetitions and modifications in the layering are considered to be the result of currents churning within the hot magma, or pulses of

new magma injected into the intrusion. As crystallization proceeds, volatile elements such as water, carbon dioxide, and sulfur gradually become concentrated in the remaining magma. Sulfur has the ability to scavenge many metals, including iron, copper, nickel, and platinum group metals. Laboratory studies have shown that if the sulfur concentration is high enough, metallic sulfide droplets can form a separate, immiscible liquid. Like water in oil, the denser sulfide magma droplets sink and accumulate toward the base of the intrusion. Many geologists believe the layers and masses of metallic sulfide ore found in the large mafic igneous complexes formed in this manner.

Detailed studies of the chemical composition of the Stillwater complex, however, suggest that the crystallization sequence was interrupted at about the level of the J-M Reef by an influx of new, somewhat different magma. The evidence suggests that the magma was sulfur-saturated, and its influx is believed to have triggered the precipitation of the platinum minerals. Research on the Bushveld complex has also suggested multiple episodes of magma injection, with the Merensky Reef forming at the base of a magma pulse. It should be emphasized, however, that even after a century of investigation, the origin of the ore at the Bushveld, Stillwater, and Sudbury complexes is still the subject of considerable debate.

Hypotheses for ore formation at Sudbury include the separation of droplets of immiscible sulfide liquid from a mafic magma, but a lively debate exists regarding the mechanism by which the mafic magma was produced. It has long been noted that a distinctive zone of broken and shattered rock many kilometers wide underlies and surrounds the Sudbury igneous complex. Overlying the complex is a thick sequence of fragmentary rocks, originally interpreted as being volcanic in origin. In 1964, Robert Dietz suggested that Sudbury was the site of a tremendous meteorite impact that formed a large crater and shattered the surrounding rock. It was later proposed that the impact caused the melting that produced the mafic igneous rock and that the supposed volcanic rock was actually material that had been ejected during the meteorite impact and had fallen back into the crater. In this view, the igneous rocks are not intrusions, as was long believed, but a great sheet of impact melt, or molten rock formed at the surface by the heat generated by impact. This theory continues to cause controversy. The evidence for a meteorite impact is strong, but some geologists consider any impact to be unrelated to the Sudbury deposit. Others not only believe the impact theory but also suggest that the Bushveld magma was triggered by a meteorite impact. The debate continues, and its outcome has implications for the presence or absence of metallic ore deposits beneath the large lunar craters.

While field and laboratory work on the great platinum deposits of the world continues, so does experimental work aimed at understanding the conditions under which these deposits formed. Laboratory scientists are duplicating conditions found in nature in order to increase their understanding of the behavior of platinum group elements during crystallization from magmas, and during the formation of immiscible liquids, as well as their mobility at submagmatic temperatures in water-rich solutions.

INDUSTRIAL VALUE

The platinum group metals are used extensively in modern industrial society because of their chemical inertness, high melting points, and extraordinary catalytic properties. Platinum group metals are important parts of the automotive, chemical, petroleum, glass, and electrical industries. Other important uses are found in dentistry, medicine, pollution control, and jewelry. The automotive industry is the single largest consumer of platinum group metals. Since 1974, platinum-palladium catalysts have been used in the United States to reduce the emission of pollutants from automobiles and light-duty trucks. A typical catalytic converter contains 0.057 ounce of platinum, 0.015 ounce of palladium, and 0.006 ounce of rhodium. In the European Economic Community, all cars with engines larger than 2 liters produced after October, 1988, must have converters.

The electrical industry is the second largest consumer of platinum group metals. Palladium is used in low-voltage electrical contacts, and platinum electrical contacts protect ships' hulls from the corrosive activity of seawater. The dental and medical professions utilize nearly as much platinum group metals as the electrical industry. Palladium is alloyed with silver, gold, and copper to produce hard, tarnish-resistant dental crowns and bridges. Other medical uses include treatment for arthritis and some forms of cancer. Platinum group metals are also used internally in cardiac pacemakers and in a variety of pin, plate, and hinge devices used for securing human bones.

The chemical industry uses platinum and palladium as catalysts for a variety of reactions involving hydrogen and oxygen. Molecules of either of these gases are readily adsorbed onto the surface of the metals, where they dissociate into a layer of reactive atoms. Oxygen atoms on platinum, for example, increase the rate at which sulfur dioxide, a common industrial pollutant, is converted into sulfur trioxide, a component of sulfuric acid, the most widely used industrial chemical. Other pollution-control devices are aimed at the control of ozone levels in the cabins of commercial jet airplanes and the oxidation of noxious organic fumes from factories and sewage treatment plants. Platinum group catalysts are also used in the production of insecticides, some plastics, paint, adhesives, polyester and nylon fibers, pharmaceuticals, fertilizers, and explosives. In the petroleum industry, platinum group metals are used by refineries both to increase the gasoline yield from crude oil and to upgrade its octane level. Because material used as a catalyst is not consumed in the chemical reactions (although small amounts are lost), the many important chemical uses actually consume only a small amount of the platinum group metals.

Platinum group metals' ability to withstand high temperatures and corrosive environments has led to their use in the ceramics and glass industry. Thin strands of glass are extruded through platinum sieves to make glass fibers for insulation, textiles, and fiber-reinforced plastics. High-quality optical glass for television picture tubes and eyeglasses is also melted in pots lined with nonreactive platinum alloys. Crystals for computer memory devices and solid-state lasers are grown in platinum and iridium crucibles. As ingots and bars, platinum group metals are sold to investors, and platinum and palladium alloys are commonly used for jewelry. Brilliant rhodium is electroplated on silver or white gold to increase whiteness, wear, and resistance to tarnishing.

SUPPLY VERSUS DEMAND

The world's reserves of platinum group metals are large, but distribution is concentrated in a relatively few locations. South Africa is the largest producer of platinum. Japan is the largest consumer nation, and the United States is second.

Historically, U.S. production has been extremely small and has consisted almost entirely of platinum and palladium extracted during the refining of copper. Stillwater's J-M Reef is a significant discovery, but it is expected to supply only about 7 percent of the nation's projected needs. Therefore, the U.S. State Department has added the platinum group metals to a list of strategic materials that are considered essential for the economy and for the defense of the United States, and that are unavailable in adequate quantities from reliable and secure suppliers.

Consequently, the search continues for new sources of platinum group metals. A potential source may be in the incrustations of iron and manganese found on the submerged slopes of islands and seamounts throughout the world's oceans. These metallic crusts and nodules are believed to have formed by extremely slow precipitation from seawater. Although they are composed mostly of iron and manganese, they contain many metals, including those of the platinum group, and the volume of these deposits is staggering. While commercial development is unlikely before the early part of the twenty-first century, these ferromanganese crusts are considered to be an attractive, long-term resource.

Eric R. Swanson

FURTHER READING

CPM Group. *The CPM Platinum Group Metals Yearbook 2011.* Hoboken, N.J.: John Wiley & Sons, 2011. Written by a precious metals and commodities research and consulting company for the platinum group metals market. Includes charts and tables covering supply and demand. Discusses inventory statistics, marketing trends, and physical characteristics and structure of platinum group metals.

Crundwell, Frank K., Michael S. Moats, Venkoba Ramachandran, et al. *Extractive Metallurgy of Nickel, Cobalt, and Platinum-Group Metals.* Amsterdam: Elsevier, 2011. Discusses nickel and cobalt metallurgy, the platinum-group metals, as well as smelting, refining, and extraction processes. A final chapter covers the recycling of these metals. Includes appendices with data used in the extraction and refining processes. Well indexed.

Evans, Anthony M. *An Introduction to Economic Geology and Its Environmental Impact.* Oxford: Blackwell Scientific Publications, 1997. Provides an excellent introduction into the field of economic geology.

Emphasis is on types of deposits, environments of formation, and economic value, along with the impact those deposits have on their environments. Written for undergraduate students. Well illustrated and includes an extensive bibliography.

Gee, George E. *Recovering Precious Metals.* Palm Springs, Calif.: Wexford College Press, 2002. Focuses on gold and silver recovery and recycling processes. Discusses many precious metal sources, including silver from photography solutions, and mirror maker's solutions. Discusses separating gold and silver from platinum.

Gray, Theodore. *The Elements: A Visual Exploration of Every Known Atom in the Universe.* New York: Black Dog & Leventhal Publishers, 2009. An easily accessible overview of the periodic table. Provides a useful introduction of each element. Those studying nonchemistry disciplines may find this useful as well. Appropriate for general readers. Includes excellent images and complete with indexing.

Guilbert, John M., and Charles F. Park, Jr. *The Geology of Ore Deposits.* Long Grove, Ill.: Waveland Press, Inc., 1986. The best available college-level text on ore deposits. Covers the geology of the Sudbury and Bushveld complexes in the chapter on deposits related to mafic igneous rocks.

Hartley, Frank R., ed. *Chemistry of the Platinum Group Metals: Recent Developments.* New York: Elsevier, 1991. Provides information on the procedures used to determine the geochemical makeup and properties of platinum group metals.

Laznicka, Peter. *Giant Metallic Deposits.* 2d ed. Berlin: Springer-Verlag, 2010. Discusses the location, extraction, and future use of many metals; geodynamics that result in large metal ore deposits and the composition of metals within deposits; and hydrothermal deposits, metamorphic associations, and sedimentary associations. Organized by topic rather than metal.

Loferski, Patricia J. "Platinum-Group Metals." In *Minerals Yearbook.* Washington, D.C.: Department of the Interior, issued annually. Published annually in three volumes by the United States Department of the Interior. Contains general information on nonmetallic resources and all metals, including platinum; information on U.S. resources by state; and comparable information for other countries of the world.

Mertie, J. B., Jr. *Economic Geology of the Platinum Metals.* U.S. Geological Survey Professional Paper 630. Washington, D.C.: Government Printing Office, 1969. A comprehensive survey and extensive bibliography of all platinum deposits.

Peterson, Jocelyn A. *Platinum-Group Elements in Sedimentary Environments in the Conterminous United States.* Washington, D.C.: Government Printing Office, 1994. Contains detailed information on platinum ores and deposits, as well as sedimentary basins. Filled with color and black-and-white maps throughout, and also includes a fold-out map.

St. John, Jeffrey. *Noble Metals.* Alexandria, Va.: Time-Life Books, 1984. Part of Time-Life's Planet Earth series. Appropriate for a general audience.

Vermaak, C. Frank. *The Platinum-Group Metals: A Global Perspective.* Randburg, South Africa: Mintek, 1995. Offers a look into platinum ores and mines, platinum mining, and the platinum group industry. Color illustrations and bibliographical references.

Weast, Robert C., ed. *CRC Handbook of Chemistry and Physics.* 92d ed. Boca Raton, Fla.: CRC Press, 2011. Contains in condensed form an immense amount of information from the fields of chemistry and physics, and alphabetized description of all the known elements. Written in a manner accessible to those with little scientific background. New editions are published at frequent intervals, ensuring that the information is up to date.

Young, Gordon. "The Miracle Metal: Platinum." *National Geographic* 164 (November, 1983): 686-706. Offers excellent pictures and covers many of the uses of platinum.

See also: Aluminum Deposits; Building Stone; Cement; Coal; Diamonds; Dolomite; Earth Resources; Fertilizers; Gold and Silver; Hydrothermal Mineralization; Industrial Metals; Industrial Nonmetals; Iron Deposits; Manganese Nodules; Pegmatites; Salt Domes and Salt Diapirs; Sedimentary Mineral Deposits; Strategic Resources; Uranium Deposits.

PLUTONIC ROCKS

Plutonic rocks crystallize from molten magma that is intruded deep below the earth's surface. Exposed at the surface by erosion, they occur mainly in mountain belts and ancient "shield" areas of continents. Many of the world's principal ore deposits are associated with plutonic rock bodies; the rocks themselves may be also exploited economically, as in the production of granite for building stone.

PRINCIPAL TERMS

- **anorthosite:** a light-colored, coarse-grained plutonic rock composed mostly of plagioclase feldspar
- **batholith:** the largest type of granite/diorite pluton, with an exposure area in excess of 100 square kilometers
- **feldspar:** an essential aluminum-rich mineral in most igneous rocks; two types are plagioclase feldspar and alkali feldspar
- **gabbro:** a coarse-grained, dark-colored plutonic igneous rock composed of plagioclase feldspar and pyroxene
- **granite:** a coarse-grained, commonly light-colored plutonic igneous rock composed primarily of two feldspars (plagioclase and orthoclase) and quartz, with variable amounts of dark minerals
- **magma:** a molten silicate liquid that upon cooling crystallizes to make igneous rocks
- **pluton:** generic term for "intrusion"
- **stock:** a granite or diorite intrusion, smaller than a batholith, with an exposure area between 10 and 100 square kilometers

PLUTONIC ROCK OCCURRENCE

Plutonic rocks crystallize from molten silicate magmas that intrude deep into the earth's crust. This same magma (molten silicate liquid) may eventually flow out onto the surface of the earth as lava that crystallizes to produce volcanic rocks. Volcanic rocks can be distinguished from plutonic rocks by the relative size of their mineral grains. Rapidly cooled volcanic rocks have nearly invisible crystals (or glass), while the slowly cooled plutonic rocks have larger crystals clearly visible to the naked eye.

Because they form deep underground, plutonic rocks require special circumstances to become exposed at the surface. Uplift of the crust in a mountain range or high plateau accompanied by nearly constant erosion by streams or glaciers may eventually uncover once-buried plutonic rocks. The best places to see excellent exposures of these rocks are in mountain ranges such as the Rockies, Sierra Nevada, and Appalachian ranges of North America, the Alps of Europe, and the Himalayas of Asia, among many others. Another major plutonic rock terrain exists in "shield" areas found on all the world's major continents. These areas comprise the ancient cores of the continents and consist of rocks that are billions of years old, most of the rocks having once existed in ancient mountain ranges now eroded down to relatively flat plains. In North America, this area is called the "Canadian Shield" and covers most of Canada and the northern portions of the states of Minnesota, Wisconsin, Michigan, and New York; buried extensions of the shield underlie much of the rest of the eastern and central United States. In certain places, these rocks are exposed on the surface, as in the center of the Ozark Plateau of southeastern Missouri.

PLUTONIC ROCK VARIETIES

Plutonic rocks come in many varieties depending upon the chemistry of the parent magma and their mode of emplacement in the crust. Igneous magmas vary chemically between two major extremes: "felsic" magma, in which the concentration of dissolved silica (silicon dioxide) is high and the concentrations of iron and magnesium are relatively low, and "mafic" magma, in which the concentration of silica is low and the concentrations of iron and magnesium are relatively high. Granite and related rocks (generally light-colored) are produced by crystallization of felsic magmas. Their light color and the presence of the mineral quartz (silicon dioxide) distinguish granitic rocks from the dark plutonic rock gabbro, which crystallizes from mafic magmas. Other rocks, such as diorite, crystallize from magma that is intermediate in composition between felsic and mafic extremes. Diorite and its relatives are generally gray-colored and are commonly mistaken for granite. For example, much of the Sierra Nevada range in eastern California is composed of granodiorite, although it is popularly known as a "granite" mountain range.

GRANITIC OR DIORITE BODIES

Plutonic rocks are emplaced in the crust in a variety of geometric forms, collectively called "plutons." By far the largest plutons are batholiths, huge masses of granite, diorite, or both with surface exposures exceeding 100 square kilometers. In western North America, some batholiths are exposed over a considerable portion of whole states, such as the Boulder and Idaho batholiths of Montana and Idaho, and the Sierra Nevada and Southern California batholiths of California. Batholiths also occur in the Appalachian Mountains, particularly in the White Mountains of New Hampshire and in parts of Maine. Most batholiths attain their large size by the successive addition of smaller plutons called "stocks." Stocks are generally exposed over tens or hundreds of square kilometers.

Minor plutonic bodies include dikes and sills, tabular intrusions that commonly represent magma that has filled in fractures that either cut across layers in country rock (dikes) or that intruded parallel to rock layers (sills). Some sills fill up with so much magma that they expand and force the overlying layers of rock to bow upward. Such plutons are called "laccoliths," and some, like the Henry Mountains of southeastern Utah, have attained the scale of small mountains.

GABBROIC BODIES

Batholiths, stocks, and laccoliths are predominantly granitic or diorite bodies. Mafic (gabbroic) intrusions occur in their own particular geometric forms, mostly as dikes and sills. Some of these bodies reach enormous size and are exposed over areas that rival large stocks or even granitic batholiths. Many of these bodies show evidence of having concentrated layers of crystals that gravitationally settled after crystallization. Known variously as "gravity-stratified complexes" or "layered mafic-ultramafic complexes," these bodies are commonly rich sources of economically important metallic ores, particularly those of chromium and platinum. The best examples of these bodies are in South Africa (the Bushveld and the Great Dyke), Greenland (the Skaergaard Complex), and North America (the Stillwater complex in Montana, the Muskox complex and Kiglapait complex in Canada, and the Duluth Gabbro north of Lake Superior).

Other mafic-ultramafic complexes may produce massive layers of metallic sulfides rich in copper, nickel, and variable amounts of gold and silver, among other metals. In North America, the Sudbury "nickel irruptive" in Ontario, Canada, is a well-known example of a rich sulfide ore body associated with gabbroic magma. Melting to produce this magma has been attributed to the ancient impact of a large meteoroid. Gravitational segregation of metallic oxides is also known from gabbroic magmas. The rich iron deposits in Kiruna, Sweden, are believed to form from the settling of large blobs of liquid iron oxide that later crystallized to the mineral magnetite. Titanium deposits in anorthosites (feldspar-rich gabbro) at Allard Lake, Quebec, may have formed by gravitational concentration of iron-titanium-rich fluids within the gabbroic magmas that also produced the associated anorthosite rock.

PLUTON-PRODUCING MAGMAS

The origin of the magmas that create plutons depends in large part on their chemical compositions. Most granite-composition magma probably arises by partial melting of siliceous metamorphic rocks in the deeper parts of the continental crust. Rocks in these regions have compositions that are already close to granitic, so melting them leads inevitably to the production of granitic liquids. These liquids (magma), being less dense than the surrounding, cooler rocks, rise through the crust and join with other bodies to make stocks and, possibly, batholiths.

Mafic magmas of the kind that crystallize gabbro originate in the upper mantle, where they arise by the partial melting of mantle peridotite. Peridotite is the major constituent of the upper mantle and consists of the mineral olivine (iron-magnesium silicate) with minor pyroxene (calcium iron-magnesium silicate) and other minerals. Laboratory experiments have shown that partially melting peridotite at pressures like those in the mantle produces mafic liquids capable of crystallizing gabbro (or basalt at the surface).

A special kind of gabbro, anorthosite, is produced by the separation and concentration of plagioclase feldspar from the gabbroic magma subsequent to its production in the mantle. Because plagioclase commonly is less dense than the surrounding iron-rich gabbroic liquid from which it crystallized, plagioclase crystals may literally float to the top of the magma chamber to form a concentrated mass of feldspar. Anorthosite makes up most of the light-colored regions of the moon (the lunar highlands) but also occurs as large intrusions on Earth, as in the Duluth Gabbro complex of northern Minnesota; in the Adirondack Mountains of New York State; the Laramie Anorthosite of Wyoming; and the Kiglapait, Nain, and Allard Lake complexes in Labrador. Mysteriously, most of Earth's anorthosite bodies were generated during only one restricted period in geologic time, between about 1 and 1.5 billion years ago.

The origin of diorite magma is complicated by the fact that it can arise in a variety of ways. Granodiorite plutons, like those in the Sierra Nevada range, represent magma reservoirs under now-vanished volcanoes. The magma that produced these intrusions was generated by melting along a subduction zone, where the Pacific Ocean lithospheric plate (crust plus uppermost mantle) was sinking under the North American plate in a zone roughly parallel to the West Coast. A similar process is currently producing the active volcanoes in the Cascade Range of the Pacific Northwest. Any number of parent materials being pulled down to great depth (and, thus, greater temperature) in a subduction zone may mix with subducted ocean water and subsequently melt to produce diorite-type magma. These materials include hydrated (water-saturated) oceanic basalt lava, and even some hydrated mantle materials (serpentine). Ample evidence suggests that some diorite is also produced by the physical mixing of granitic magma produced by the melting of continental crustal materials, and gabbroic magma produced in the mantle. These contrasting liquids intermingle on their way upward, finally intruding as intermediate-composition diorite magmas.

ALKALINE PLUTONS

One other group of plutonic rock is worthy of mention, even though surface exposures are relatively rare. These are the "alkaline" intrusives, so named because they tend to be enriched in the alkali elements potassium and sodium and relatively depleted in silica. Alkaline magmas originate by partial melting of mantle peridotite at extreme depths and high pressures (20 kilobars or greater). Because high pressure acts to discourage melting (even at high temperatures), special conditions are required

(commonly tectonic rifting) to generate alkaline magmas, making them among the rarest of igneous rocks.

A fairly familiar variety of alkaline-type volcano-intrusive body is the kimberlite "pipe" that in some areas contains diamond. Diamond-bearing kimberlites are especially common in South Africa and other African countries, India, and Russia. North American localities include Murfreesboro, Arkansas, and the Arctic of Canada. Kimberlite rock itself is a complex mixture of mantle and crustal fragments, carbonate minerals, and silicate minerals crystallized from the alkaline magma. Diamonds, if present, were swept up by the magma from high-pressure areas of the mantle and propelled by expanding carbon-dioxide gas to the surface at speeds estimated, in some cases, to exceed supersonic velocities. Surface deposits (diatremes) of kimberlite consist of volcanic ash ejected during powerful volcanic explosions.

Other important alkaline plutons include the nepheline syenites, light-colored, coarse-grained rocks consisting mostly of the mineral nepheline (sodium aluminum silicate). Important North American localities include Magnet Cove, Arkansas, and Bancroft, Ontario. Depending on the particular locality, these rocks are potential repositories of rare elements, and thus unusual minerals. Commercial exploitation of nepheline syenites has produced the metals beryllium, cesium, thorium, uranium, niobium, tantalum, and zirconium, to name just a few. These rock bodies may also be a fertile source of apatite (hydrated calcium phosphate), an ore of phosphorus, and the dark-blue mineral sodalite (sodium aluminum silicate chloride), used as a semiprecious gemstone and for sculpted carvings. Another alkaline plutonic rock rich in rare elements and minerals is carbonatite, a magmatic rock composed mostly of calcium, magnesium, and sodium carbonates. Like nepheline syenites, carbonatites are highly prized for their mineral treasures, although they are relatively rare, and their individual surface exposures are limited in size. The best examples are in the East African rift and in South Africa.

STUDY METHODS

Plutonic rocks are igneous rocks; thus, their study entails the same methods that might also apply to volcanic rocks. Field studies of plutons include the construction of geologic maps showing spatial distribution and structure of the various rock types in the pluton. Most plutons contain more than one type of igneous rock or at least show some chemical and mineralogical variation from one locality to the next. During mapping or other field surveys, samples are normally collected to be chemically analyzed by various techniques, including atomic absorption analysis, X-ray fluorescence, or neutron activation analysis. Individual minerals in the rocks may be chemically analyzed using an electron microprobe, a machine that gives a full chemical analysis of a 1-micron spot on a single mineral in a matter of minutes. More sophisticated analyses include isotopic abundance ratios and radiometric ages determined from mass spectrometry. Collected chemical data are normally plotted on diagrams that help show how and why the plutons may have changed over time, and may lead to an understanding of the ultimate origin of their parent magmas.

The identities, textures, and spatial orientations of minerals in rocks are assessed by preparing microscopic slides of thin slices of the rocks called "thin sections." A skilled igneous petrologist (geologist who specializes in igneous rocks) can determine much about the history of a plutonic rock merely by studying a thin section under a microscope. For example, the identification of constituent minerals under the microscope serves to classify the rock, and the relative volume of individual minerals provides hints about the rock's chemistry. For example, a rock with a high volume of dark, mafic (iron and magnesium-rich) minerals would suggest a high concentration of iron and magnesium in that rock compared to one with fewer dark minerals.

ECONOMIC VALUE

Plutonic rocks are the source of many of the raw materials that are used in industrial society. Granite and diorite are used as construction stones in buildings and monuments and as crushed stone for roadways and concrete. Gabbro, however, is generally shunned for decorative purposes because its high iron content causes it to oxidize (rust) over time. Conversely, the special kind of light-colored gabbro, called "anorthosite" (mostly plagioclase feldspar), is prized as a polished building facing stone; it is also used to grace floors or countertops in banks and office buildings.

Additionally, many of the richest metallic ore deposits originate in or adjacent to plutonic bodies. The extensive list of metals from granitic deposits includes copper, gold, silver, lead, zinc, molybdenum, tin, boron, beryllium, lithium, and uranium. Gabbroic and related deposits contribute nickel, iron, titanium, chromium, platinum, copper, gold, and silver.

Plutons are also the source of some of the most precious and semiprecious gemstones. For example, the precious gems topaz and emerald occur in ultra-coarse grained granitic deposits called "pegmatites." Pegmatites also provide the semiprecious gems aquamarine (a blue-green form of emerald), tourmaline (elbaite), rose quartz, citrine (yellow quartz), amethyst, amazonite (aqua-colored microcline feldspar), and zircon (zirconium silicate). Also, large mica crystals from pegmatites are used as electrical and thermal insulators, and feldspar minerals are powdered to make porcelain products and potassium-rich fertilizer. The principal source of the element lithium, used in lubricants and psychoactive drugs, is granitic pegmatite. Lithium is obtained from spodumene (a lithium pyroxene) and lepidolite (a lithium mica), minerals that occur exclusively in pegmatites.

In addition to providing a highly desirable construction stone, anorthosite plutons may also be the source of the semiprecious gemstone labradorite. Labradorite is a high-calcium form of plagioclase feldspar (the major mineral in anorthosites) that displays a green, blue, and violet iridescence similar to the play of colors in the tail of a male peacock. Gem-quality labradorite crystals may be made into jewelry, and polished slabs of labradorite anorthosite are used as countertops, desktops, and building facings.

GEOLOGIC VALUE

In a larger sense, plutonic rocks (particularly granitic and diorite plutons) form the bulk of the world's continents. During the early evolution of the earth, low-density granitic plutons rising from the early, primitive mantle coalesced to form the cores of the first continents. In later eras, continents have continued to grow, as plutons and overlying volcanic rocks have added new material to the margins. This process is well illustrated by the plutonic terrain of the Sierra Nevada range, the andesite (the volcanic equivalent of diorite) volcanoes of the Cascades of the Pacific Northwest, and the Andes Mountains of South America. Were it not for the formation and expansion of continents by the continuing addition of plutons over geologic time, life on Earth would be considerably different. Without continents and the dry land they provide, life would be confined to the oceans, with obvious implications for the evolution of human beings.

Finally, plutonic rocks in natural settings enhance the beauty and general aesthetic value of the landscape. Deeply eroded plutons have produced some of the most striking landscapes in North America, Europe, and Asia, many of which have been set aside as parks and recreation areas. In North America, especially magnificent landscapes in eroded plutons occur at Yosemite National Park in the Sierra Nevada range of California, the Boulder batholith area of Montana, the high peaks area of the Adirondack Dome of New York State, and the White Mountains of New Hampshire.

John L. Berkley

FURTHER READING

Ballard, Robert D. *Exploring Our Living Earth.* Washington, D.C.: National Geographic, 1983. Aimed at the general reader and available in most public libraries. Covers the earth as an "energy machine," with sections on volcanism, mountain-building, earthquakes, and related phenomena. Richly illustrated with colored photographs and drawings. An excellent source of information on plate tectonics, the unifying theory that seeks to explain the origin of major Earth forces and surface features. Because plutonic rocks are involved in both mountain-building and volcanic activity, an understanding of plate tectonics is essential to understanding the origin of plutons. Glossary, bibliography, and comprehensive index.

Bédard, Jean H. "Parental Magmas of the Nain Plutonic Suite Anorthosites and Mafic Cumulates: A Trace Element Modelling Approach." *Contributions to Mineralogy and Petrology* 141 (2001): 747–771. Discusses the trace elements found within mafic and anorthositic rock formations. Draws conclusions on the melt origins of these cumulates and the details of crystallization.

Best, Myron G. *Igneous and Metamorphic Petrology.* 2d ed. Malden, Mass.: Blackwell Science Ltd., 2003. A college-level textbook that should be accessible to most readers. Part 1 contains comprehensive treatments of all major plutonic rock bodies, complete with drawings, diagrams, and photographs detailing the essential features of each pluton type. Chapter 1 contains a section on how petrologists study rocks. The appendix contains chemical analyses of plutonic rocks and descriptions of important rock-forming minerals. One of the best books available for the serious student of igneous and metamorphic rocks.

Beus, S. S., ed. *Rocky Mountain Section of the Geological Society of America: Centennial Field Guide.* Vol. 2. Denver, Colo.: Geological Society of America, 1987. Produced for professional geologists but also useful to the general reader. Describes automobile field trips to geological localities in the Rocky Mountain states. Illustrated with photographs, geological cross-sections, various diagrams, and maps. Descriptions and explanations of the various localities allow an appreciation for the details of plutons and their relationships to associated rocks.

Decker, Robert, and Barbara Decker. *Volcanoes.* 4th ed. San Francisco: W. H. Freeman, 2005. Suitable for the general reader, text provides useful background for the understanding of igneous rock bodies.

Grotzinger, John, et al. *Understanding Earth.* 5th ed. New York: W. H. Freeman, 2006. One of many introductory textbooks in geology for college students.

Hacker, Bradley R., Ju G. Liou, et al., eds. *When Continents Collide: Geodynamics and Geochemistry of Ultra-High Pressure Rocks.* Boston: Kluwer, 2010. An excellent introduction to the study of geologic metamorphism. Examines the geochemical and geophysical effects of high pressures on metamorphic rocks. Intended for readers with some background in Earth sciences. Bibliography and index.

Jensen, Mead L., and Alan M. Bateman. *Economic Mineral Deposits.* New York: John Wiley & Sons, 1981. One of the most comprehensive single-volume treatments of world ore deposits available. Illustrated with abundant monochrome diagrams and photographs; includes tables listing statistics about major ore-bearing regions. Important ore deposits associated with plutonic intrusions are described in detail. Extensive index. Available in most college or metropolitan public libraries.

Jerram, Dougal, and Nick Petford. *The Field Description of Igneous Rocks.* 2d ed. Hoboken, N.J.: Wiley-Blackwell, 2011. Begins with a description of field skills and methodology. Contains chapters on lava flow and pyroclastic rocks. Designed for student and scientist use in the field.

Prinz, Martin, George Harlow, and Joseph Peters. *Simon & Schuster's Guide to Rocks and Minerals.* New York: Simon & Schuster, 1978. Readily available in most bookstores. Contains beautiful color photographs of museum-quality minerals and rocks, along with comprehensive descriptions. Provides a concise description of igneous processes, including plutonic bodies and intrusive processes. Explains how magmas are generated and emplaced in the crust and how they become differentiated. Includes a glossary and index. Highly recommended both for "rock hounds" and professionals interested in exploring plutonic rocks in the field.

Smith, David G., ed. *The Cambridge Encyclopedia of Earth Sciences.* New York: Cambridge University Press, 1981. One of the best and most comprehensive resources on Earth processes available for the nonspecialist. Richly illustrated with colored photographs, drawings, and diagrams. Provides in-depth explorations of plate-tectonic concepts relevant to plutonic and volcanic processes, as well as topics concerning plutonic rocks and minerals. Includes an extensive glossary and further reading section.

Webster, J. D. *Melt Inclusions in Plutonic Rocks: Short Course Series 36.* Ottawa: Mineralogical Associations of Canada, 2006. Discusses melt inclusion research as a tool for determining melt composition. Developed as the text for a short course, it covers research methods and recent results. Multiple chapters discuss plutonic rocks.

Will, Thomas M. *Phase Equilibria in Metamorphic Rocks: Thermodynamic Background and Petrological Applications.* New York: Springer, 1998. Offers a clear description of phase rule and its effects on the equilibrium of metamorphic rocks. Offers an in-depth examination of the geochemical makeup of metamorphic rocks and their properties. Bibliographic references.

Wilson, B. M. *Igneous Petrogenesis: A Global Tectonic Approach.* Dordrecht: Springer, 2007. Provides fundamental information on igneous petrology. Discusses a broad range of topics related to igneous rocks, including geophysics and geochemistry, trace elements, radiogenic isotopes, and global tectonic processes.

See also: Andesitic Rocks; Anorthosites; Basaltic Rocks; Batholiths; Carbonatites; Feldspars; Granitic Rocks; Igneous Rock Bodies; Kimberlites; Komatiites; Metamorphic Rock Classification; Orbicular Rocks; Pyroclastic Rocks; Rocks: Physical Properties; Xenoliths.

PYROCLASTIC FLOWS

A pyroclastic flow is a fast-moving avalanche of hot gas, fluidized lava fragments, and ash that is ejected from an erupting volcano. In some cases reaching speeds of 160 kilometers (100 miles) per hour or higher, pyroclastic flows can reach temperatures of more than 540 degrees Celsius (1,000 degrees Fahrenheit), are extremely destructive, and can cover large areas surrounding an eruption.

PRINCIPAL TERMS

- **caldera:** an underground magma chamber
- **fluidization:** process whereby the rocks and minerals contained in a volcanic eruption are vaporized within a pyroclastic flow
- **ignimbrite:** a pumice-filled pyroclastic flow caused by the collapse of a volcanic column and accelerated by gravity
- **lahar:** combination of water, ash, and mud that creates tremendous mudslides in a pyroclastic flow
- **nuée ardente:** a type of dense pyroclastic flow that occurs after a volcanic dome collapse; also known as block and ash and Peleen flow
- **Plinian:** volcanic eruption that sends ash and debris high above the volcano in a column
- **pumice:** light, frothy volcanic rock formed when lava is cooled above ground following a violent eruption and pyroclastic flow
- **vesiculated:** gas-infused rock within a pyroclastic flow

BASIC PRINCIPLES

A pyroclastic flow occurs when superheated gas and rock are ejected forcefully from a volcano during eruption. Pyroclastic flows may occur when, after a volcanic eruption sends a plume of ash and rock fragments into the sky, gravity causes the cloud to collapse and the superheated material to flow to the surface. In other cases, pyroclastic flows can occur when a volcano explodes outward, sending a pyroclastic flow along the side of the mountain and along the ground.

Pyroclastic flows comprise hot gas, rock, and ash. When such a flow is released from an active volcano, it travels downhill at extreme speed, causing enormous destruction along its path. Pyroclastic flows have been the most destructive components of some of the most violent eruptions in history, including the eruptions of Mount Vesuvius, Krakatoa, and Mount St. Helens.

Pyroclastic flows combine with a great deal of other material as they travel after an eruption. Flows are so destructive that they can, for example, disrupt lakes and ponds that are located on the volcano's surface. Additionally, the intense heat that accompanies the flows is enough to instantly melt the snow and ice packs at summits. Furthermore, volcanic eruptions release a tremendous amount of water vapor into the sky, along with ash and soot, causing intense rainstorms. All of this water combines with mud and falling ash to form a slurry called a "lahar." Lahars add to the pyroclastic flow's overall destructiveness.

There are two general components of a pyroclastic flow. One is basal flow, which comprises dense rock fragments and material. Basal flow moves along the contours of the land as it travels down and away from the volcano. Another component of a pyroclastic flow is the cloud of ash, gas, and other lighter material, which surges above the basal flow.

In addition to causing extreme damage to the area surrounding a volcano, pyroclastic flows als deposit a large volume of volcanic material. Pyroclastic flow deposits, which are sometimes found many miles from the eruption site, occur in three general layers. The bottom layer (the basal or sillar level) contains fine grains of pumice and glass. The middle level usually comprises welded pumice containing a combination of unsorted fragments of pumice and other volcanic minerals (such as silicon). Atop the welded middle level is a layer of ash. In many cases, the pyroclastic flow deposit is mixed with the water and debris of lahars. Scientists often look to flow deposits for information about the type of activity that has occurred in a volcanic area.

FLUIDIZATION AND FLOW VELOCITY

A central element of the velocity of a pyroclastic flow is its fluidization. A pyroclastic flow contains large quantities of rock and ash in addition to gas. However, substances therein are superheated to the point in which the particles have little friction. Combined with the water and gas from the eruption and with the vaporized water and snow that the flow encounters as it moves downslope, the flow travels rapidly and unobstructed, moving along the contours of the mountain.

Fluidization is, in turn, dependent on the types of material contained in the flow. The granular minerals and rocks are major contributors to how the flow moves, playing a role in the flow's density. The more dense the flow, the faster and farther it travels. Pyroclastic flows have been known to travel great distances (sometimes over large bodies of water). In Japan, for example, an eruption in 4000 B.C.E. sent a pyroclastic flow more than 60 km (37 mi) from its source, including 9.5 km (6 mi) over water. Researchers found that the flow deposit was about 2 meters (6.5 feet) thick only, a testament to the flow's low density.

Scientists are still attempting to study the natural process of fluidization and how pyroclastic flows carry fluidized minerals from volcanic sites. Increased research on the natural process of fluidization will help researchers understand the nature of pyroclastic flows and how they transport minerals across the earth's surface.

IGNIMBRITES

There are two general classifications of pyroclastic flows. The first of these is the ignimbrite or "pumice flow." Ignimbrites are volcanic clouds that consist largely of pumice, a light, gaseous rock that is formed from frothy lava. These clouds are vesiculated (the rock contained within the cloud contains a large amount of gas), which means that they have a low density. The grains of pumice contained in the flow vary in size, particularly as they proceed from their volcanic source.

Because they are low in density, ignimbrites are expelled rapidly. During a Plinian eruption, wherein the eruption is vertical, a cloud is expelled high into the air above the volcano. However, as the eruption continues to send denser debris into the sky, the debris causes the cloud to collapse. The main component of ignimbrite flow, containing large concentrations of pumice, stays on the ground; the cloud of ash hovers above it. Meanwhile, the fluidized pumice flow rushes down the slope and outward, accelerated by its fall from high above the volcano.

Pumice flows emanating from another type of eruption—the caldera collapse—can be some of the most devastating types of pyroclastic flows. Here, a caldera (an underground magma chamber) is drained in a volcanic eruption. The collapse sends a sheet of pumice outward rather than upward, filling adjacent valleys and causing widespread destruction. In some cases, ignimbrites emanating from a caldera collapse have been known to wipe out an area hundreds of kilometers wide, leaving extremely thick deposits of pumice far from the volcano.

NUÉE ARDENTE FLOW

The second type of pyroclastic flow is called the "nuée ardente" ("glowing cloud"), so named by French geologist Antoine Lacroix. Nuées ardentes are frequently associated with Peleen eruptions, after the deadly 1902 eruption of Mount Pelée in Martinique. That eruption was marked by two massive, fiery pyroclastic flow clouds. The vertical column's intimidating luminescence captured observers' attention, although the surface level of the flow was far more destructive.

Nuées ardentes are considerably denser in composition than are ignimbrites. Nuées ardentes occur when the dome (a mass of viscous lava that cools and forms a bulbous mass around the volcano's vents) of a volcano collapses explosively in an eruption. The release of fluidized ash and rock in the pyroclastic flow is too dense to travel into a volcanic column; it therefore spills rapidly from the vents and along the contours of the mountain at incredible speeds. Nuées ardentes are nonvesiculated because they contain blocks of fragments from the dome and ash and other particles (hence, the more accurate name provided to these occurrences: "block and ash" flows).

When a dome collapses, the flow surges from the vent, upward or outward, or both. The dense cloud then bends down as the pyroclastic flow begins to travel along the surface of the volcano. A second component of the nuée ardente, the ash cloud surge, moves above the flow, masking it. Because they are so dense, nuées ardentes are not as extensive as ignimbrites. Still, they move with tremendous speed and, as was the case of St. Pierre, Martinique, in 1902, can cause widespread damage to the nearby environment.

HISTORIC EXAMPLES OF PYROCLASTIC FLOWS

Pyroclastic flows have been the main force behind some of the most awesome and destructive volcanic eruptions in recorded history. For example, 7,700 years ago, a 3.6-km (12,000-foot) volcano dubbed

Mount Mazama erupted in what is now central Oregon. The caldera feeding the volcano collapsed with such force that the resulting ignimbrite flow sent a sheet outward for nearly 32 km (20 mi) in every direction. The eruption destroyed a large portion of the mountain itself and created what is now Crater Lake, the deepest freshwater lake in North America.

The origin of the term "Plinian" (named for Pliny the Younger, who recorded his observations of the subsequent devastation) comes from the eruption of Mount Vesuvius in 79 C.E., one of the most infamous volcanic eruptions in human history. This Plinian eruption created a devastating ignimbrite pyroclastic flow, which quickly engulfed the nearby Roman cities of Pompeii and Herculaneum, leaving both communities (and the people and animals who did not escape the rapidly advancing flow) buried in pumice.

On May 7, 1902, Mount Pelée erupted near St. Pierre in Martinique. Witnesses report having seen two clouds emitted from atop the 1,400-m (4,600-ft) mountain: A black, glowing column spewed vertically and another rushed down the sides of the mountain. Only two of the thirty thousand people living in St. Pierre survived the nuée ardente flow.

One of the most famous examples of pyroclastic flows in modern history is the 1980 eruption of Mount St. Helens in Washington State. This event featured both types of pyroclastic flow. On May 18, the mountain erupted violently, sending a Plinian column more than 19 km (12 mi) into the sky. When the massive column collapsed, the resulting ignimbrite flows, coupled with a significant lahar, destroyed millions of trees and caused massive steam explosions in nearby lakes and rivers.

In the following few weeks, a lava dome formed at the mountain's vent. During subsequent eruptions in June, the dome's collapse produced several nuée ardente flows, triggering more avalanches and mudslides that contributed to the major alteration of the area around the mountain.

EFFECTS OF PYROCLASTIC FLOWS

Pyroclastic flows are major events that can substantially alter the landscape and environment in and around a volcano. Many of the effects of pyroclastic flows are destructive in nature, devastating all forms of life in their paths.

Pyroclastic flows also can significantly alter the topography of the mountain and surrounding areas, creating new crevasses and surface features along the volcano's slopes and, therefore, new routes for descending lava and debris in future eruptions. In the long term, nuées ardentes and ignimbrites also can alter the shapes and courses of rivers, lakes, and streams.

STUDYING PYROCLASTIC FLOWS

One of the difficulties scientists encountered while attempting to study the volcanic activity at Mount St. Helens in 1980 was that so much of the mountain was obscured by the massive ash clouds that the eruption produced. Indeed, even when dormant a volcano presents a number of challenges for scientists, not the least of which includes traveling up steep, dangerous terrain to the cone.

Modern technology has greatly enhanced scientists' ability to study pyroclastic flows and other aspects of volcanism. For example, the use of computer models has enabled scientists to analyze potential flow courses and velocities. Such data enable them to isolate hazard zones and potentially save lives in the event of an eruption.

In other cases, scientists may utilize radar systems, based on the ground and mounted on aircraft and satellites, to study the paths of pyroclastic flow surges. For example, in a recent study, scientists successfully analyzed flow deposits and mapped flow paths on an active volcano by using synthetic aperture radar. The use of such technology precludes the need to climb up a potentially dangerous slope in unpredictable weather and atmospheric conditions. Other scientists have used ground-penetrating radar systems to map flow paths and layers in the topography of a volcano, making it easier to create models of volcanic activity.

In addition to radar and computer modeling, another invaluable technological tool in the study of pyroclastic flows is the global positioning system. Utilizing this satellite-based technology, scientists can analyze pyroclastic flow patterns, movement, and composition despite cloud cover and other visual obstructions.

Michael P. Auerbach

FURTHER READING

Constantinescu, Robert, Jean-Claude Thouret, and Ioan-Aurel Irimus. "Computer Modeling as a Tool for Volcanic Hazards Assessment: An Example of Pyroclastic Flow Modeling at El Misti Volcano, Southern Peru." *Geographia Technia* 11, no. 2 (2011): 1-14. Describes the need to study pyroclastic flows and other areas of volcanic activity to develop systems that detect volcanic hazards. The authors suggest computer modeling as a means toward this end.

Dale, Virginia H., et al., eds. *Ecological Responses to the 1980 Eruption of Mount St. Helens.* New York: Springer, 2005. Discusses the tremendous destruction that occurred at Mount St. Helens during the eruptions and pyroclastic flows in 1980. The editors review the ecologic rebound that has occurred since the eruption.

Oppenheimer, Clive. *Eruptions That Shook the World.* New York: Cambridge University Press, 2011. Describes some of the most significant volcanic eruptions that have occurred through Earth's history. Discusses how these events have shaped (and reshaped) the planet's geography and topography.

Petersen, R. E. "Geological Hazards to Petroleum Exploration and Development in Lower Cook Inlet." *Environmental Assessment of the Alaskan Continental Shelf: Lower Cook Inlet Interim Synthesis Report* (1979): 51-70. Discusses the risks of petroleum exploration projects in Augustine Island in the western Gulf of Alaska. Describes the 1976 eruption of that island's volcano, which included nuée ardente pyroclastic flows and major lava flows and ash clouds.

Saucedo, R., et al. "Modeling of Pyroclastic Flows of Colima Volcano, Mexico: Implications for Hazard Assessment." *Journal of Volcanology and Geothermal Research* 139, nos. 1/2 (2005): 103-115. Examines a week-long series of eruptions that occurred at a Mexican volcano in 1913 that included pyroclastic flows. Attempts to reconstruct the flow paths to better understand the volcano's activity and the hazards it presents.

Scarth, Alwyn. *La Catastrophe: The Eruption of Mount Pelée, the Worst Volcanic Disaster of the Twentieth Century.* New York: Oxford University Press, 2002. A detailed look at the 1902 eruption of Mount Pelée in Martinique, including the 300-mph nuée ardente that killed nearly thirty thousand people in the nearby town of St. Pierre.

See also: Geologic Settings of Resources; High-Pressure Minerals; Hydrothermal Mineralization; Pyroclastic Rocks; Sulfur.

PYROCLASTIC ROCKS

Pyroclastic rocks form from the accumulation of fragmental debris ejected during explosive volcanic eruptions. Pyroclastic debris may accumulate either on land or under water. Volcanic eruptions that generate pyroclastic debris are extremely high-energy events and are potentially dangerous if they occur near populated areas.

PRINCIPAL TERMS

- **ash:** fine-grained pyroclastic material less than 2 millimeters in diameter
- **ignimbrite:** pyroclastic rock formed from the consolidation of pyroclastic-flow deposits
- **lapilli:** pyroclastic fragments between 2 and 64 millimeters in diameter
- **pumice:** a vesicular glassy rock commonly having the composition of rhyolite; a common constituent of silica-rich explosive volcanic eruptions
- **pyroclastic fall:** the settling of debris under the influence of gravity from an explosively produced plume of material
- **pyroclastic flow:** a highly heated mixture of volcanic gases and ash that travels down the flank of a volcano; the relative concentration of particles is high
- **pyroclastic surge:** a turbulent, low-particle concentration mixture of volcanic gases and ash that travels down the flank of a volcano
- **stratovolcano:** a volcanic cone consisting of lava, mudflows, and pyroclastic rocks
- **tephra:** fragmentary volcanic rock materials ejected into the air during an eruption; also called pyroclasts
- **tuff:** a general term for all consolidated pyroclastic rocks
- **volatiles:** fluid components, either liquid or gas, dissolved in a magma that, upon rapid expansion, may contribute to explosive fragmentation

FORMATION OF PYROCLASTIC ROCKS

Pyroclastic rocks form as a result of violent volcanic eruptions such as that of Mount St. Helens in 1980 or Mount Vesuvius in 79 C.E. Molten rock, or magma, within the earth sometimes makes its way to the surface in the form of volcanic eruptions. These eruptions may produce lava or, if the eruption is highly explosive, fragmental debris called tephra or pyroclasts. The term "pyroclastic" is from the Greek roots *pyros* ("fire") and *klastos* ("broken"). Dissolved water and gases (volatiles) are the source of energy for these explosive eruptions. All molten rock contains dissolved fluids such as water and carbon dioxide. When the molten rock is still deep within the earth, the confining pressure of the overlying rock keeps these volatiles from being released. When the magma rises to the surface during an eruption, the pressure is lowered, and the gases and water may be violently released, causing fragmentation of the molten rock and some of the rock surrounding the magma. This type of explosive eruption is more common in rocks rich in silica, which are more viscous (flow more thickly) than those that are silica-poor. External sources of water, such as a lake or groundwater reservoir, may also provide the necessary volatiles for an explosive eruption.

Pyroclastic debris can be produced from any of three different types of volcanic eruptions: magmatic explosions; phreatic, or steam, explosions; and phreatomagmatic explosions. Magmatic explosions occur when magma rich in dissolved volatiles undergoes a decrease in pressure such that the volatiles are rapidly released or exsolved. The solubility of volatiles in magma is partially controlled by confining pressure, which is a function of depth. Solubility decreases as the magma rises toward the surface. At a certain depth, carbon dioxide and water begin exsolving and become separate fluid phases. At this point, the magma may undergo explosive fragmentation either through an open vent or by destroying the overlying rock in a major eruptive event.

As a magma rises toward the surface, it may encounter a groundwater reservoir or, in a subaqueous vent, interact with surface water. In both cases, the superheating and boiling of water followed by its explosive expansion to gas may fragment the magma and the surrounding country rock. The ratio between the mass of water and the mass of magma controls the type of eruption. If there is little water in relation to magma, the explosive activity may be confined to the eruption of steam and is called a phreatic explosion. If the magma contains significant quantities of dissolved volatiles and encounters a large amount of water, the resulting explosion is termed phreatomagmatic.

COMPOSITION OF PYROCLASTIC DEPOSITS

Pyroclastic deposits are composed of tephra, or pyroclasts. These fragments can have a wide range of sizes. Particles less than 2 millimeters in diameter are termed ash; those between 2 and 64 millimeters are called lapilli; and those greater than 64 millimeters are called blocks, or bombs. There are three principal components that make up pyroclastic debris: lithic fragments, crystals, and vitric, or juvenile, fragments. Lithic fragments can be subdivided into pieces of the surrounding rock explosively fragmented during an eruption (accessory lithics), pieces of already solidified magma (cognate lithics), and particles picked up during transport of eruptive clouds down the flanks of a volcano (accidental lithics). Crystal fragments are whole or fragmented crystals that had solidified in the magma before eruption. Vitric or juvenile fragments represent samples of the erupting, still molten, magma. They may be either partly crystallized or uncrystallized (glass). Pumice is a type of juvenile fragment that contains many vesicles, or holes, as a result of the rapid exsolution of gases during eruption. Small, very angular glass fragments are called shards.

TYPES OF PYROCLASTIC DEPOSITS

Three types of pyroclastic deposit can be distinguished based on the type of process that forms the deposit: pyroclastic-fall deposits, pyroclastic-flow deposits, and pyroclastic-surge deposits. These types can all be formed by any of the previously described different types of volcanic eruption. Any of these deposits may be termed a tuff if the predominant grain size is less than 2 millimeters.

Pyroclastic-fall deposits form from the settling of particles out of a plume of volcanic ash and gases erupted into the atmosphere that form an eruption column. Tuffs formed in this way are coarsest near the eruption center and become progressively finer farther away. Ash falls can also be derived from the top of more dense pyroclastic flows, as the finer-grained material is turbulently removed from the upper portion of the pyroclastic flow and then settles to the ground.

Two types of pyroclastic deposits result from the formation of dense clouds of ash during an eruption and the subsequent transport of debris, in the form of a hot cloud of ash, lapilli, and gases. Pyroclastic flows have a relatively high particle concentration and in some areas—the western United States, for example—these flows form enormous deposits with volumes as large as 3,000 cubic kilometers. Pyroclastic-flow deposits rich in pumice are termed ignimbrites. Pyroclastic surges are generally turbulent, low-particle concentration density currents. Surge deposits are volumetrically less important than those of pyroclastic flows but can be very destructive. Both flows and surges may have emplacement temperatures of up to 800 degrees Celsius. Because of their lower density and turbulent nature, pyroclastic surges may attain velocities up to 700 kilometers per hour. It is this combination of speed and temperature that makes pyroclastic surges so dangerous.

STUDY OF PYROCLASTIC ROCKS

Geologists study pyroclastic rocks using field techniques, laboratory analyses, and theoretical considerations of eruption processes. Observations of deposits in the field remain a significant cornerstone of geologic interpretation. Pyroclastic deposits form essentially as sedimentary material—that is, as fragments or clasts moved in air or water and deposited in layers. As such, many of the techniques used by sedimentologists (geologists interested in the formation and history of sedimentary rocks) are employed in the study of pyroclastic deposits. Careful examination of a variety of sedimentary features within pyroclastic deposits can aid in the interpretation of the processes of transport and deposition. This information, studied over as wide a geographic area as possible to ascertain systematic changes in the deposits, will assist in understanding the geologic history of the region.

Much can be learned through analysis of the composition of pyroclastic rocks, which is generally done using a variety of laboratory techniques. The use of specialized microscopes allows geologists to examine very thin sections of rock in order to observe textures and to discern mineral composition. Geochemical techniques have become very popular and powerful in the study of all kinds of rocks, including pyroclastic rocks. By looking at the amounts of certain elements that occur in extremely low abundances and also at relative proportions of certain types of isotopes (naturally occurring forms of the same element that differ only in the number of neutrons in the nucleus), scientists can understand more about the processes taking place deep within the earth that lead to the formation of magma and eventually to the eruption of pyroclastic debris. Scanning electron microscopes have been used to study in detail the surface features and textures of fine volcanic ash particles. This information can lead to better understanding of eruptive and transport processes that formed and deposited the pyroclastic particles.

Theoretical studies associated with pyroclastic rocks revolve primarily around considerations of the mechanics of high-temperature, high-velocity eruption clouds and their transport and deposition. This type of reasoning allows a geologist to infer certain conditions of eruption from an analysis of the deposits. The geologic rock record has abundant pyroclastic deposits, and it is through this type of inference that geologists interpret the geologic history of a region. A comprehensive understanding of pyroclastic deposits must include a thorough understanding of the processes by which the deposit forms.

Association with Violent Eruptions

Most pyroclastic rocks are associated with stratovolcanoes, also called composite volcanoes. These volcanic edifices are built by a combination of extrusive and explosive processes and thus are formed of both pyroclastic debris and lava. Well-known examples of stratovolcanoes include Mount St. Helens, Mount Fuji, and Mount Vesuvius. Volcanoes such as those found on the Hawaiian Islands are of a less energetic variety (with less explosive eruptions) called shield volcanoes; this variety produces insignificant quantities of pyroclastic material. Stratovolcanoes are located around the world and are associated with the global process of plate tectonics, the slow motion of large slabs of crust in response to flow in the earth's mantle. It is at the zone of plate destruction that rock is melted and makes its way to the surface, sometimes producing pyroclastic eruptions.

Violent volcanic eruptions that may produce pyroclastic deposits are among the most powerful events occurring on Earth. Historically, many of the most destructive volcanic eruptions have involved pyroclastic surges. The eruption in 79 C.E. of Mount Vesuvius generated pyroclastic debris that buried the towns of Pompeii and Herculaneum, killing large numbers of people. In 1902, on the island of Martinique in the Caribbean, the violent eruption of Mount Pelée produced a pyroclastic surge that swept down on the city of St. Pierre, killing all but a handful of a population of about thirty thousand. The eruption of Mount St. Helens in 1980 and of El Chichón in Mexico in 1982 both produced pyroclastic surges. Two thousand people were killed as a result of the El Chichón eruption. Pyroclastic surges and flows generally do not present a hazard beyond a radius of about 20 kilometers. In 1991, Mount Pinatubo in the Philippines unleashed the largest eruption in nearly a century.

Pyroclastic deposits form a major portion of some volcanic terrains. Some of these deposits are enormously extensive, indicating that the eruptions that produced them were much larger than any witnessed in modern times. It is not clear whether these deposits reflect an overall increase in volcanic activity in Earth's

A pyroclastic flow deposit. (U.S. Geological Survey)

past or whether this type of titanic eruption occurs sporadically throughout geologic time. Titanic eruptions inject so much debris into the upper atmosphere that global weather can be affected. Earth experienced brilliant red sunsets and lowered temperatures because dust blocked the sun for several years following the 1883 eruption of Krakatau in the strait between Java and Sumatra, which completely destroyed an island and discharged nearly 20 cubic kilometers of debris into the air. The explosion was heard nearly 5,000 kilometers away in Australia, and darkness fell over Jakarta, 150 miles away. In 1815, an even larger eruption of Tambora, Indonesia, vented so much ash that global climate was significantly cooled. Some geologists speculate that enormous volcanic eruptions in Earth's past even led to extinctions, including that of the dinosaurs, by producing so much ash that the amount of sunlight Earth received was reduced and the entire food chain disrupted. The largest pyroclastic eruptions of all are those in which the roof of a magma chamber simply caves in and hundreds, even thousands, of cubic kilometers of ash are erupted. Such eruptions, sometimes called "supervolcanoes," include several in Yellowstone National Park (in the last 2 million years); Toba, Sumatra (70,000 years ago); Long Valley, California (700,000 years ago); and the Jemez Caldera, New Mexico (1.2 million years ago).

Bruce W. Nocita

FURTHER READING

Blong, R. J. *Volcanic Hazards: A Sourcebook on the Effects of Eruptions.* Sydney, Australia: Academic Press, 1984. Discusses the nature of volcanic hazards with case histories. Suitable for college-level students.

Branney, Michael J., and Peter Kokelaar. *Pyroclastic Density Currents and the Sedimentation of Ignimbrites.* London: Geological Society of London, 2002. Describes ignimbrite deposits and pyroclastic currents. Presents field studies, experiments, and research methods. Includes photographs, tables, references, and extensive indexing.

Cas, R. A. F., and J. V. Wright. *Volcanic Successions: Modern and Ancient.* Winchester, Mass.: Unwin Hyman, 1987. Eleven of the fifteen chapters deal wholly or in part with pyroclastic rocks. Takes a sedimentological approach to the study and interpretation of pyroclastic deposits. Indispensable for geologists interested in pyroclastic rocks. Suitable for college-level students.

Cattermole, Peter John. *Planetary Volcanism: A Study of Volcanic Activity in the Solar System.* 2d ed. New York: John Wiley & Sons, 1996. Examines volcanism and geology on Earth and other planets in the solar system. A good introduction for the layperson without much background in astronomy or Earth sciences.

Decker, Robert, and Barbara Decker. *Volcanoes.* 4th ed. San Francisco: W. H. Freeman, 2005. Provides the reader with a good overview of different types of volcanoes and volcanic processes, including those that produce pyroclastic debris. Suitable for high school students.

Faure, Gunter. *Origin of Igneous Rocks: The Isotopic Evidence.* New York: Springer-Verlag, 2010. Discusses chemical properties of igneous rocks, and isotopes within these rock formations. Provides specific locations of igneous rock formations and rock origins. Includes many diagrams and drawings along with the overview of isotope geochemistry in the first chapter to make this accessible to undergraduate students as well as professionals.

Fisher, R. V., and H. U. Schmincke. *Pyroclastic Rocks.* New York: Springer-Verlag, 1984. Provides one of the most comprehensive treatments of pyroclastic deposits and the processes that form them. Suitable for college-level students.

Francis, Peter, and Clive Oppenheimer. *Volcanoes.* 2d ed. New York: Oxford University Press. 2004. Covers the volcanic activity from lava flows to pyroclastic currents. Provides many examples from Hawaii, Italy, Ethiopia, Japan, Mount St. Helens, and Krakatau. Written for the layperson, but some knowledge of geology is helpful.

Grotzinger, John, et al. *Understanding Earth.* 5th ed. New York: W. H. Freeman, 2006. One of many introductory textbooks in geology for college students. Offers a good chapter on igneous rocks.

Jerram, Dougal, and Nick Petford. *The Field Description of Igneous Rocks.* 2d ed. Hoboken, N.J.: Wiley-Blackwell, 2011. Begins with a description of field skills and methodology. Contains chapters on lava flow and pyroclastic rocks. Designed for student and scientist use in the field.

Sigurdsson, Haraldur, ed. *Encyclopedia of Volcanoes.* San Diego, Calif.: Academic Press, 2000. Contains a complete summary of the scientific knowledge of volcanoes. Eighty-two well-illustrated overview articles are accompanied by a glossary of key terms. Written in a clear style that makes it generally accessible. Cross-references and index.

Simkin, Tom, L. Siebert, L. McClelland, et al. *Volcanoes of the World: A Regional Directory, Gazetteer, and Chronology of Volcanism During the Last Ten Thousand Years*. Berkeley: University of California Press, 2011. Suitable for all readers. Provides recent studies and describes new eruption events. Includes photographs of common rocks, updated data on volcanoes, and new references.

Sutherland, Lin. *The Volcanic Earth: Volcanoes and Plate Tectonics, Past, Present, and Future*. Sydney, Australia: University of New South Wales Press, 1995. Provides an easily understood overview of volcanic and tectonic processes, including the role of igneous rocks. Includes color maps and illustrations, as well as a bibliography.

See also: Andesitic Rocks; Anorthosites; Basaltic Rocks; Batholiths; Building Stone; Carbonatites; Earth Resources; Granitic Rocks; Igneous Rock Bodies; Kimberlites; Komatiites; Orbicular Rocks; Pegmatites; Plutonic Rocks; Rocks: Physical Properties; Xenoliths.

Q

QUARTZ AND SILICA

Silica is one of the most common substances in the earth's crust and mantle. It is composed of silicon dioxide tetrahedra that bond according to unique atomic characteristics to form crystalline structures. Quartz is a crystalline form of silica that forms under conditions of high temperatures and pressures in liquid environments. Silica and quartz are used for a variety of industrial applications, including chemistry and construction, and in the development of fiber-optic technologies.

PRINCIPAL TERMS

- **amethyst:** a variety of quartz noted for its violet to purple color produced by irradiated molecules of iron contained within its crystalline structure
- **chalcedony:** a form of cryptocrystalline quartz formed from microcrystals of quartz and other included minerals
- **conchoidal fracture:** fracturing of the type exhibited by materials that do not break along planes of separation
- **feldspar:** a silicate rock that forms in igneous environments and is the most common mineral in the earth's crust
- **fiber optics:** flexible transparent fiber made of fused quartz or silica glass; used in computer and telecommunications applications
- **flint:** a type of microcrystalline quartz that exists in large quantities within the earth's crust and is notable for its glass-like characteristics
- **igneous rock:** rocks formed from magma that cools either on the earth's surface or in pockets within the mineral layers of the earth's crust
- **quartz:** crystalline form of silica that develops when silica is heated in a liquid or semiliquid environment and then slowly cooled
- **silicates:** compounds originating from tetrahedra of silicon dioxide
- **tetrahedron:** a polyhedron composed of four triangular sides organized into pyramidal shape

SILICON DIOXIDE

Silicon dioxide (SiO_2), also called silica, is an abundant chemical substance in hundreds of different types of minerals and rocks. It forms a majority of Earth's lithosphere.

The upper layers of the earth consist of the rocky crust, which comprises the outer 30 to 90 kilometers (19 to 56 miles) of the terrestrial environment and 5 to 8 km (3 to 5 mi) of the ocean floor. Below the crust is the mantle, a layer of ductile rock that reaches from the outer core of the earth to the bottom of the crust and comprises more than two-thirds of Earth's mass. Silica accounts for more than 80 percent of the crust and approximately 50 percent of the mineral contained within the mantle, making it the most abundant single mineral compound on the planet.

Within the earth's crust, silica usually appears as quartz, a crystalline form of silicon dioxide. Quartz is the second most abundant mineral in the earth's crust, second only to feldspar, which is another mineral composed almost entirely of silica and included minerals. Quartz is composed of silica molecules organized into a tetrahedral pattern, which lends quartz its distinct six-sided crystalline shape, topped with a six-sided pyramid at the leading end of each crystal. Sand found on beaches is composed almost entirely of fragmented quartz crystals, blended with other elements in smaller quantities.

Both silica and quartz have been used in a variety of industrial and technological applications. Silica is a component of many building materials and has been used in the manufacture of microelectronic components. Quartz crystals are used in the manufacture of watches. Some of the more attractive varieties of quartz are prized for their aesthetic properties and are used in the manufacture of jewelry.

Silica is also melted and polished to make a type of glass called fused quartz or fused silica, which is used in the manufacture of eyeglasses and other optical lenses. Fused silica is also used in the construction of optical fibers for telecommunications applications.

Optical fibers are superior to metal fibers at transmitting electrical signals because of the chemical properties of silica, which make the material largely immune to the effects of electromagnetic interference. Fiber optics have been used in a variety of applications, from lighting and telecommunications to computing. Quartz also is the ore used in the development of silicon computer chips.

CHEMISTRY OF SILICA AND QUARTZ

Silicon dioxide forms spontaneously at a variety of temperatures and pressures when silicon atoms encounter atomic oxygen. Oxygen atoms form covalent bonds to a silicon atom such that four atoms of oxygen are organized around each silicon atom in a tetrahedron or pyramid-shaped array.

The connections between silicon and oxygen and between molecules of silicon dioxide create flexible bonds that allow silicon dioxide to organize into a variety of shapes. By controlling the temperature and pressures at which silicon and oxygen interact, it is possible to develop silicate forms with a variety of structures.

Each tetrahedral molecule of silica then bonds to other silica molecules to form a three-dimensional structure. The nature of the crystal structure depends on the angle at which the silica molecules bond. Typically, molecules bond at an angle of between 140 to 160 degrees. Quartz crystals form when silica molecules bond at angles between 143.6 and 153 degrees.

Once formed, silica is highly stable and will resist further chemical reactions because of the strength of the covalent silicon-oxygen bonds within the crystal structure. Silica glass, which is a pure form of quartz glass containing more than 99 percent silica, is therefore used for a number of chemical applications, providing an inert environment that can contain chemical substances without becoming reactive.

CRYSTAL STRUCTURE AND FORMATION

Quartz comes in two basic varieties, based on the size of the crystals within a deposit. Macrocrystalline quartz contains crystals that can be viewed with the naked eye, while crypto- or microcrystaline quartz are a form in which the constituent crystals can be viewed only with a microscopic viewing aid.

Microcrystaline quartz can further be broken down into two basic types, fibrous and grainy, based on the appearance of thin layers of material viewed through a polarizing microscope. The difference in structure between macro- and microcrystalline quartz is related to the environment in which the crystals form.

Most macrocrystalline quartz forms either in heated, silica-rich water solutions or within igneous rock deposits from silica-rich magma. Macrocrystalline quartz is a primary component of many rock types, meaning that it was one of the basic components present during the formation of the rock. Individual crystals of quartz are generally considered a secondary product, forming within quartz rocks or within other rocks that contain quartz as part of their overall structure.

Macrocrystalline quartz grows in hydrothermal environments, where water temperature is between 100 and 450 degrees Celsius (212 to 842 degrees Fahrenheit) with correspondingly high pressure. In this environment, rocks containing silica dissolve to produce orthosilicic acid (H_4SiO_4), which reacts to form layers of crystalline silica, giving off water in the process. If quartz grains are already present in the solution, these new layers of crystallized silica will add to existing layers, building into larger crystals in the process.

If no preformed crystals are present, then crystalline silica can bond together to form new crystals, though this occurs only in environments in which the temperature and pressure slowly decrease. This is largely because silica tetrahedra are continually being both added and removed from the aggregating crystals. For crystals to form, there must be sufficient dissolved silica in the environment, but macrocrystalline structures will not form if silica saturation rises above a certain critical level. At this point, crystal aggregation occurs so frequently throughout the entire medium that only tiny, microcrystalline structures will form.

In environments closer to the surface of the earth, the medium will cool more rapidly, and fewer crystals will form. In some situations, crystal-forming aqueous environments develop within a solid host rock, which may lead to large crystals as the heat gradually escapes, buffered by the surrounding rock layer. Extremely large quartz deposits result from tectonic collisions, which can drive sediment from deep within the mantle or crust toward the surface through millions of years. In this situation, temperatures and pressures slowly change, and pockets of heated water may be able to generate quartz crystals for thousands of years before crystal formation ceases.

In molten rock, silica does not exist as an individual tetrahedral but forms instead into long chains of tetrahedra that move together through the molten rock. Silica chains of this type cause the viscosity of flowing magma and lava, and magma with higher proportions of silica will be more viscous. Whether cooling magma will form quartz crystals depends on the chemical composition of the magma and the conditions under which the magma cools.

Magma that erupts onto the surface and cools rapidly will not form crystals because insufficient time exists for the bonds between tetrahedra in the silica chains to break and then re-form into crystallized structures. Lava therefore gives rise to either porous volcanic rock or, when the lava cools extremely rapidly, glassy structures like obsidian. Similarly, high-viscosity magma, which is rich in silica chains, will not generally form into crystals and usually gives rise to other types of silicate mineral. Alternatively, when molten magma cools beneath the surface, cooling occurs gradually, forming into a variety of silica-based rock types. In granite, for instance, mica, feldspar, and quartz are the primary types of silicate crystals that form as the granitic magma cools. Quartz is usually the last type of crystal to form in a sample of granite and therefore occurs in pockets within the stone.

Microcrystalline structures, sometimes called chalcedony, form in aqueous environments through a similar mechanism as macrocrystalline quartz, but they tend to form under different basic conditions. Higher saturation of silica tetrahedra favors the development of microcrystalline structures. Microcrystalline quartz also tends to form in environments where the overall temperature is below 150 degrees Celsius (320 degrees Fahrenheit); microcrystal growth is highly inhibited at temperatures above 200 degrees Celsius (395 degrees Fahrenheit).

In addition, microcrystalline structures tend to grow only in aqueous environments and are inhibited by environments in which there is insufficient water for the formation of dissolved silica tetrahedra. For this reason, microcrystalline structures are largely absent from magmatic rocks and other types of igneous rock deposits.

VARIETIES OF QUARTZ

There are many varieties of quartz, all based on the size and shape of the crystals and on the inclusion of other minerals and elements that cause variations in color and texture. Common types of macrocrystalline quartz include rock crystal, amethyst, milky quartz, and smoky quartz.

Rock crystal is the name for basic, colorless macrocrystalline quartz, consisting of 99.5 percent or more silica. Rock crystal quartz can be found around the world and in many types of geological environments.

A valuable and highly prized type of quartz is amethyst, a variety of macrocrystalline quartz with included pockets of purple, red, or violet formed by the presence of embedded iron molecules. When these iron molecules are exposed to radiation during formation, the oxygen that is bonded to silicon oxidizes in such a way to produce the distinctive color. Amethyst remains highly sensitive to radiation; the violet hues will fade in amethyst exposed to sunlight and other forms of radiation. The color in most amethyst is concentrated at the pyramidal heads of the crystals, though in some cases iron can form into waves of color called phantoms within the body of the crystals.

Iron present in quartz also can produce ametrine, a type that varies in colors and often includes both yellow and violet sections within a single crystal. The yellow areas reflect yellow light because of the higher concentration of iron atoms within the lattice.

Smoky quartz is named for its gray to brown color formed by the irradiation of aluminum molecules within the lattice. The color in smoky quartz can occur in isolated areas or can affect the entire crystal structure. The color in smoky quartz forms long after the initial formation of the crystal, after extended exposure to irradiation chemically transforms the included aluminum ions through oxidation, thereby altering their structure and interaction with light. Smoky quartz can be artificially created by exposing quartz with included aluminum to radiation.

Milky quartz has a translucent white color that forms from the inclusion of numerous microscopic chambers containing liquid or gas molecules that alter the way the crystal reacts to light. Milky quartz has been discovered in a variety of environments and ranges in color from slightly translucent to specimens that are nearly solid white and opaque. Because inclusions tend to distort crystal growth, milky quartz exhibits unusual morphological shapes, with crystals often shifting in their growth patterns as the crystal develops.

One of the most common types of microcrystalline quartz is agate, which is a translucent, multicolored rock that exhibits bands of different colors organized parallel to the surface of the rock. Agate does not appear to have a crystalline structure unless examined with a high-powered electron or other type of microscope. Agate is not composed of pure silica, and the colored bands result from inclusions of various types of minerals and other rock types. Agates can have a variety of colors, from greens and reds to shades of brown and gray.

One of the most common types of microcrystalline quartz is flint, which is translucent to opaque and exhibits no obvious crystalline structure. Flint is usually dark brown, gray, or tan. Color varieties are caused by included minerals and oxides, which typically include iron, aluminum, and other metals.

One of the most unusual properties of flint is that it exhibits conchoidal fracturing, which is fracturing that does not follow a standard plane of separation. Conchoidal fractures and flint fractures, which are formed similar to glass fractures, have made flint useful in a variety of applications. Because it is fine grained and can be sharpened because of its fracturing pattern, flint was used to manufacture arrowheads and other types of weaponry, making it one of the most anthropologically important rocks on Earth.

Micah L. Issitt

FURTHER READING

Duff, Peter McLaren D. *Holmes' Principles of Physical Geology.* 4th ed. London: Chapman & Hall, 1998. Introduction of geological processes and research. Includes a discussion of igneous rocks, including silica and quartz in Chapter 5.

Grotzinger, John, and Thomas H. Jordan. *Understanding Earth.* 6th ed. New York: W. H. Freeman, 2010. Detailed coverage of Earth systems science and the formation of the lithosphere. Includes a discussion of igneous rock formation and descriptions of many types of igneous rocks.

Monroe, James S., Reed Wicander, and Richard Hazlett. *Physical Geology.* 6th ed. Belmont, Calif.: Thompson Higher Education, 2007. General-interest text discussing processes, methodology, and theoretical principles of modern geological research. Includes a discussion of igneous minerals and their formation and importance to industrial applications.

Montgomery, Carla W. *Environmental Geology.* 4th ed. Columbus, Ohio: McGraw-Hill, 2008. Basic text covers geological research and industrial geology from an environmental standpoint. Discusses environmental impact of igneous rock mining and the use of igneous rocks in construction.

Stanley, Steven M. *Earth System History.* New York: W. H. Freeman, 2004. Detailed text places the formation of rock types within the context of Earth's overall evolution and the development of the environmental spheres. Chapter 6 discusses the development of igneous rocks.

Woodhead, James A., ed. *Geology.* Pasadena, Calif.: Salem Press, 1999. Introductory text to the geologic sciences. Contains detailed discussions of igneous rocks, their categorization, and their chemical analysis.

See also: Crystals; Feldspars; High-Pressure Minerals; Hydrothermal Mineralization; Igneous Rock Bodies; Silicates; Siliciclastic Rocks; Toxic Minerals; Ultra-High-Pressure Metamorphism; Ultrapotassic Rocks.

R

RADIOACTIVE MINERALS

Radioactive minerals contain unstable atoms, such as uranium and thorium, within their crystal structure. Primary radioactive minerals form within magma pockets in the earth's mantle while secondary radioactive minerals form when radioactive substances derived from magmitic rocks are dissolved into solution and later combine with other materials to solidify into other types of rock. Radioactive minerals are used in radiometric dating; the manufacture of metals, industrial gases, and filaments; and the development of fission reactors to derive nuclear fuel. Researchers are working on ways to derive fuel from fission involving thorium.

PRINCIPAL TERMS

- **alpha rays:** combination of two protons and two neutrons released from a radionuclide during the radioactive decay process
- **beta rays:** electrons or positrons released from a nucleus during the radioactive decay process
- **gamma rays:** electromagnetic rays released from a variety of sources including the decay of unstable radionuclides, which is a form of ionizing radiation and is therefore potentially harmful to biological tissues
- **metamict rock:** mineral that has lost its defined crystal structure during the process of radioactive decay, generally forming into an amorphous mineral
- **radioactive decay:** process by which an unstable nucleus releases excess energy to reach a more stable state, transforming a radionuclide into a daughter nuclide
- **radionuclide:** atom with an unstable nucleus due to a lack of balance among protons and neutrons within the atomic nucleus
- **thorium:** naturally occurring atomic element characterized by its mildly radioactive characteristics and a decay pattern that involves the release of alpha rays into the environment
- **uranium:** naturally occurring atomic element characterized by its highly radioactive nature and the emission of alpha, beta, and gamma rays during the decay process

RADIONUCLIDES AND RADIOACTIVE DECAY

Atomic elements found in nature may be classified as either stable or unstable based on the energy contained within the element's atomic nucleus. If the particles that make up the nucleus (protons and neutrons) of the atom are balanced in regard to energy, the atom is stable. If the nucleus contains excess energy it is said to be unstable or radioactive.

Unstable atoms, called radionuclides, have either excess protons or neutrons; in this case the atom will eventually emit the excess energy either by ejecting a proton or neutron into the environment; by emitting beta particles or positrons to convert protons to neutrons, or vice versa; or by emitting excess energy in the form of light radiation or photons. This stabilization process, called radioactive decay, converts the radionuclide into a more stable atom called a daughter nuclide.

Three basic types of radioactive decay exist: alpha, beta, and gamma. Each type releases a different type of radiation. Alpha decay occurs when a radionuclide emits two protons and two neutrons into the environment, referred to as alpha rays. Beta decay involves the emission of an electron or positron into the environment, which converts a proton to a neutron, or vice versa. Both alpha and beta decay contribute to the stabilization of the radionuclide and its conversion to a daughter nuclide. Gamma decay occurs in conjunction with both alpha and beta decay and involves the release of photons from the extra energy produced by alpha or beta decay.

Gamma radiation is the most damaging form of radiation to biological structures because it moves easily through the atmosphere and can be blocked only by heavy shielding. Beta and alpha radiation, conversely, travel only a short distance through the atmosphere before they are absorbed by atmospheric atoms and molecules. Therefore, most dangerous radiation is in the form of gamma radiation, though direct exposure to alpha and beta radiation, at extremely close range, can also damage biological tissues.

As unstable atoms decay, they transition into a different type of atom. The resulting material, therefore, is distinct in many characteristics from the parent material. Physicists have found that different radioactive materials decay at certain rates and that these rates can be used to determine the age of the material. Radioactive isotopes of certain atoms, like carbon, will decay into nonradioactive isotopes of carbon in a predictable time, so measuring the proportion of stable to radioactive isotopes within a sample of carbon-rich rock can be used to estimate the time that has elapsed since the rock formed.

RADIOACTIVE MINERAL
FORMATION

Radioactive minerals generally contain either thorium or uranium within their crystal structure. More than two hundred minerals have been discovered containing one of these elements. Occasionally, radioactive isotopes of phosphorus may occur in certain minerals.

Radioactive minerals can be divided broadly into primary and secondary minerals, based on their geologic origin. Primary materials are those that form within magma pockets in the earth's mantle. Radioactive minerals are a natural product of magma and form into solid rock when magma erupts through the crust as lava or cools beneath the surface to form igneous rock deposits.

Another source of radioactive minerals are hydrothermal vents, which are areas in which magma beneath the surface heats pockets of water in the deep ocean. As this superheated water erupts through the oceanic crust, it carries a variety of minerals derived from the deeper crust and the mantle, including radioactive minerals, which later solidify into rock.

Secondary radioactive minerals are those that form when primary radioactive minerals dissolve into a solution within the soil or some aqueous environment and then re-form into solid rock of a different type. Secondary radioactive minerals are generally sedimentary rock, formed when solutions containing radioactive minerals mobilize and then become solidified and compacted through evaporation and concentration within the crust.

Radioactive uranium is well known because of its use in nuclear fission technology. The most common primary minerals containing uranium include uraninite, the silicate mineral coffinite, and the complex oxide mineral brannerite. There also exist a wide variety of secondary uranium minerals, including carbonates and silicate-based

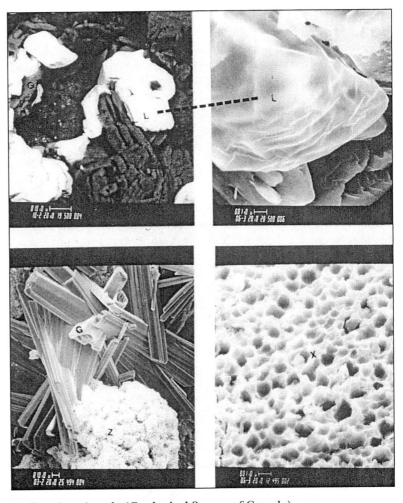

Radioactive minerals. (Geological Survey of Canada)

minerals. Thorium is the second most common radioactive element found in naturally occurring minerals. The most common primary minerals of thorium include thorianite, thorouraninite, and the silicate mineral uranothorite. Thorium also is included in a variety of secondary radioactive minerals and can be found in trace amounts in several minerals, including monazite, which is a phosphate mineral.

IDENTIFICATION AND CLASSIFICATION OF RADIOACTIVE MINERALS

As radioactive minerals decay, the chemical structure of the mineral is altered, leading to subsequent changes in the structure of the crystal lattice. This process is called metamictization and leads to the formation of what geologists call metamict minerals.

Metamictization generally takes thousands of years because radioactive decay occurs at a gradual pace. In most cases, metamictization results in an amorphous mineral shape, which is a state in which the shape and structure of the crystal lattice is unclear, resulting in smoothed edges and an overall globular appearance.

Radioactive minerals containing both thorium and uranium are difficult to distinguish based on morphological characteristics and also are similar in appearance to many other types of nonradioactive minerals. This presents a difficulty because the handling and collecting of radioactive minerals can be dangerous; handling and collecting is therefore prohibited or restricted in many locations. The exact identification of radioactive materials often requires the use of X-ray diffraction, which utilizes the pattern created by X rays focused through thin sections of a crystal to illustrate the atomic and chemical structure of the substance in question.

Many radioactive minerals appear in two general varieties: bright green or yellow minerals and amorphous and dark or cloudy colored minerals. Primary radioactive minerals are more likely to be yellow or light green because they have not yet transitioned to their eventual form through metamictization. As mentioned, radioactive minerals that have undergone significant metamictization will appear amorphous and are generally dull and cloudy in color and luster. Primary radioactive minerals are therefore easier to identify than are metamict minerals, and

diffraction analysis may be required to achieve a full identification. Both primary radioactive minerals and metamict minerals may emit harmful radiation. Geologic organizations recommend caution when collecting minerals that are potentially radioactive in origin.

USES OF RADIOACTIVE MINERALS

The most familiar use of radioactive minerals is in nuclear fission technology, which utilizes uranium as a source of fuel. Uranium-rich minerals such as uraninite and tobernite constitute one of the earliest material stages of the fission process, as these minerals yield sufficient uranium to produce fuel. Uranium ore is then crushed into a fine power called yellowcake, which can be further processed to derive the raw fuel for fission.

Thorium extracted from thorium-rich radioactive minerals is mixed with magnesium alloy to create stronger metallic compounds for industrial uses. This process has yielded strong metals used in aircraft construction and other applications. Because of their high melting point, thorium metals also are used as filaments in the construction of lamps and lighting technology.

Plans exist to use thorium, which is less radioactive than uranium, as a fuel for nuclear reactors. The most recent plan calls for the development of a liquid-fluoride thorium reactor (LFTR), which contains uranium and thorium in separate compartments filled with liquid fluoride salts. The uranium donates neutrons to the thorium, transforming the thorium into an isotope of uranium and producing heat energy from fission in the process. While thorium fission is still in its infancy, significant funding has been dedicated to research because of the substantial advantages of thorium versus uranium fuel.

Early research indicates that LFTRs may increase the efficiency of nuclear reactions by more than 20 percent and will require far less investment in terms of initial materials than conventional uranium reactors. In addition, thorium reactors are being designed in such a way that most of the by-products of the process can be utilized in another stage of the reactor, thereby greatly reducing the amount of nuclear waste.

Another major benefit to thorium fission is that thorium is far more abundant than uranium in the earth's crust. Thorium is a basic component of the crust in many locations and is more abundant than many other common minerals, such as tin, silver, and mercury.

Micah L. Issitt

FURTHER READING

Duff, Peter McLaren D. *Holmes' Principles of Physical Geology.* 4th ed. London: Chapman & Hall, 1998. Detailed college-level text describing a variety of subjects in the Earth sciences. Chapter 14 contains a discussion of radioactive minerals in their naturally occurring state and examines safely handling and identifying radioactive minerals.

Eisenbud, Merril, and Thomas F. Gesell. *Environmental Radioactivity: From Natural, Industrial, and Military Sources.* Waltham, Mass.: Academic Press, 1997. Advanced text examines sources of radiation found within natural environments. Includes a description of naturally occurring radioactive elements and radioactive materials and contaminants produced by human activity.

Lauf, Robert. *Introduction to Radioactive Minerals.* Atglen, Pa.: Schiffer, 2007. Concise introduction to radioactive minerals written for the general collector, amateur mineralogist, and rock collector. Explains the major types of radioactive minerals, their standard locations, and how to safely handle them.

Malley, Marjorie C. *Radioactivity: A History of a Mysterious Science.* New York: Oxford University Press, 2011. Popular science book traces the history of human discovery and knowledge of radioactive minerals. Contains a discussion of naturally occurring radioactive minerals and their uses, including radioisometric dating of geologic samples.

Schumann, Walter. *Minerals of the World.* 2d ed. New York: Sterling, 2008. General-interest text describing mineral formation, appearance, classification, and collection for the amateur and general interest geology student. Contains a description of radioactive materials, including a description of classification, collection, and identification.

Stanley, Steven M. *Earth System History.* 2d ed. New York: W. H. Freeman, 2004. Comprehensive introductory text describing geologic processes and the development of current geologic formations. Chapter 6 examines the use of radioactive isotopes in dating geologic samples. Also describes the formation of radioactive minerals through igneous processes.

See also: Earth Science and the Environment; Geothermal Power; Hydrothermal Mineralization; Igneous Rock Bodies; Lithium; Ultra-High-Pressure Mineralization; Unconventional Energy Resources; Uranium Deposits.

REGIONAL METAMORPHISM

Regional metamorphism, which takes place in the roots of actively forming mountain belts, is a process by which increased temperatures and pressures cause a rock to undergo recrystallization, new mineral growth, and deformation. These changes diagnostically cause minerals to develop a preferred orientation, or foliation, in the rock.

PRINCIPAL TERMS

- **equilibrium:** a situation in which a mineral is stable at a given set of temperature-pressure conditions
- **facies:** a part of a rock or group of rocks that differs from the whole formation in one or more properties, such as composition, age, or fossil content
- **igneous rock:** a rock formed from the cooling of molten material
- **mica:** a platy silicate mineral (one silicon atom surrounded by four oxygen atoms) that readily splits into thin, flexible sheets
- **mineral:** a naturally occurring chemical compound that has an orderly internal arrangement of atoms and a definite formula
- **sedimentary rock:** a rock formed from the physical breakdown of preexisting rock material or from the precipitation, chemically or biologically, of minerals
- **shale:** a sedimentary rock composed of fine-grained products derived from the physical breakdown of preexisting rock material
- **strain:** change in volume or size in response to stress
- **stress:** force per unit of area
- **texture:** the size, shape, and arrangement of crystals or particles in a rock
- **zeolites:** members of a mineral group with very complex compositions: aluminosilicates with variable amounts of calcium, sodium, and water and a very open atomic structure that permits atoms to move easily

ROCK TEXTURE AND COMPOSITION

Metamorphism is a process whereby igneous or sedimentary rocks undergo change in response to some combination of increased temperature, pressure, and chemically active fluids. Several types of metamorphism are recognized, but regional metamorphism is the most widespread. Regionally metamorphosed rocks in association with igneous rocks make up about 85 percent of the continents. In order to describe regionally metamorphosed rocks

and to understand the processes by which they form, one must understand two properties of these rocks: texture and composition. Texture refers to the size, shape, and arrangement of crystals or particles in a rock. Composition refers to the minerals present in a rock.

The effect of temperature and pressure on texture and composition is aided by fluids. Fluids transport chemical constituents and facilitate chemical reactions that take place in the rocks. As metamorphic conditions increase, metamorphic reactions readily occur if fluids are present; however, fluids are gradually driven off as recrystallization reduces open spaces. Most of the chemical reactions that occur during metamorphism result in the loss of water or carbon dioxide. Once fluids are eliminated and temperatures and pressures decrease, the reactions generally occur very slowly. Therefore, in most metamorphic rocks, the minerals preserved are those that represent the highest temperatures and pressures attained, assuming equilibrium when adequate fluids were still available.

Regionally metamorphosed rocks characteristically develop distinctive textures, largely because of the role of two types of pressure in concert with elevated temperatures: confining pressure, or the force caused by the weight of the material overlying a rock, and a stronger directed pressure, or a horizontal force created by mountain-building processes.

FOLIATION, SCHISTOSITY, AND GNEISSIC TEXTURE

As metamorphism commences, rocks become deformed, and original features are strongly distorted or obliterated. Original minerals either recrystallize or react to form new minerals and develop a pronounced preferred orientation, or foliation, in response to the directed pressure. Preferentially aligned platy minerals (such as the micas), formed during the early stages of metamorphism, enhance the development of foliation. Foliation assumes different forms depending on the degree of metamorphism.

Although the minerals in most rock types characteristically develop preferred orientations, shales and other fine-grained rock types display more distinct changes of texture. Shales undergo textural changes as they pass from conditions of low-grade metamorphism (low temperatures and pressures) at shallow depths to high grades (higher temperatures and pressures) deeper in the earth. At low grades of regional metamorphism, microscopic platy minerals grow and align themselves perpendicular to the orientation of the directed pressure. This realignment allows this fine-grained rock, called a slate, to split or cleave easily along this preferred direction. Under somewhat higher metamorphic grades deeper in the earth, the platy minerals continue to develop parallel arrangements in response to the directed pressures, but the resultant mineral grains, typically the micas, are now large enough to be identified with the unaided eye. This texture, a schistosity, forms in a rock called a schist. At higher grades, the platy minerals react to form new minerals, and a gneissic texture develops. This texture is distinguished by alternating layers of light and dark minerals that give the rock, called a gneiss, a banded appearance.

INDEX MINERALS

Just as textural changes in metamorphic rocks are generally predictable as grade increases, minerals also appear or disappear in a systematic sequence and become even better gages of the conditions of metamorphism. A progression of key minerals in shales, called index minerals, forms as the grade of metamorphism increases. These index minerals, therefore, occur in rocks metamorphosed at a given range of temperature-pressure conditions. When mapped in the field, areas containing index minerals are called metamorphic zones. As the metamorphic grades increase, the typical sequence of the index minerals is chlorite, biotite, garnet, staurolite, kyanite, and sillimanite.

Because metamorphic zonation based on index minerals utilizes minerals most likely to form from shales, zonation can be limited if shales are not abundant. A more precise definition of metamorphic change is recognized, which is not based on single minerals in one rock type but on sequences of mineral assemblages in groups of associated rocks within ranges of temperature-pressure conditions. These ranges of temperatures and pressures, in which diagnostic mineral assemblages in different rock types may exist, are called metamorphic facies. The zeolite

facies represents the lowest conditions of regional metamorphism. At the lowest extremes of this facies, conditions considered sedimentary merge with those that are metamorphic. This facies is named for the zeolite group of minerals that forms within this facies. As metamorphic conditions increase, the assemblages of minerals distinctive of the zeolite facies break down, and new assemblages form that are distinctive of either the blueschist or the greenschist facies. As the names of these facies imply, the textures formed under these conditions are commonly schistose, and a number of the minerals impart either blue or green tints to the rocks in their respective facies. In a given area of regional metamorphism, it is typical to find rocks that represent a sequence of metamorphic facies that follow one of two pathways. The greenschist path has temperature increasing more rapidly than pressure, relatively speaking, so the combination of facies that it crosses is referred to as a low pressure/high temperature sequence. The less common blueschist path, in which temperature increases more slowly relative to pressure, represents a high pressure, low temperature sequence.

UNDERSTANDING REGIONAL METAMORPHISM PROCESSES

By understanding how mountains are formed, scientists can learn how regionally metamorphosed rocks are generated. Regional metamorphism occurs in the roots of actively forming mountain belts. Linear zones thousands of kilometers long and hundreds of kilometers wide can be involved. It will take millions of years before the products of regional metamorphism are exposed in the Andes or Cascades that are forming now. Among the youngest exposed metamorphic rocks in the world are those in the Olympic Mountains of Washington, some of which are only ten million years old. On all the continents, however, broad areas of exposed rocks, called shields, now reveal the products of numerous ancient mountain-building events.

In the 1960's, the concept of plate tectonics revolutionized geology and scientists' understanding of the processes in regional metamorphism. For decades, geologists have known that the earth is divided into several concentric layers. The outermost zone, the crust, consists of relatively low-density rocks and averages 35 kilometers thick on the continents and 10 kilometers thick under the oceans. Beneath the crust is the mantle (nearly 3,000 kilometers thick), and then the core.

With the advent of plate tectonics, another important subdivision was recognized. The upper 60 to 100 kilometers, which includes the crust and a portion of the upper mantle, behaves in a relatively rigid fashion and is composed of about twenty pieces that move independently of one another. This rigid zone is called the lithosphere, and the pieces are called lithospheric plates. Because each plate is moving relative to its neighbors, three types of boundaries with other plates exist: those in collision, those pulling apart, and those sliding by one another. The rigid lithosphere apparently floats on a soft plastic layer called the asthenosphere, which extends from 60 to 100 kilometers to at least 250 kilometers into the mantle. Plate tectonics allows us to understand the processes that occur during mountain building.

CONVERGENT BOUNDARIES

Regional metamorphism is restricted to margins in collision, called convergent boundaries. The collision of two lithospheric plates produces several profound effects. One plate subducts, or sinks, under the other but manages to crumple the edge of the continent into a mountain belt. An oceanic trench forms where the descending plate sinks beneath a continent. As one plate plunges downward, melting occurs where temperatures sufficiently rise. The molten material rises through the lithosphere to produce volcanoes at the surface and masses of igneous rock at depth. It is in this realm—extending from the zone of collision to well beyond the igneous bodies—that regional metamorphism occurs. Within this area, two distinct sequences of regional metamorphism develop. The high pressure/low temperature sequence forms in the region nearest the trench. As sediments and volcanic rocks are carried rapidly (in geological terms) downward into the subduction zone, they attain relatively high pressures but remain relatively cold because they have not had the time to heat up. The minerals that form in these high-pressure conditions are those of the zeolite and blueschist facies. If the downward movement continued uninterrupted, the rocks would eventually heat up. Before that can happen, however, thin, cold slabs with high-pressure mineral assemblages within them are chaotically pushed or thrust back toward the surface. Why this thrusting occurs is not entirely clear, but it is not

uncommon to find a region 100 kilometers wide, adjacent to a trench, composed entirely of highly deformed high pressure/low temperature facies rocks. This region is called an accretionary wedge because of its inferred shape in a vertical slice through the earth. It is this material that allows the continents to grow larger through time. The California Coast Ranges are a good example of rocks that formed in such a setting.

STUDY OF REGIONALLY METAMORPHOSED ROCKS

Geologists study regionally metamorphosed rocks from several different perspectives. At the core of these studies are field observations. Descriptions of rock units and structures provide information on how these rocks have formed. Much of this fieldwork is synthesized into a geological map.

In the laboratory, samples from the field are studied by a variety of physical and chemical methods. Studies with the petrographic microscope yield information on mineral compositions and textural relationships that provide clues for classifying and determining modes of origin for these rocks. X-ray powder diffraction techniques are commonly used to identify metamorphic minerals not readily identified by visual inspection. Since metamorphic rocks are markedly changed texturally and mineralogically from their original state, chemical analysis provides data that can be compared to probable premetamorphic rock types. Standard classical chemical methods of analysis may be used, but a number of more sophisticated, faster, and simpler spectrographic methods are popular. Atomic absorption and X-ray fluorescence spectroscopy are two of the more widely applied techniques to determine elemental abundances, although emission spectrographic and neutron activation analyses are also used. Another important tool is the electron microprobe. Most published mineral analyses are generated by electron microprobe analyses. Mass spectroscopic analyses also provide information on the distribution of isotopes in metamorphic rocks and minerals. These data are useful in determining age and conditions of formation.

Rock deformation experiments provide information about the mechanical properties of metamorphic rocks. Rock samples subjected to stress and strain tests yield data on properties such as plasticity, strength, and viscosity. These data are particularly

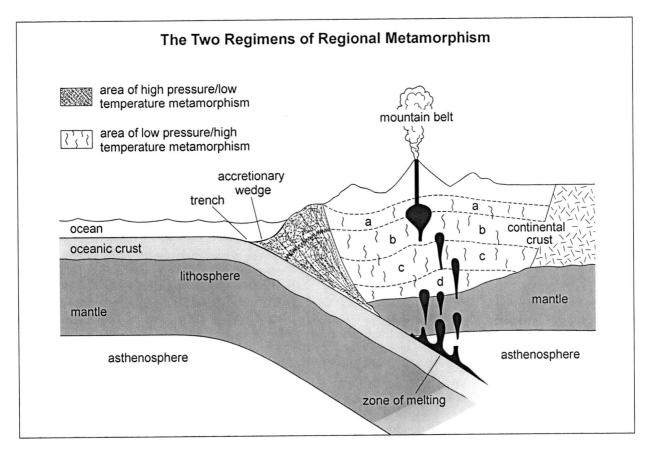

The Two Regimens of Regional Metamorphism

important in understanding how directed pressures influence textures. Another phase of the study of metamorphic rocks is experimental petrology. Here, metamorphic minerals are synthesized under controlled-equilibrium conditions. From these studies, geologists gain knowledge about the actual ranges of stability of these minerals in naturally occurring environments. The study of the crystal chemistry and thermodynamics provides information regarding actual conditions of formation and potential reactions. The roles of variables—such as entropy, volume change, and heat of reaction—provide clues to the behavior of mineral assemblages at varying temperature and pressure conditions. Theoretical petrology—the treatment of data on metamorphic rocks by mathematical models or the principles of theoretical physics—also provides important information in the study of regional metamorphism.

In addition to analyzing the samples collected, scientists analyze field data in the laboratory. Statistical analysis of orientations of foliations and other structural features is greatly facilitated by the computer. Laboratory analyses are recast to generate a variety of graphical treatments to check for chemical trends or metamorphic facies relationships.

APPLICATIONS FOR INDUSTRY AND ENGINEERING

The products of regional metamorphism touch people's lives in many ways. Marbles, gneisses, and slates are used as building and cut ornamental stones, and some quartzites, marbles, and gneisses are used as aggregate. The minerals graphite, talc, vermiculite, and asbestos are four of the more commonly used mineral products of regional metamorphism, whereas wollastonite, garnet, kyanite, emery, and pyrophyllite are mineral products of limited use. Graphite is most commonly used in the metallurgical industry, with lesser amounts used as lubricants, paints, batteries, and pencil "leads," and in electrodes. Talc is also a product with diverse uses. The ceramics industry is the largest consumer, followed by paint and paper manufacturers. Vermiculite is a common product in thermal and acoustic insulation. Asbestos, despite concerns about its health risks, remains an important fire-retardant material for appropriate uses.

Regionally metamorphosed rocks provide certain advantages in construction but also present unique engineering problems. A scan of the skyline of the island of Manhattan in New York City reveals that the largest skyscrapers are restricted to several areas of the island, surrounded by expanses of much shorter buildings. The reason that buildings such as the Empire State Building occupy only certain areas is that they sit directly on structurally strong regionally metamorphosed rocks that occur at or near the surface. In other areas of the island, these rocks are too deeply buried, and the overlying sediments are too weak to support the larger structures. In other cases, inclined foliations paralleling slopes in hilly or mountainous regions represent potential planes of slippage that can give way and produce massive rock slides, particularly when road, mine, or dam construction modifies the landscape. In areas where foliations are not adequately taken into account, the cost can be millions of dollars in repairs and/or great loss of life. Special techniques and restrictive measures need to be applied when construction occurs in areas where foliations present potential problems.

Ronald D. Tyler

FURTHER READING

Augustithis, S. S. *Atlas of Metamorphic-Metasomatic Textures and Processes.* Amsterdam: Elsevier, 1990. A thoughtful treatment of metamorphic rocks, mineralogy, and metasomatism. Excellent illustrations support concepts and processes. Appropriate for the college student or layperson with an interest in Earth sciences.

Bates, Robert L. *Geology of the Industrial Rocks and Minerals.* Mineola, N.Y.: Dover, 1969. Concentrates on the occurrences and uses of rocks and minerals. Deals with metamorphic rocks, metamorphic minerals, and minor industrial minerals, respectively. Provides excellent summaries of how the products of regional metamorphism are utilized. Includes an index and good bibliography.

Best, Myron G. *Igneous and Metamorphic Petrology.* 2d ed. Malden, Mass.: Blackwell Science Ltd., 2003. A readable advanced college-level text for the general reader who wants to learn more about metamorphic textures. Includes the topics of mineralogy, chemistry, and the structure of metamorphic rocks. Discusses both field relationships and global-scale tectonic associations.

Blatt, Harvey, Robert J. Tracy, and Brent Owens. *Petrology: Igneous, Sedimentary, and Metamorphic.* 3d ed. New York: W. H. Freeman, 2005. Deals with the descriptions, origins, and distribution of igneous, sedimentary, and metamorphic rocks. Includes an index and bibliographies for each chapter.

Bucher, Kurt, and Martin Frey. *Petrogenesis of Metamorphic Rocks.* 7th ed. New York: Springer-Verlag, 2002. Provides an excellent overview of the principles of metamorphism. Contains a section on geothermometry and geobarometry.

Chernicoff, Stanley. *Geology: An Introduction to Physical Geology.* 4th ed. Upper Saddle River, N.J.: Prentice Hall, 2006. Offers an overview of the geology of Earth and surface processes. Includes a Web address that provides regular updates on geological events around the globe.

Dietrich, R. V., and B. J. Skinner. *Rocks and Rock Minerals.* New York: John Wiley & Sons, 1979. Describes rocks and minerals in very clear, concise terms, emphasizing simple identification techniques. Gives an excellent but succinct overview of metamorphic rocks and their conditions of formation, notable occurrences, and uses.

Ernst, W. G. *Earth Materials.* Englewood Cliffs, N.J.: Prentice-Hall, 1969. A short, highly regarded introductory-level book that uses crystal structures, the concepts of chemical equilibria, physical chemistry, and thermodynamic principles to explain the formation of minerals in the three major rock types. Considers the stability of metamorphic mineral assemblages and textures in terms of temperature-pressure conditions.

Fettes, Douglas, and Jacqueline Desmons, eds. *Metamorphic Rocks: A Classification and Glossary of Terms.* New York: Cambridge University Press, 2007. Discusses the classification of metamorphic rocks. Mentions feldspars throughout the book in reference to the various types of metamorphic rocks and basic classifications.

Gillen, Cornelius. *Metamorphic Geology: An Introduction to Tectonic and Metamorphic Processes.* 2d ed. Berlin: Springer, 2011. Provides an introduction to metamorphism and indicates the strong relationships among mountain building, plate tectonics, and metamorphic processes. Regional examples are largely European, but this is one of the few introductory-level texts fully devoted to metamorphism. Includes glossary and short bibliography.

Mason, R. *Petrology of the Metamorphic Rocks*. London: Allen & Unwin, 1978. In addition to experimental and theoretical petrology, isotope and electron-microprobe analyses, this second-year college-level text covers descriptions of metamorphic rocks in the field and under the microscope. Written from the perspective of a European geologist.

Miyashiro, A. *Metamorphism and Metamorphic Belts*. New York: Springer, 1978. Deals with the basic concepts, characteristics, and problems of metamorphic geology. Focuses on the use of shales and specific igneous rocks to help explain metamorphism. Summarizes classic metamorphic belts throughout the world.

Newton, Robert C. "The Three Partners of Metamorphic Petrology." American Mineralogist 96 (2011): 457-469. Discusses field observations, experimental petrology, and theoretical analysis of the earth's crust. Highlights the importance of collaboration to petrology.

Philpotts, Anthony, and Jay Aque. *Principles of Igneous and Metamorphic Petrology*. 2d ed. New York: Cambridge University Press, 2009. Easily accessible for students of geology. Discusses igneous rock formations of flood basalts and calderas. Covers processes and characteristics of igneous and metamorphic rock.

Tennisen, A. C. *Nature of Earth Materials*. 2d ed. Englewood Cliffs, N.J.: Prentice-Hall, 1983. Written for the nonscientist. Covers the nature of atoms and minerals as well as igneous, sedimentary, and metamorphic rocks. Also includes a section on the uses of these materials. Well illustrated, with a bibliography and subject index.

Vernon, Ron H. *A Practical Guide to Rock Microstructure*. New York: Cambridge University Press, 2004. Investigates the microscopic structures of sedimentary, igneous, metamorphic and deformed rocks. Discusses techniques and common difficulties with microstructure interpretations and classifications. Includes color illustrations, references, and indexing.

Winter, John D. *An Introduction to Igneous and Metamorphic Petrology*. Upper Saddle River, N.J.: Prentice Hall, 2001. Provides a comprehensive overview of igneous and metamorphic rock formation. Discusses volcanism, metamorphism, thermodynamics, and trace elements; focuses on the theories and chemistry involved in petrology. Written for students taking a university-level course on igneous petrology, metamorphic petrology, or a combined course.

See also: Blueschists; Contact Metamorphism; Metamorphic Rock Classification; Metamorphic Textures; Metasomatism; Pelitic Schists; Sub-Seafloor Metamorphism.

REMOTE SENSING

Remote sensing makes use of the interaction of energy waves to measure objects or materials from a distance, without physical contact. It is an analytic technique with wide-ranging applications in many fields, including meteorology, resource management, national security, environmental monitoring, agriculture, and mapping.

PRINCIPAL TERMS

- **backscattering:** a reflection of energy waves by an object
- **bathymetry:** the study of underwater depth
- **depth sounding:** determining the depth of a specific point underwater; the data are used in bathymetry
- **electromagnetic radiation:** energy emitted and absorbed by charge particles; travels in a wave-like manner
- **hyperspectral imaging:** a remote sensing technique that detects electromagnetic radiation across a wide spectrum of wavelengths, including visible light
- **inverse problem:** determining information about an object or phenomenon by analyzing observed measurements about the object or phenomenon
- **light detection and ranging:** a remote sensing technique involving the emission of beams of light
- **multispectral scanning:** a remote sensing technique that detects discrete bands of electromagnetic radiation
- **radar:** a remote sensing technique involving the emission of radio waves or microwaves
- **sonar:** a remote sensing technique for object detection and navigation that utilizes the propagation of sound waves
- **spectral signature:** a unique identifier of a material based on how the material absorbs or reflects electromagnetic radiation wavelengths

WHAT IS REMOTE SENSING?

In most cases, remote sensing involves the active or passive reading of a variety of radiations from the electromagnetic spectrum as they are emitted, reflected, or scattered from the object or phenomenon being observed. Other propagated signals (such as sound waves) also can be used.

Because remote sensing can be performed from a large distance, it has wide-reaching applications in many different areas of study. It is particularly useful in applications involving the environment, agriculture, meteorology, mapping, and national security.

Environmental remote sensing is aimed at understanding natural and human-made changes to Earth's ecosystems. One major environmental application of remote sensing is land conservation, which involves monitoring deforestation and land usage, tracking glacial decline, overseeing urban growth, and keeping hazardous waste in check. Environmental remote sensing also can be used in the conservation of natural resources, both renewable (oceans, soil, forests, wetlands) and nonrenewable (natural gas, minerals, oil).

Remote sensing also has direct applications to agriculture. It can be used to monitor crop conditions, predict crop yield, and measure soil erosion. (The monitoring of soil erosion is crucial; preventive measures do exist and can be implemented when needed.) Remote sensing also is useful for monitoring crop damage from weeds, insects, wind, hail, and herbicides, and the resulting information can help farmers know which sections of a farm, for example, need fertilizer or pesticides.

In meteorology, remote sensing can be used to predict weather patterns and track short-term and long-term weather trends, such as hurricanes and tropical storms. It also is an essential technique for monitoring ozone depletion and global warming.

Remote sensing is widely used in map creation, particularly in terms of topography, the study of Earth's surface features (such as mountain ranges and other landforms). In addition to mapping surface features, remote sensing can help to map coastal and ocean depths through a technique known as depth sounding. The study of these depths is called bathymetry. Because remote sensing does not require physical contact, it is useful in the exploration of dangerous or inaccessible areas.

Finally, as national security often requires the detailed knowledge of inaccessible and hazardous areas, remote sensing is an essential tool for military surveillance and strategy. Security-related remote sensing often requires particularly powerful tools that have both a high temporal resolution (timely results) and a high spatial resolution (finely detailed results).

THE HISTORY OF REMOTE SENSING

The modern history of remote sensing began in 1858, when French photographer and balloonist Gaspard-Félix Tournachon (also known by the pseudonym Félix Nadar) took photos of Paris from his hot-air balloon, in which he had a makeshift darkroom. In the next century, camera photography was the only method of overhead remote sensing, although the technology developed alongside the development of flight technology. In addition to balloons, other platforms were used to carry a camera, including messenger pigeons, kites, and early rockets.

Modern remote sensing began in the 1950's with the growing availability of satellite platforms, electro-optical sensor systems, and quantitative analytical tools for measuring and interpreting the resulting photographic and electro-optical images. The year 1959 marked the first time photos of Earth were taken from space, from a nonorbital space flight of *Explorer 6*. In 1961 an unpiloted Mercury-Atlas flight (MA-4) took the first color photos of Earth from space. The Television Infrared Observation Satellite (TIROS-1) began monitoring weather and the global environment from space in 1960. The Cold War also was a driving factor in the development of remote sensing during this time; systematic aerial photography became a necessary defensive tool.

The early 1970's brought about the further development of space-based remote sensing. The first Landsat multispectral scanner system, for example, launched in 1972 to begin the systematic acquisition of satellite imagery of Earth. Although quite advanced at the time, its features and resolution are modest by twenty-first century standards. Groups such as the National Aeronautics and Space Administration (NASA), the Jet Propulsion Laboratory (a NASA program), and the U.S. Geological Survey have continued to drive the development of remote sensing.

REMOTE SENSING AND THE ELECTROMAGNETIC SPECTRAL RANGE

Remote sensing depends on the interaction of energy and the matter that is being measured; the observed energy, more specifically, is radiation that is present in the electromagnetic spectrum (including the visible spectrum; that is, what humans can see).

Electromagnetic radiation refers to energy waves traveling at the speed of light; the spectrum is divided into regions by wavelength.

The visible spectrum, for example, has wavelengths between 380 nanometers (nm) and 760 nm (790-400 terahertz). Radio waves, though, are much larger, with wavelengths reaching hundreds of meters. Remote sensing instruments can measure electromagnetic radiation that is emitted, reflected, or scattered by an object. Many materials found on Earth and in the atmosphere have a unique spectral signature—a "fingerprint" that allows conclusions to be drawn from the measurements taken by a remote sensing instrument.

Remote sensing can be divided into three classes with respect to the wavelength region being detected. These classes are visible and reflective infrared, thermal infrared, and microwave. Visible and reflective infrared remote sensing is mainly limited to passive, not active, remote sensing because the energy source is the sun (and not the sensing instrument). This type of remote sensing is particularly useful for materials at Earth's surface, where reflected energy from the sun exceeds the earth's own emitted energy. This region of the spectrum includes wavelengths between 0.4 and 3 micrometers (μm), encompassing the visible, near infrared, and short-wave infrared regions of the electromagnetic spectrum.

Thermal (long-wave) infrared remote sensing measures energy radiated from objects, beginning at a wavelength around 5 μm and peaking at 10 μm. Because this type of remote sensing does not depend on the reflectance of solar energy, thermal infrared remote-sensing can be done at night. Forward-looking infrared cameras (found on military planes) use this region of the spectrum to aid operators in steering at night.

Microwave remote sensing deals with wavelengths on a larger order, typically between 1 millimeter and 30 centimeters. While some passive remote sensing systems can detect microwaves, active remote sensing also can be used by measuring the way the observed material backscatters the radiation emitted by the sensing system.

PASSIVE REMOTE SENSING TECHNIQUES AND TOOLS

Remote sensing can be described as passive or active. Passive remote sensing involves the detection of radiation emitted or reflected by the observed object, material, or phenomenon; the sensing system measures but does not actively emit radiation itself. Passive remote sensing can encompass visible and reflective infrared, thermal infrared, and some microwave remote sensing. Radiation sources include reflected solar energy and energy inherent in the ground, atmosphere, and clouds.

Passive remote sensing techniques and sensor types include film and digital photography, multispectral scanners, hyperspectral imaging tools, passive sonar, radiometers, photometers, and seismometers. For the purposes of remote sensing, film photography is most often black and white (panchromatic) and aerial (shot from aircraft or satellites). The use of film photography for remote sensing skyrocketed after World War I as aircraft and photographic technology were both progressing rapidly. Color photography became more popular after World War II, but color photography is more expensive than panchromatic.

Even more expensive is infrared photography, which has clear advantages. Like color film, infrared photography detects three distinct colors. Where color film detects blue, green, and red, infrared film detects green, red, and infrared. The advantage to infrared is that this particular set of wavelengths shows more clearly the distinction between land and water and between healthy and unhealthy vegetation.

Digital photography also has become increasingly useful in remote sensing. Many digital imaging techniques rely on sensors called charge-coupled devices, which allow for the movement of a charge to a format that can be digitally analyzed.

NASA's Landsat program, which began in 1972 with the launch of remote sensing satellite Landsat 1, made use of a multispectral scanning system that could detect four spectral bands of reflected sunlight. One satellite that launched later in the program, Landsat 3, could detect an additional band: thermal infrared radiation.

While multispectral imaging covers a number of discrete spectral bands, hyperspectral imaging is similar but covers a whole spectral range without breaks between bands. NASA's Airborne Visible/Infrared Imaging Spectrometer (AVIRIS) uses this technology to detect more than 220 contiguous spectral bands between wavelengths of 400 and 2,500 nm.

Other useful remote sensing tools include radiometers, which measure radiant flux, and photometers, which measure luminous flux. While radiant flux describes the total power of electromagnetic radiation

emitted from or landing on an object, luminous flux describes only the power of the visible light emitted from the object. (Radiant flux includes infrared and ultraviolet light, not just visible light.)

The term *remote sensing* is often defined as being dependent on the electromagnetic spectrum. However, the more general idea of measuring something without physical contact also encompasses techniques such as passive sonar (using the propagation of sound waves to navigate or detect objects) and seismography (using seismic waves to determine information about earthquakes, volcanic eruptions, and other ground movements).

ACTIVE REMOTE SENSING TECHNIQUES AND TOOLS

Unlike passive remote sensing systems, active remote sensing systems actually create an artificial radiation source as a probe, emitting energy toward the object to be observed and then measuring the reflection or backscattering. Radar (originally an acronym for "radio detection and ranging") and LIDAR (light detection and ranging) are two of the most common forms of active remote sensing, but the definition can be expanded to include active sonar, which uses sound waves rather than electromagnetic radiation.

Radar techniques make use of radio waves or microwaves to gather information about objects, including their position, speed, and direction. The radar dish or antenna transmits energy wave pulses, which bounce off objects; a sensor at the dish or antenna detects the energy that is returned. Radar signals work especially well with highly conductive materials such as metals and wetlands.

Common uses of radar techniques involve weather monitoring, air traffic control, and security (antimissile systems, for example). It is important to note that although radar has a wide variety of uses, it has a number of limitations: signal noise, interference and clutter from irrelevant objects, intentional or unintentional jamming of the signal, and the curvature of a radar beam path from atmospheric effects.

One common type of radar is Doppler radar, a specialized tool that focuses on the velocity of objects. For example, police can use Doppler radar to detect speeding cars. This technology is based on the Doppler effect, the shift between an emitted wave frequency and the wave frequency observed by someone or something moving relative to the wave source. Consider an ambulance speeding past with siren blaring. The siren sounds differently as it approaches, passes, and moves farther away; the siren itself is emitting a stable wave frequency. Doppler radar puts this into practice by detecting the change in the frequency of the energy reflected back by the moving object in question and then using this information to calculate the object's velocity.

Another common type of radar is synthetic aperture radar (SAR), which is often used for terrain mapping, environmental monitoring, and military purposes. SARs emit a radiation beam from a moving source, enabling the generation of extraordinarily high-resolution images by utilizing the movement of the radar platform in a way that simulates a large antenna or aperture. Because of the movement required, these radars are most commonly found on aircraft or spacecraft. SAR Earth-observation satellites include RADARSAT from Canada (a satellite pair), *TerraSAR-X* from Germany, and NASA's *Magellan*.

LIDAR works in much the same way as radar except that it uses infrared, visible, or ultraviolet light instead of microwaves or radio waves. LADAR is more accurate than radar when it comes to the detection of object heights and features because its light beam reacts with a wider range of materials than does a radar beam.

LIDAR applications are varied and include agriculture, archaeology, meteorology, and military. In agriculture, for example, LIDAR can be used to make detailed topographic maps of farmland, giving farmers a detailed look at which sections of their crops get less sun and need more fertilizer. In the military, LIDAR can be used to detect landmines and applications of biologic warfare, among other uses. Beyond the electromagnetic spectrum, active sonar is useful for underwater measurements, particularly the distance to an object.

THE RESOLUTIONS OF REMOTE SENSING SYSTEMS

Remote sensing systems work on the premise of what is known as the inverse problem: determining information about an object or phenomenon by analyzing observed measurements about the object or phenomenon. (Consider, for example, figuring out a type of animal based on its observed footprints.) The quality of the observations of a remote sensing system depends on four main types of resolution: spatial, spectral, radiometric, and temporal.

Spatial resolution refers to the pixel size in an image generated by a remote sensing system and its relationship to the sample size it represents in the area being observed. Low resolution is good for meteorology, for instance, because meteorologists are interested in weather patterns occurring over large sections of land. Meteorology would not need a resolution that can make buildings clear, for example. The equipment in the Landsat program, for instance, tends to have moderate resolution, enough to show large human-made structures (like the Great Wall of China) but not enough to show houses, because that information is barely relevant to satellite images of the earth as a whole. For mapping and military purposes, however, it is generally important for a remote sensing system to have high resolution to provide an accurate and detailed look at small pieces of a large area.

Spectral resolution refers to the wavelength width of the frequency bands recorded by a remote sensing system. Spectral resolution is usually discussed in terms of how many bands a system records, particularly in the case of multispectral systems, which measure discrete bands (as opposed to hyperspectral imaging, which looks at contiguous bands of a spectrum without breaks between the bands). For example, NASA'S MODIS, a sensing instrument aboard two Earth-observation satellites, detects thirty-six spectral bands. Band 1 detects wavelengths between 620 and 670 nm, which corresponds to chlorophyll in vegetation, so this band contributes to information about land boundaries. Meanwhile, bands 17 through 19 cover three discrete bands between wavelengths of 890 and 965 nm, corresponding to cloud and atmospheric properties.

Similarly, radiometric resolution refers to the system's ability to differentiate between different magnitudes of energy, described in binary data format. A remote sensing system with high radiometric resolution can detect small differences in energy magnitudes.

Finally, temporal resolution refers to the frequency at which a plane or satellite flies over the object or area being observed, when applicable. In meteorology and military applications, high temporal resolution is necessary to provide up-to-date information. Low resolution is fine for mapping, however, because mapping does not generally require time-sensitive information.

INFORMATION EXTRACTION FROM REMOTE SENSING SYSTEMS

Remote sensing systems can provide a wide range of data. Depending on the goal of the remote sensing, the extraction of information takes different forms. For mapping, for example, information extraction tends to be image-centered, focusing on the spatial relationships among images on the ground to create a larger picture. Image-centered information extraction, called photointerpretation, still relies heavily on the human eye, even with modern technology automating part of the process. Photointerpretation involves the analysis of shapes, sizes, tones, textures, patterns, and shadows.

For purposes other than mapping, a data-centered approach is often appropriate, involving analysis of the spectral absorption features of the observed object or area and the fractional abundances of the isotopes in the materials. Long-term monitoring of the environment tends to combine these two approaches. Although scientists need data referring to the composition of the objects in question, it is necessary to put the information in a spatial framework.

Rachel Leah Blumenthal

FURTHER READING

Lillesand, Thomas M., Ralph W. Kiefer, and Jonathan W. Chipman. *Remote Sensing and Image Interpretation.* Hoboken, N.J.: John Wiley & Sons, 2008. Starting with the foundations of remote sensing, this text covers the general elements of the techniques and then gives detailed explanations of how remote sensing images can be analyzed and interpreted.

Liu, Jian Guo, and Philippa J. Mason. *Essential Image Processing and GIS for Remote Sensing.* Hoboken, N.J.: Wiley-Blackwell, 2009. This text focuses on the interpretation of data provided by remote sensing techniques, making it useful for professionals in the field. Case studies augment the text, which is otherwise quite technical.

Rencz, Andrew N., and Susan L. Ustin. *Manual of Remote Sensing.* Hoboken, N.J.: John Wiley & Sons, 2004. Provides a wide overview of the field of remote sensing and its applications to the fields of natural resource management and environmental monitoring.

Schott, John R. *Remote Sensing: The Image Chain Approach.* New York: Oxford University Press, 2007. Although focused mostly on passive, not active, remote sensing, this text still gives a broad overview of the topic, including a historical perspective. Suitable for professionals in the field and students alike.

Schowengerdt, Robert A. *Remote Sensing: Models and Methods for Image Processing.* Burlington, Mass.: Academic Press, 2007. Provides college-level students with a detailed and technical look at remote sensing techniques, applications, and interpretation. Specific to remote sensing so not recommended for general Earth science students.

Skidmore, Andrew. *Environmental Modeling with GIS and Remote Sensing.* New York: Taylor & Francis, 2003. Focusing on the environmental applications of remote sensing techniques, this text is accessible and suitable for general environmental sciences students and those specializing in remote sensing.

See also: Earth Resources; Earth Science and the Environment; Earth System Science; Geographic Information Systems; Land Management; Land-Use Planning.

ROCKS: PHYSICAL PROPERTIES

Rocks and rock products are used so widely in everyday life that physical properties of rocks affect everyone. Major properties of rocks fall into two categories: those properties that compose the exterior nature of the rock, such as color and texture, and those properties that make up the rock's internal nature, such as strength, resistance to waves, and toughness.

PRINCIPAL TERMS

- **coefficient of thermal expansion:** the linear expansion ratio (per unit length) for any particular material as the temperature is increased
- **compressive strength:** the ability to withstand a pushing stress or pressure, usually given in pascals or pounds per square inch
- **density:** mass per unit volume
- **elasticity:** the maximum stress that can be sustained without suffering permanent deformation
- **hardness:** the resistance to abrasion or surface deformation
- **shear, or shearing, strength:** the ability to withstand a lateral or tangential stress
- **toughness:** the degree of resistance to fragmentation or resistance to plastic deformation

COLOR

Physical properties of rocks are important not only to geologists and geophysicists but also to construction engineers, technicians, architects, builders, and highway planners. In fact, rock properties affect everyone. Major physical properties include color, hardness, toughness, density, compressive strength, shear strength, tensional and bending (transverse) strength, elasticity, coefficient of thermal expansion, electrical resistivity, absorption rate of fluids, rates of weathering, chemical activity, response to freeze-thaw tests, texture, spacing between fractures, ability to propagate waves, radioactivity, and melting point.

Color is one of the most interesting physical properties of rocks. It is determined by the colors of the component minerals, arrangement of these minerals, and weathering pigments. Color and patterns of colors are important in assessing the aesthetic qualities of ornamental and building stone. The unweathered colors must be pleasing to the eye, and staining that is detrimental to the appearance must not occur in appreciable amounts. The quality and beauty of many marbles are renowned when used in ornamental stone and sculpture. Building limestone must have a pleasing gray or tan color and not contain iron or manganese minerals, which would stain the rock brown or black during weathering. The pink-and-black or white-and-black mottling of attractive granites, diorites, and syenites is a familiar sight in business, school, university, and other public buildings. It is not, therefore, necessary or even always desirable that a color be uniform throughout.

HARDNESS, TOUGHNESS, AND DENSITY

Hardness, or resistance to abrasion, can be found by using the Mohs scale, which was derived by Friedrich Mohs in 1822 to measure relative resistance to abrasion in minerals and rocks. Any substance will be able to scratch any substance softer than or as hard as itself. Ten standard minerals are used: Talc is the softest and designated as having a hardness of 1, and diamond is the hardest and designated as having a hardness of 10. Rock hardness is of particular importance to those people who use rocks or rock products in building façades, monuments, tombstones, patios, and other structures. A variety of tests are used to determine resistance to abrasion over a time interval.

Toughness differs from hardness. This property is defined as the degree of resistance to brittle fracture or plastic deformation of a particular substance. Toughness is determined by a test using repeated impacts of a heavy object. A hammer is dropped from a specified height upon a sample, and this height is continually increased. The height of the fall in centimeters upon breakage is then defined as toughness. An example of a very tough material is jade, whereas brittle substances include rock salt. The so-called French coefficient of wear is also used to measure toughness. This test measures the amount of material worn off a sample by tumbling in a drum under standard conditions. Toughness is a very important property in rocks that are used as crushed rock in building roads and airstrips, which are subject to repeated stresses.

The density of a rock is its mass or weight divided by unit volume (amount of space). More frequently, the rock is compared with an equal volume of water that has a density of almost exactly 1 gram per cubic centimeter. This number, which is dimensionless, is known as specific gravity. There is a considerable variation in specific gravity among rock types and even within them. Most limestone and dolomite rocks range from about 2.2 to 2.7 in specific gravity, whereas basalt and traprock are considerably heavier, at 2.8 to 3.0. Density is not synonymous with strength and toughness. Some low-density materials are strong, and some very dense materials cut by fractures are weak.

COMPRESSIVE, SHEARING, AND BENDING STRENGTHS

Compressive strength, or bulk modulus, is measured on the basis of the highest pressure or stress that a rock can withstand per square unit, usually measured in pascals or pounds per square inch. One pound per square inch, or psi, is about 7,000 pascals, and 1,000 psi is about 7 million pascals or 7 megapascals, abbreviated MPa. This strength is generally greater than transverse, tensional, or shearing strength. Limestones average about 100 MPa (15,000 psi), granites about 175 MPa (25,000 psi), quartzite about 200 MPa (30,000 psi), and basalts and traprock about 350 MPa (48,000 psi). This strength will vary considerably from rock to rock of the same composition because of variation in structural properties, as most rocks have fractures and voids and differ in grain size and shape.

Shear strength, or modulus of rigidity, is measured on the basis of the highest lateral stress in pounds a rock can withstand per square inch. An example of shear stress is the stress exerted by a car sideswiping another car. Values for limestones average about 14 MPa (2,000 psi) and granites about 20 MPa (3,000 psi). Shear strength is generally much less than compressive strength.

Bending, or transverse, strength is defined as the strength of a slab loaded at the center and supported only by adjustable knife edges. This strength is determined by the "modulus of rupture," which is a function of the rupture load in pounds, the length of a slab, the width or breadth of a slab, and the thickness of a slab.

ELASTICITY AND LINEAR EXPANSION

The so-called modulus of torsional rigidity, or elasticity (Young's modulus), is also measured in pascals or pounds per square inch. This property is a measurement of how difficult it is to deform a material. Young's modulus is the force that would have to be applied to deform a material by 100 percent of its original size. The amount that a material deforms under pressure is the ratio of the applied force to Young's modulus. This modulus is extremely variable, ranging typically from about 200,000 to 400,000 MPa (3×10^6 psi to 6×10^6 psi) for limestone, and about 400,000 to 560,000 MPa (6×10^6 psi to 8×10^6 psi) for granite; that is, a stress of 20 MPa applied to limestone will deform it by about 1/10,000 of its length. Elasticity refers to the ability of a material to spring back to its original shape without plastic deformation or rupture. Beyond a certain value of stress, rupture or flow will occur.

The coefficient of thermal expansion measures expansion along a line through the rock with increase in temperature. It is also termed linear expansion. This property is expressed as a ratio of change in length divided by unit length times change in temperature. Typical values for common rocks used in crushed stone or building stone range from about 4×10^{-6} (four millionths) to 12×10^{-6} (twelve millionths) per degree Celsius. Knowledge of the coefficient of thermal expansion is very important to engineers who design structures such as highways and bridges. On bridges, expansion-contraction joints are commonly used in consideration of this property.

RESISTIVITY AND FLUID ABSORPTION RATE

A measurement important to petroleum geologists, geophysicists, and engineers is electrical resistivity of rocks. The resistivity may be defined as the reciprocal of electrical conductivity. Resistivity is measured in ohm-centimeters—that is, electrical resistance along a centimeter's length. It will vary greatly depending on whether a rock is dry or contains saline water, organic compounds such as petroleum or natural gas, or fresh water in its pores or cracks. Igneous rocks such as basalts and traprock will vary from about 1×10^4 to 4×10^5 ohm-centimeters, and granites about 10^7 to 10^9 ohm-centimeters. Basalts have less resistivity and greater conductivity than granite because they generally contain more abundant amounts of dark, nonmetallic minerals than granites. Limestones and sandstones commonly range from 10^3 to 10^5 ohm-centimeters, but those containing much saline water will have much smaller values, because water with dissolved salts is a good conductor of electricity. Metallic ore deposits will show very low resistivities because of the high conductivity of metals.

The absorption rate of fluids by rocks is another property often measured. A dry rock sample of known weight (or dry aggregate of rock chips) is soaked in water for twenty-four hours, dried under surface conditions, and weighed. The weight of water absorbed is given as a percentage of the dry weight of the rock. This property is extremely variable in rocks. For example, limestones may vary from 0.03 percent to 12 percent absorption rate. Very fine-grained

limestones will commonly have a higher absorption rate than coarse-grained limestones.

SOUNDNESS, TEXTURE, AND SPACING BETWEEN FRACTURES

Soundness, or resistance to weathering, is important because rock materials are exposed to outside conditions, where temperature and moisture conditions may vary considerably. Chemicals may also attack the rock, and in arctic or temperate climates, freezing and thawing may occur. In a climate where much freezing and thawing occur, the breakage and

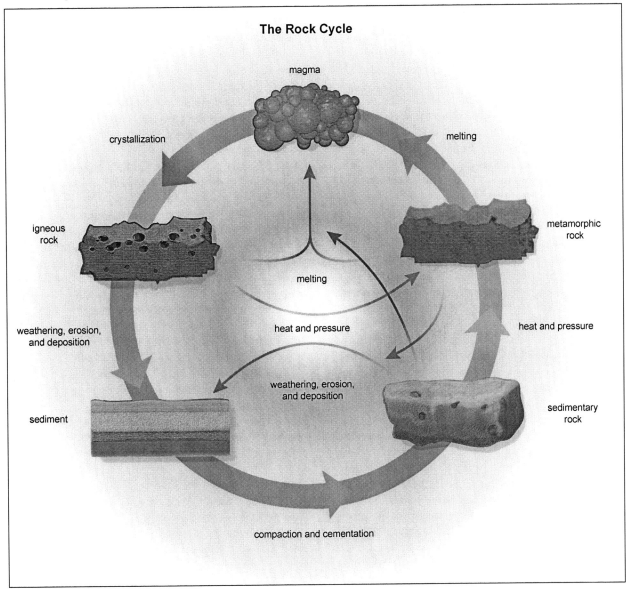

The Rock Cycle

magma

crystallization

melting

igneous rock

metamorphic rock

melting

weathering, erosion, and deposition

heat and pressure

heat and pressure

weathering, erosion, and deposition

sediment

sedimentary rock

compaction and cementation

deterioration of cement and rock materials caused by this phenomenon will be noticed. Engine-driven vehicles and equipment would not last very long in winter if antifreeze were not added to the engine block and radiator. Examples of chemical weathering can be found in an old cemetery, where gravestone inscriptions of the same age on chemically resistant rocks such as quartzite are much clearer than those on chemically reactive rocks such as marble.

Texture relates to grain size and grain shape, as well as fabric, which is the relationship of the grains to one another. Appearance, resistance to weathering, absorption rate characteristics, and strength are all related to texture.

Spacing between fractures is a significant structural property of rocks. For building stone, it is usually desirable to have massive rock or rock separated at intervals of meters by layering or bedding planes or by vertical cracks or joints. Preferably, the distance between planes will be fairly constant and the cracks flat and even. Other faces may be sawed out by means of wire saws and silica grit. For some uses of building stone, such as roofing or facing slates and paving stones, smooth, closely spaced fractures are desirable, provided the stone is relatively strong and free from fractures in between.

WAVE TRANSMISSION ABILITY

The ability to transmit waves is an important rock property to petroleum geologists and engineers, as well as construction engineers regarding the design of buildings, roads, bridges, and dams in earthquake-prone areas. This property depends on the density and the strength of the rock, especially its compressibility and its rigidity (shear) modulus. The property is useful to engineers in predicting how strongly rocks will shake in an earthquake, and to petroleum geologists for interpreting the results of seismic surveys. Basically, there are two types of waves that result from earthquakes or human-made explosions: longitudinal and transverse waves. Longitudinal waves, similar to sound waves, create pushing and pulling effects on molecules while traveling along a straight line. Transverse waves move sideways as they travel. Compressive strength (pushing strength) is important in determining the speed of longitudinal waves, whereas modulus of rigidity or shear strength is more important in determining the speed of transverse waves. In general, the greater the strength,

the greater the wave speed through a material; the greater the density, the less the wave speed through a material or rock under a given temperature and pressure. The density of darker igneous rocks is generally greater than that of lighter-colored igneous and sedimentary rocks. Also, the density of rocks increases toward the earth's center, but velocity is usually greater, because strength tends also to increase. This factor generally outweighs the density factor and produces a net increase in velocity or speed.

RADIOACTIVITY AND MELTING POINT

Radioactivity, another physical-chemical property, is demonstrated by those rocks that spontaneously emit radiant energy (which results from the disintegration of unstable atomic nuclei). Scientists use the heat generated by absorption of the radiation emitted by these rocks to establish the temperature of the earth's interior and its thermal evolution. Radioactivity has been of particular interest since the early 1980's because of the health hazard posed by radon. Radon is an intermediate product of radioactive decay, and it can seep into basements and cause lung cancer. Proper designing or repair of basement interiors may be necessary to reduce rates of radon infiltration.

Rock melting point is a rock property that is generally of interest to petrologists, engineers on deep-drilling projects, and engineers designing furnaces or kilns that require rock or brick that has a melting point higher than that of the materials being processed in the kiln. Except in the case of dry, completely one-mineral rocks, melting will occur over a range of temperatures rather than at merely one temperature. This type of melting occurs for three reasons. First of all, several types of minerals, each with a different melting point when pure, usually occur in rocks. Second, each mineral mutually affects the melting points of the other minerals, usually lowering the melting points and extending them over a range of temperatures. Third, water and other fluids may mutually lower the melting point of a rock.

PHYSICAL PROPERTY TESTS

A great variety of tests are used to determine the degree or types of physical properties of rocks. Hardness may be determined by comparing a rock with known hardness points made from minerals of known hardness or from ceramic cones of various known hardnesses, either on the Mohs ten-point scale or on a more elaborate 1,000-point industrial scale. Another test is called the Dorry hardness test. Rock cones 2.5 centimeters in diameter are loaded with a total weight of about 1,000 or 1,250 grams and then subjected to the abrasive action of a fine quartz of known size fed upon a rotating cast-iron disk. The loss in weight of the core after 1,000 revolutions is used to compute value of hardness. One of the most widely used machines to test hardness is the Los Angeles Rattler (abrasion testing machine). It has a steel drum 70 centimeters in diameter and 50 centimeters long. The sample is inserted with a certain number of steel balls, and the drum is rotated five hundred revolutions at thirty revolutions per minute. The material is then sized into further grades, and the coarser sets of remaining fragments are subjected to one thousand further revolutions using various numbers of steel balls weighing about 5,000 grams. Another test, the Brinell test, determines hardness by pressing a small ball of hard material on the sample and measuring the size of the depression.

Toughness is tested on samples with a diameter of about 2.5 centimeters. The sample is held in a test cylinder on an anvil and subjected to the fall of a steel hammer or plunger weighing 2 kilograms. The first fall is from a 1-centimeter height; this height is progressively increased by 1 centimeter until breakage occurs. Height of fall is then expressed as toughness.

Specific gravity is measured by placing the sample in water. It is measured, dried to obtain dry weight, then immersed in water for twenty-four hours, surface dried, and weighed. Finally, it is weighed again in water. Loss of weight from dried weight initially is true specific gravity. The second step, involving the twenty-four-hour immersion, is necessary to ascertain that pore spaces in the sample are filled with water so that a true measure of buoyancy can be taken in the third step.

Compressive strength is measured on test rock cylinders 5 centimeters in diameter and 5.6 centimeters long. The sample is placed in a cylindrical casing in a press, and the press is lowered at a specified rate. Amount of compression in pounds per square inch at failure is the compressive strength.

The so-called soundness test measures resistance to chemical weathering or, alternatively, resistance to a major type of mechanical weathering, called ice or frost wedging. In the first case, the sample (usually about 1,000 grams) is covered with a saturated solution of sodium or magnesium sulfate for eighteen hours. Then it is oven-dried. The test is repeated usually five times. If the sample decomposes, it is called unsound; if it shows signs of decomposition but is still intact, it is called questionable. If it shows no signs or extremely minor signs of decomposition, it is called sound. In place of ice, the test uses expansion of the sulfate to measure resistance to wedging. In the case of freeze-thaw tests, different-sized materials are tested by freezing in water to an air temperature of −22 degrees Celsius, then thawed at room temperature (about 20 degrees Celsius). A first cycle lasts twenty-four hours and may be repeated. Damage to the sample is then noted.

SEISMOGRAPHIC STUDY

Wave velocities, or speeds through rocks, are dependent on density and strength. These velocities can be determined theoretically or in the laboratory if density and strength can be measured. Conversely, compressive and shearing strength may be measured knowing rock density and velocity of wave transfer.

Geophysicists and engineering geologists may determine wave speeds through the earth's crust by setting up a seismograph station in the field. Explosives are set off, and travel time and velocity of waves are monitored by a system of geophones that relay the wave energies to the seismograph, which then measures the waves on a recording drum with a stylus. Characteristic wave reflections occur at certain depths. By knowing the times of explosive detonation and the time of arrival of the two kinds of earth body waves (primary and secondary waves), and by understanding that distance traveled for the two waves is the same, scientists can determine the wave speeds. This information is vital when drilling for oil or gas and digging deep foundations, especially in limestone sinkhole areas or in areas prone to earthquakes.

ENGINEERING AND CONSTRUCTION APPLICATIONS

Knowledge of particular rock physical properties is important to geologists, engineers, geophysicists, hydrologists, builders, architects, industrial city and regional planners, and the general public. Rock properties are especially important in regard to the construction of buildings, dams, roads, airstrips, human-made lakes, and tunnels; drilling for petroleum and natural gas; mining and quarrying; monitoring earthquake hazards; measuring properties of aggregate and construction materials; determining rates of heat flow and mechanical strain in the earth; monitoring radioactive hazards; studying chemical makeup and physical structure of the earth's interior; and evaluating aesthetic properties of ornamental construction materials.

Within earthquake-prone areas (as in California) or areas affected by wind shear (as in the tops of high buildings), the utmost importance is granted to the nature and strength of materials surrounding foundations, the foundations themselves, groundwater behavior and its effects on building strength, and the ability of building materials to withstand vibrations. Engineering codes have been enacted within some of these areas so that earthquake-resistant or wind-shear-resistant buildings, dams, tunnels, and highways may be constructed.

David F. Hess

FURTHER READING

Birch, Francis, J. F. Schairer, and H. Cecil Spicer. *Handbook of Physical Constants.* Geological Society of America Special Paper 36. Baltimore: Waverly Press, 1942. Covers physical and chemical properties of rocks and minerals. Offers prefatory discussion with most of the tables. Especially important are discussions and tables concerning rock strength and wave propagation. Contains an extensive section on chemical properties of rocks and minerals. Suitable for college-level students.

Carmichael, Robert S., et al. *Practical Handbook of Physical Properties of Rocks and Minerals.* Boca Raton, Fla.: CRC Press, 1989. A compilation of major physical and chemical properties of rocks and minerals. Of special interest are the sections on radioactive and electrical properties of rocks. Most of the tables are preceded by discussion of the particular property illustrated. Indispensable for geologists and engineers.

Dorn, Ronald I. *Rock Coatings.* New York: Elsevier, 1998. Covers the physical properties of rocks and their distribution patterns on the surface of the earth. Illustrations and maps are particularly useful, as is the extensive, detailed bibliography.

Fettes, Douglas, and Jacqueline Desmons, eds. *Metamorphic Rocks: A Classification and Glossary of Terms.* New York: Cambridge University Press, 2007. Discusses the classification of metamorphic rocks. Feldspars are mentioned throughout in reference to the various types of metamorphic rocks and basic classifications.

Haussühl, Siegfried. *Physical Properties of Crystals: An Introduction.* Weinheim, Germany: Wiley-VCH, 2007. Begins with foundational information and discusses tensors. Contains some mathematical equations. Best suited for more advanced students, professionals, and academics.

Jerram, Dougal, and Nick Petford. *The Field Description of Igneous Rocks.* 2d ed. Hoboken, N.J.: Wiley-Blackwell, 2011. Begins with a description of field skills and methodology. Contains chapters on lava flow and pyroclastic rocks. Designed for student and scientist use in the field.

Pellant, Chris. *Smithsonian Handbooks: Rocks and Minerals.* New York: Dorling Kindersley, 2002. An excellent resource for identifying minerals and rocks. Contains colorful images and diagrams, a glossary, and an index.

Potts, P. J., A. G. Tindle, and P. C. Webb. *Geochemical Reference Material Compositions: Rocks, Minerals, Sediments, Soils, Carbonates, Refractories, and Ore Used in Research and Industry.* Boca Raton, Fla.: CRC Press, 1992. Focuses on the composition of rocks, determinative mineralogy, and analytical geochemistry as they apply to scientific research and industrial use. Includes a useful bibliography.

Schon, J. H. *Physical Properties of Rocks.* Amsterdam: Elsevier, 2004. Contains information on geophysics useful to a number of professions. Discusses the physical properties of rocks, geophysics theories, and experiments. Suited for engineers, geologists, geophysicists, and well loggers.

Vernon, Ron H. *A Practical Guide to Rock Microstructure.* New York: Cambridge University Press, 2004. Investigates the microscopic structures of sedimentary, igneous, metamorphic, and deformed rocks. Discusses techniques and common difficulties with microstructure interpretations and classifications. Includes color illustrations, references, and indexing.

Wenk, Hans-Rudolf, and Andrei Bulakh. *Minerals: Their Constitution and Origin.* Cambridge, England: Cambridge University Press, 2004. Covers the structure of minerals, physical characteristics, processes, and mineral systematic. Multiple chapters devoted to non-silicates, followed by chapters discussing silicates.

See also: Andesitic Rocks; Anorthosites; Basaltic Rocks; Batholiths; Carbonatites; Granitic Rocks; Igneous Rock Bodies; Kimberlites; Komatiites; Orbicular Rocks; Plutonic Rocks; Pyroclastic Rocks; Xenoliths.

S

SALT DOMES AND SALT DIAPIRS

In addition to their importance in the supply of minerals for the food processing and chemical industries, salt domes and salt diapirs are associated with economic reserves of crude oil, natural gas, and sulfur. They are also utilized as repositories for valuable and strategic materials and, potentially, for nuclear waste.

PRINCIPAL TERMS

- **cap rock:** an impervious rock unit, generally composed of anhydrite, gypsum, and occasionally sulfur, overlying or capping any salt dome or salt diapir
- **evaporite minerals:** that group of minerals produced by evaporation from a saline solution; for example, rock salt
- **precipitation:** a process whereby a solid substance is separated from a solution
- **rim syncline:** a circular depression found at the base of many salt domes and diapirs
- **sedimentary rock:** any rock formed by the accumulation and consolidation of loose sediment, as in the precipitation of salt crystals from seawater
- **shale sheath:** a variable thickness of ground-up sedimentary rock found along the flanks of many salt domes and diapirs

BEDDED SALT DEPOSITS

Salt domes and salt diapirs are found throughout the world in association with bedded salt, an evaporative sedimentary rock known to be part of the rock record of every geologic time period. While salt deposits underlie at least twenty-six of the United States, salt domes and diapirs in North America are confined to the Gulf coast basin, extending from central Mississippi and Louisiana and coastal Texas south into the Isthmus of Tehuantepec in Mexico, and to some interior salt deposits in Colorado and Utah. The Gulf coast basin contains more than five hundred salt domes and diapirs, located within at least nine separate sub-basins. Other significant concentrations of salt domes or diapirs occur throughout European countries and in the following areas: the Zechstein Basin of Germany and Poland; the Donets Basin of Russia and the northern regions of Siberia; along

the Mediterranean coast of Algeria and Morocco; in Gabon and Senegal; Peru; and the Iran, Aden, and Yemen sections of the Arabian Peninsula.

Bedded salt deposits result from geological processes that have been in operation for several hundred million years. Precipitation of salt deposits occurs within modified marine environments whenever evaporation exceeds surface runoff and rainfall, and wherever restriction of the free circulation of seawater exists. The evaporation of a 300-meter column of seawater will theoretically result in the precipitation of 4.6 meters of evaporate minerals, of which 3.5 meters will be salt (chemically composed of sodium chloride). Bedded salt units are generally confined to a structural basin, a downwarped region of the earth's crust containing thick sequences of sedimentary rock. In North America, the Louann Salt of the Middle Mesozoic age (approximately 175 million years old) underlies more than 2 million square kilometers of the Gulf Coast basin and is several thousand meters thick.

FORMATION OF SALT STRUCTURES

The mechanism whereby bedded salt units are transformed into salt domes and salt diapirs has been generally known for decades. After precipitation from an evaporating body of saline water, pure, bedded rock salt has a specific gravity of 2.16 grams per cubic centimeter. When that salt bed is initially overlain by younger sedimentary rocks composed of sandstone, siltstone, shale, and clastic limestone—normal rock associations found wherever salt domes occur—the salt maintains density equilibrium with the younger rocks that have abundant open pores. Their specific gravities range from 1.7 to 2.0 grams per cubic centimeter. Upon burial at depths of several thousand meters, however, the pores in the sedimentary rock are compressed and the specific gravity of the overburden will increase to levels ranging from 2.4 to 2.8 grams per cubic centimeter, while the salt, being incompressible, remains at the 2.16 specific gravity level. Under these conditions, buoyancy will act upon the salt, encouraging movement toward the surface of the earth. This potential for movement is accelerated below depths of 3,000 meters by conditions of increasing temperature and pressure, altering the visco-elastic properties of the salt as it changes from a solid to a slowly flowing plastic material. Through this combination of density imbalance and altered visco-elastic condition, bedded salt will begin to arch overlying rocks at those sites marked by slight salt-surface irregularities. As long as the overlying rocks are arched or folded and thus remain mechanically intact, the resulting structure is termed a salt "dome." Should the overlying strata be broken and penetrated by the rising column of salt, the structure becomes a salt "diapir" (meaning "through-piercing"), a form of intrusive sedimentary rock. Studies of the thickness and age of salt basins suggest that the rate of salt movement and penetration, while variable, generally averages less than 2 millimeters per year. During periods of very active sedimentation and crustal folding, domal activity is most active. In the Aquitaine basin of France, there are seven mapped phases of salt upwelling that relate directly to periods of crustal deformation in the nearby Alps.

Once initiated, upward flow will continue until the causative forces of buoyancy and flow properties are altered by diminished pressure and temperature, or until the base of the salt dome or diapir is cut off from the bedded unit acting as the salt source. In general, the thicker the bedded salt source and the deeper it is buried, the higher the developing salt structure. A salt dome or diapir will usually not reach to the surface of the earth; occasionally, though, surface penetration occurs, as in the case of the salt domes of the Persian Gulf region. There, 40 percent of the nearly two hundred salt domes rise above the surrounding desert, often displaying salt "glaciers" that flow laterally away from the center of the exposed salt mass. One exposed dome, Kuh-e Namak, has a surface elevation of 1,310 meters and a diameter of 9.6 kilometers.

INTEGRAL FEATURES

Through the process of arching or penetrating overlying sedimentary rock, all salt domes and diapirs take on certain identifiable characteristics. These features are found within the interior of the salt structure, forming the top of the dome or diapir and associated with the sedimentary rock adjacent to the salt structure.

Studies of salt mines within near-surface diapirs of the Gulf Coast basin give direct evidence of vertical flowage. This flowage is demonstrated by stretched, elongated halite crystals and vertically oriented, tightly folded layers of salt accentuated by color changes between alternating pure and impure halite beds. Further evidence of vertical intrusion into overlying rock is seen in the occasional sighting of inclusions of sandstone and shale within salt structures; such foreign rock is identical in composition to adjacent intruded rock.

Topping most shallow salt domes and diapirs is an impervious mantle of minerals, on average 90 to 120 meters thick, termed the "cap rock." While found on domes up to 3,000 meters deep, cap rock thickness is greatest over the center of the dome, generally decreases with dome depth, and is absent along dome flanks. A typical, well-developed cap rock is mineralogically composed of anhydrite grading upward into calcite (calcium carbonate). Often, a transition zone containing varying amounts of barite (barium sulfate), gypsum, and sulfur separates the two principal layers. Some salt domes within the Poland section of the Zechstein basin of Europe have a cap rock composed of clay minerals. It is generally agreed that cap rock is gradually formed by upward-migrating salt structures encountering waters of decreasing salinity contained within the sedimentary rock being penetrated. Incomplete solution of the impure rock salt results in the accumulation of insoluble residues, principally anhydrite, along the leading edge of the salt intrusion. Transition minerals are believed to result from later alteration of the anhydrite.

ADJACENT FEATURES

The most common features external, or adjacent, to salt domes and salt diapirs include a shale sheath, an overlying domed structure accentuated by a central graben, an overall complex pattern of faulting, and a peripheral rim syncline.

Where encountered by drilling in the Gulf Coast basin, the shale (clay) sheath covers the flanks of a dome or diapir much like a cap rock covers the top. This sheath is of variable thickness and is composed of finely ground rock formed by the frictional drag of the upward movement of the salt structure against adjacent sedimentary sandstones and shales.

As bedded salt beds become mobile under conditions of deep burial, they form circular structures created by arching of the overlying sedimentary rock. Diapiric movement is accomplished by salt, and developing cap rock, penetration of the rock cover along a complex system of fractures (faults) that radiate outward from the circular structure. As such movement actually extends (stretches) the overlying rock cover, the structure center is commonly interrupted by a graben, a depressed block bounded by faults. This combination of a regional arched structure interrupted by complex faulting and a localized central graben is what complicates the geologic interpretation of the upper portions of any salt dome or diapir.

The most significant aspect of the exterior base of a salt structure is the presence of a rim syncline, a structural depression that partially or completely encircles most domes and diapirs. This feature is caused by the downward movement of overlying sedimentary rock into the void formed by the lateral flow of bedded salt into the developing dome. Details of rim synclines are little known as a consequence of their association with the base of salt domes and diapirs.

DISCOVERY OF SALT STRUCTURES

Under normal conditions, salt domes and salt diapirs are not exposed at the surface of the earth and therefore are not accessible for direct study. An exception would include the diapiric salt structures of the Persian Gulf region, which are surface exposed and, because of the arid climate, preserved for long-term analysis. Salt domes and diapirs can also be directly studied where salt has been extracted from the upper levels of the structure, either by brine-well production, such as at the Bryan Mound Dome in Brazoria County, Texas, or by direct salt mining, as conducted at the Avery Island Dome in Iberia Parish, Louisiana. Another salt dome, the Jefferson Island Dome, was the location of a bizarre disaster in 1980, when an oil drilling rig in nearby Lake Peigneur accidentally drilled into the salt mine. As salt dissolved, the lake drained into the mine, causing extensive property damage along the shore and destroying both the mine and the drill rig. Analyses of the walls of such mines give evidence pertaining to the layering and folding of salt beds, permitting the development of theories of salt mobility and flowage.

The formation of a salt dome or diapir by long-term upward salt movement usually does not manifest itself in the development of surface relief effects, which identify the presence of the shallow salt structure. Therefore, the discovery of the great majority of intrusive salt structures is accomplished by instrumentation designed to measure contrasting physical characteristics of the salt and adjacent sedimentary rock indirectly. In 1917, a buried salt dome was identified near Hanigsen, Germany, by a torsion balance, an early prospecting instrument designed to measure alterations in the gravitational field of the earth. Subsequent drilling into this dome verified the accuracy of the torsion-balance method, which was soon being employed worldwide in the discovery of salt structures. By 1935, the cumbersome torsion balance had been successfully modified in the form of the first practical gravimeter, a modern field instrument that directly reads gravity differences with an accuracy of less than one millionth of Earth's total gravitational field. Because of density differences between the salt and adjacent rock material, such accuracy permits easy identification by gravimeter of a salt dome or salt diapir, as well as identification and thickness analyses of the cap rock.

After World War I, the invention of the refraction seismology process, which measures the change in velocity and deflection of an elastic wave moving from one rock formation to another, opened a new and very successful era of salt dome and salt diapir discovery in the Gulf Coast basin. By the second half of the twentieth century, improvements in the reflection seismology process using technology related to refraction seismology were being employed worldwide in the continued discovery of intrusive salt structures, especially within marine waters off the Gulf Coast of North America and offshore Africa. The seismic reflection process makes it possible to identify the overall geometry of intrusive salt structures with considerable precision. During the 1990's, adaptation of three-dimensional reflection seismology, technology that is capable of documenting not only the geology of the base of salt domes and diapirs but also the structure underlying the source salt layer, is yet another major advancement in the identification and scientific study of intrusive salt.

MINERAL AND FOSSIL-FUEL SOURCES

Halite (the mineralogical term for salt) is the principal evaporite mineral in salt domes and diapirs. Halite has been a significant industrial material and article of world trade for thousands of years. In the United States, half of produced salt is employed by the chemical industry in the manufacture of caustic soda, polyvinyl chloride, hydrochloric acid, and household and industrial bleach. Large volumes are used in cold climates for traction control on highways. Other uses include water treatment, the manufacture of insecticides, medicines, aluminum, and steel, and food preservation and seasoning.

Extraction of halite from salt domes and diapirs is accomplished by two means: by conventional underground mining, as conducted, for example, in the Winnfield Dome in Winn Parish, Louisiana; and by means of brine (salt solution) wells, as practiced at the Barbers Hill Dome in Chambers County, Texas. Drilled wells are also used to extract economic volumes of sulfur from the cap rock of many salt domes. Production of sulfur from domes and diapirs utilizes the Frasch superheated water process to melt and dissolve the sulfur.

Although salt domes are important sources of halite and sulfur, their greatest economic value is as sites of petroleum accumulation. The upward flow of the salt creates traps into which petroleum may migrate. Traps include oil-bearing layers that are bent upward along the flanks of the salt, the dome of rocks bent upward above the salt, and the cap rock itself. The discovery and exploratory drilling of salt domes and diapirs has been a worldwide priority for decades because of their association with crude oil and natural gas. The bulk of these fossil-fuel deposits is found in a circular area up to 1,000 meters away from the salt periphery. With the dual discovery in 1901 of the Spindletop Dome in Jefferson County, Texas, and the presence of oil within its cap rock, the center of oil and gas production in the United States moved from the Ohio River Valley to west of the Mississippi River. More than several hundred million barrels of oil has been produced from the cap rock and sedimentary rock flanking this historic salt structure.

STORAGE UNITS

In addition to their association with hydrocarbon resources, solution-created and mined cavities in salt domes and diapirs are becoming increasingly important as possible depositories of volatile, strategic, toxic, and nuclear material. The Tatum Dome of Lamar County, Mississippi, is employed by the U.S. Atomic Energy Commission to house an explosion as part of the study of peaceful applications of nuclear energy. With the signing of the Energy Policy and Conservation Act of 1972, five salt domes of the Gulf Coast basin became part of the Strategic Petroleum Reserve, the United States' emergency plan for the storage of crude oil. As a typical part of this program, two levels of mined rooms approximately 21 meters high by 30 meters square within the Weeks Island Dome in Iberia Parish, Louisiana, held a capacity 72 million barrels of crude oil during the mid-1990's. Elsewhere in Texas and Louisiana, solution-developed and mined dome cavities have been used for the storage of low-pressure liquid gas, high-pressure ethylene, and refined hydrocarbon products ranging from gasoline to fuel oil.

Intrusive salt structures are particularly amenable to the storage of such products because salt is generally impervious to the passage of liquids. Impermeability combined with a relative lack of humidity makes dome and diapir cavities particularly adaptable to the long-term storage of volatile and radioactive waste materials, as well as historical, legal, and financial records vital to industrial and national security.

Albert B. Dickas

FURTHER READING

Alsop, G. Ian, Derek John Blundell, and Ian Davison, eds. *Salt Tectonics.* London: Geological Society, 1996. A collection of essays from the Geological Society discussing the composition of evaporites, salt, and diapirs. Illustrations, index, and bibliographic references.

Halbouty, M. T. *Salt Domes: Gulf Region, United States and Mexico.* 2d ed. Houston, Tex.: Gulf Publishing, 1981. A presentation written by the acknowledged salt-dome expert among American geologists and petroleum engineers. Lists pertinent data for more than four hundred salt domes and salt diapirs within the U.S. and Mexican portions of the Gulf Coast basin. Numerous full-color figures illustrate salt-dome distribution, configuration, composition, classification, and economic significance. A very readable reference written for the general public.

Jackson, M. P. A., ed. *Salt Diapirs of the Great Kavir, Central Iran.* Boulder, Colo.: Geological Society of America, 1991. Covers salt domes of the central Iranian basin, one of the most spectacular and well-exposed salt dome regions of the world, forming circular elevations 80 to 100 meters in relief. Intended for college-level readers.

Jackson, M. P. A., D. G. Roberts, and S. Snelson, eds. *Salt Tectonics: A Global Perspective.* Tulsa, Okla.: American Association of Petroleum Geologists, 1996. A comprehensive presentation of the geologic means whereby bedded salt units are transformed into a variety of geometric shapes, including domes and diapirs. Written at the college level for anyone interested in the exploration of salt structures and their common association with economic reserves of sulfur, crude oil, and natural gas.

Kupfer, D. H., ed. *Geology and Technology of Gulf Coast Salt Domes: A Symposium.* Baton Rouge, La.: School of Geoscience, Louisiana State University, 1970. Conducted by nine scientists actively engaged in Gulf Coast basin salt-dome research. Explores the origin, composition, and development of salt structures. Presents discussion of topics not commonly found in other references, such as cap rock development, salt movements, and methods of salt intrusion. Written for the college-level reader, with numerous illustrations.

Lefond, S. J. *Handbook of World Salt Resources.* New York: Plenum Press, 1969. Discusses salt resources of 196 continental and island countries. Offers documentation of the worldwide distribution of intrusive salt structures invaluable to the student of salt domes and diapirs. Discusses the history of salt-resource utilization of each country, and offers numerous charts showing chemistry, mineralogy, geographic location, and geology. Very suitable for high school and college-level readers.

Lerche, Ian, and Kenneth Peters. *Salt and Sediment Dynamics.* Boca Raton, Fla.: CRC Press, 1995. Examines the chemical composition and properties of salt, salt domes, and sedimentary basins. Includes bibliographic references and index.

Murray, G. E. *Geology of the Atlantic and Gulf Coastal Province of North America.* New York: Harper & Brothers, 1961. A general reference to the geology of the coastal regions of the eastern and southern United States. Contains a thorough presentation of the salt domes and diapirs of the Gulf coast basin. Appropriate for the professional geologist and the general college-level reader.

Stewart, S. A. "Salt Tectonics in the North Sea Basin: A Structural Style Template for Seismic Interpreters." *Special Publication of the Geological Society* 272 (2007): 361-396. Discusses the Permian salt layer in the North Sea Basin. Explores salt tectonics. Provides information useful to professionals studying salt tectonics, paleogeomorphology, or other related topics.

Usdowski, Eberhard, and Martin Dietzel. *Atlas and Data of Solid-Solution Equilibria of Marine Evaporites.* New York: Springer, 1998. Examines seawater composition. Focuses on evaporites, marine deposits, and salt deposits. Accompanied by a computer disk that reinforces the concepts discussed. Illustrations, index, and bibliography.

Warren, John K. *Evaporites: Sediment, Resources, and Hydrocarbons.* Berlin: Springer-Verlag, 2006. Discusses a number of evaporitic minerals, with certain sections focusing on salts, chemistry and hydrology, deposit locations, and mining. Includes many drawings of geological features.

See also: Aluminum Deposits; Building Stone; Cement; Coal; Diamonds; Dolomite; Fertilizers; Gold and Silver; Igneous Rock Bodies; Industrial Metals; Industrial Nonmetals; Iron Deposits; Manganese Nodules; Pegmatites; Platinum Group Metals; Sedimentary Mineral Deposits; Uranium Deposits.

SEDIMENTARY MINERAL DEPOSITS

Sedimentary mineral deposits are accumulations of economically valuable minerals that occur in sedimentary rocks ranging in age from 2.2 billion to less than 2 million years old. Such deposits have been widely exploited in the past and continue to be important sources of ores.

PRINCIPAL TERMS

- **deposition:** the physical or chemical process by which sedimentary grains come to rest after being eroded and transported
- **diagenesis:** changes that occur in sediments and sedimentary rocks after deposition caused by interaction during burial with water trapped between the sediment grains
- **evaporite:** a mineral formed by direct precipitation from water resulting from supersaturation caused by solar evaporation in an arid setting
- **hydrothermal:** characterizing any process involving hot groundwater or minerals formed by such processes
- **midocean ridge:** a large, undersea chain of volcanic mountains encircling the globe, branches of which are found in all the world's oceans
- **ore:** a concentration of valuable minerals rich enough to be profitably mined
- **placer:** an accumulation of valuable minerals formed when grains of the minerals are physically deposited along with other, nonvaluable mineral grains
- **strata:** layers of sedimentary rock

PLACERS

In its broadest sense, a sedimentary mineral deposit is an abnormal accumulation of valuable minerals in sedimentary rocks. When such an accumulation is rich enough to warrant mining, it is referred to as a sedimentary ore deposit. Most geologists would agree, however, that further restriction should be applied to the term. In this discussion, the term "sedimentary mineral deposits" refers only to those deposits that formed through sedimentary processes. For example, hydrothermal deposits would not be considered sedimentary ore deposits even if they occurred in sedimentary rocks. Such ores form by fluids circulating through the rocks long after deposition, not during the deposition or burial of the rocks.

Several types of sedimentary processes may lead to the formation of sedimentary mineral deposits.

Primary deposits form at the same time the host sediments are deposited. They can form either as chemical precipitates or as placers. Placers form when ore minerals are eroded and transported along with sand and silt by rivers. When the sand and silt are deposited as sediments, the ore minerals are deposited along with them. Most placers form in riverbeds and tend to be small. Placer accumulations are also found in some sands deposited on beaches.

The best-known placers are those of gold. Although small by comparison to lode deposits (deposits in bedrock), placer gold deposits were responsible for setting off the many gold rushes in the United States, including those to Colorado, California, and Alaska. Other placer deposits occur as paleoplacer deposits—accumulations that, millions or billions of years ago, were placer deposits and have since been hardened into sedimentary rocks. Of particular interest to geologists are large deposits of quartz pebbles with sand-sized grains of pyrite (iron sulfide) and uraninite (uranium oxide). Pyrite and uraninite are unstable in the presence of oxygen and quickly break down during erosion. They are common, however, as detrital minerals (minerals that have been eroded and transported by streams) in rocks of Archean age (older than about 2.5 billion years). Geologists have deduced that such accumulations could have formed only if oxygen were not abundant as a free gas (not combined with other elements) in the earth's atmosphere at that time.

CHEMICAL SEDIMENTS

Many chemical sediments form when minerals precipitate directly from seawater or saline lakes onto the sea floor or lake bottom. The most common type of chemical sediment forms when sea or lake water is concentrated by solar evaporation. Such deposits are known as evaporites. Large deposits of anhydrite (calcium sulfate), gypsum (hydrated calcium sulfate), and halite (sodium chloride, also known as table salt) have formed in small seas in areas with arid climates. As water is lost from these bodies because of high rates of evaporation, new seawater enters through narrow straits, connecting the sea with the open ocean. Dissolved salts in the water do not evaporate. They are concentrated until the brine becomes saturated, at which time minerals begin to precipitate. Gypsum forms first, followed by halite. Anhydrite forms when gypsum deposits are buried by later sediments and heat expels water from the gypsum. Potash (potassium chloride) can also form in this manner if arid, evaporative conditions persist for a long enough time, but most seawater does not evaporate completely enough to precipitate potassium salts.

Saline lakes also undergo the evaporative concentration of dissolved salts, from which evaporite deposits may form as well. Lake-water chemistry is controlled by the chemical composition of the surrounding bedrock, and lake water is often quite different in chemical composition from seawater. Thus, lacustrine evaporites (evaporites forming in lakes) commonly contain minerals not associated with marine evaporites. Such deposits include various borate (boron oxide) deposits in California and Turkey, and trona (hydrated sodium carbonate) deposits of the western United States.

Chemical sediments also form when hydrothermal solutions are expelled onto the sea floor. The hot, mineral-laden hydrothermal water mixes with seawater and is rapidly cooled. The solubility of most metals decreases drastically as the temperature decreases, and various sulfide minerals precipitate, including minerals of copper, lead, zinc, and iron, as well as barite. Deposits of this type occur in many regions of the world, including North America (especially Canada), and are forming today in the Red Sea and the Pacific Ocean along the midocean ridge system.

Primary deposits also form when slight changes in the physical or chemical composition of ocean water cause dissolved minerals to precipitate. For example, the mineral apatite (calcium phosphate) is soluble in cold, deep-ocean water that contains only small amounts of dissolved oxygen. Along continental margins, upwelling of this water to the surface causes the apatite to precipitate out as sediment. If the influx of other types of sediment, such as sand and silt, is very low, rich accumulations of phosphate can result.

SEDIMENTARY IRON FORMATION

Another type of chemical sediment is known as sedimentary iron formation. During the Phanerozoic eon, which spans from 544 million years ago to the current geologic eon, deposits of this type have formed when iron was leached from rocks and transported by groundwater to shallow, restricted seas, where it precipitated as hematite (iron oxide). The iron ores of northern Europe and eastern North America, which were very important during the Industrial Revolution, are of this type.

A unique type of sedimentary iron formation is found in rocks between 1.6 and 2.5 billion years old. These deposits consist of alternating bands of iron minerals and chert (a hard rock composed of silica). These deposits formed in many parts of the world as iron precipitated from seawater. The source of this iron is the subject of much debate, but one thing is clear: The deposits, though widespread, are clearly restricted to the period in Earth's history when free oxygen was becoming abundant in the atmosphere. It is believed that this change caused the solubility of iron to decrease (iron is soluble in water only if no free oxygen is present), thereby triggering the precipitation of the iron dissolved in seawater. Upwelling events similar to those envisioned for phosphates may have brought poorly oxidized, iron-rich water upward, to levels where it oxidized and precipitated.

Iron is precipitating on the ocean floor today, along with manganese, to form widespread deposits of manganese nodules. These nodules are found in all the world's oceans and contain considerable amounts of other metals, such as copper, nickel, and cobalt. It is thought that much, if not most, of the metals are derived from hydrothermal activity along midocean ridges. Because the dissolved oxygen content in the deep oceans is low, the metals can travel considerable distances before precipitating from solution.

SECONDARY DEPOSITS

Secondary deposits form after the host sediments have been deposited but during the processes related to burial. When sediments are buried beneath new sediments, water located between the various sediment grains is squeezed out. As the sediments are buried deeper and deeper, they are warmed by heat from the earth's interior. As the temperature rises, the water being expelled from the compacting sediment mass reacts with the various grains and becomes increasingly saline. Such reactions, involving pore fluids and the enclosing rocks, are collectively known as diagenesis and occur in all sedimentary environments. Brines formed in this manner are well known from oil and gas exploration. As these brines move toward the earth's surface, the temperature decreases; in many instances, they also mix with other groundwater

Several important types of secondary deposit are known. In northern Europe, Africa, Australia, and elsewhere, thin (up to several meters thick) but widespread (up to 10,000 square kilometers) layers of shale are enriched in sulfides of copper and, in some cases, other metals, such as lead and zinc. Although there is still much controversy regarding the origin of brines, most evidence indicates that brines formed during diagenesis reacted with the shale shortly after it was deposited.

Secondary deposits are found in limestones (rocks made up of calcium carbonate) and dolomites (calcium-magnesium carbonate). Brines that were formed during diagenesis migrated into cavities in the host sediment, and sulfides (minerals that contain sulfur), particularly of lead, zinc, and, in some cases, barite and fluorite, were precipitated. These deposits exhibit many similarities to shale-hosted deposits but differ in that copper is rare or absent. They are commonly called Mississippi Valley-type deposits because they are similar to the extensive accumulations in the Mississippi River drainage area. Recent research suggests that warm fluids originating deep under the Michigan and Illinois Basins traveled hundreds of kilometers before precipitating the ores. The fluids were also responsible for converting limestone in the region to dolomite.

Secondary deposits also form when oxidized groundwaters flow through sandstones located near the earth's surface. Some metals, such as uranium, are present in small quantities in the sandstone. As the water percolates downward through the sandstone, the metals dissolve and are concentrated in the water. Eventually, the dissolved oxygen reacts with organic matter in the rocks. This reaction lowers the solubilities of the metals, and they precipitate. Most of the uranium ores of the western United States formed in this manner. Certain ores of copper are thought to have formed in a similar fashion.

FIELD STUDY

Economic geologists (geologists who study ore deposits) have a number of methods available to them for studying sedimentary mineral deposits. Perhaps the most important method is detailed mapping and description of ore deposits exposed in mines. During mining, rock material is continuously removed, thereby exposing new rocks. This exposure allows the mine geologist (a geologist who works with miners to ensure efficient mining) to observe and describe the deposit from a three-dimensional perspective not readily available to geologists conducting other types of studies. (For example, a geologist examining the face of a mountain can only guess at what the rocks beneath the mountain are like; a mine geologist examining the walls of a mine need only wait until these are stripped away to see what lies behind them.) This three-dimensional viewpoint allows the geologist to determine the spatial relationships of the minerals to one another and to the enclosing strata. Drill cores of rock are normally obtained around the periphery of the deposit, and the geologist can make similar observations in the cores. These relationships can then be used to interpret how the deposit formed.

Once studies of individual sedimentary mineral deposits are completed, they can be compared with studies of other, nearby sedimentary mineral deposits. A regional interpretation of how the deposits formed can be gathered from such comparisons. Models of this type are extremely useful in deciphering the overall geologic history of an area. They are useful also in the search for similar sedimentary mineral deposits.

MICROSCOPIC STUDY

The examination of rock samples using microscopic techniques allows geologists to observe relationships not visible to the naked eye. Several types of microscopic study are available. Standard petrographic microscopy involves observation of a thin wafer or rock. This wafer, known as a thin section (0.03 millimeter thick), is mounted on a glass microscope slide and ground until it is thin enough for visible light to pass through. Polarized light is passed through the thin section and the geologist notes how the light is affected by passing through different minerals.

Some minerals, particularly those containing valuable metals and sulfur, are opaque to visible light no matter how thin they are ground. These minerals are studied by reflected light microscopy. The thin section is polished until a very smooth surface is obtained. Then, light from above is reflected through the microscope for observation. Reflected light microscopy is often combined with standard petrographic microscopy. Together, these techniques are useful for examining the boundaries between the various minerals present in the sample, and for determining the order in which they formed.

Additional microscopic techniques are often used to supplement these studies. Epifluorescence microscopy involves illuminating the sample with violet or ultraviolet ("black") light. Under these conditions, certain minerals fluoresce, like the colors on a black-light poster, and they can be easily identified through a microscope. In cathode luminescence microscopy, the sample is put in a vacuum chamber and bombarded with an electron beam. The beam excites certain minerals, causing them to emit visible light. Fluid-inclusion microthermometry involves putting thin chips of the sample into a special heating-cooling microscope stage. The sample is then heated with hot nitrogen gas or cooled with liquid nitrogen. Tiny bubbles of water trapped in the minerals are observed as they freeze, thaw, and expand as the temperature changes. This technique yields information about the temperature and chemical composition of the hydrothermal fluids from which the minerals formed.

GEOCHEMICAL STUDY

Geochemistry is the study of the chemical characteristics of rocks. Economic geologists use several geochemical techniques, including the study of isotopes. An isotope is an atom of an element that contains a greater or lesser number of neutrons than are usually present in that element. Determining the amounts of the various isotopes of an element present in the sample can yield significant information about the origin of the deposit. Radioisotopes, those isotopes involved in radioactive decay, can be used to determine the age of a deposit; radioisotopes of lead minerals can be used to determine the source of lead that has been concentrated in the deposit as well. Isotopes of sulfur (present in most sedimentary mineral deposits) can be useful in determining the source of the sulfur and the temperature and pressure under which the deposit formed. Isotopes of hydrogen and oxygen can yield information about the source of water in hydrothermal fluids (seawater, rainwater, and water released by the melting of rocks) as well as the temperature at which the deposit formed.

CONTINUING VALUE

During the last century, exploitation of the earth's mineral wealth has expanded at unprecedented rates as advances in technology have led to improvements in mining and extraction methods and the development of new and more expansive uses of minerals and other natural resources. Sedimentary mineral deposits have been, and continue to be, the source of a major portion of the world's mineral commodities. At one time or another, the largest lead, zinc, gold, uranium, barite, boron, and iron mines in the world have exploited deposits in sedimentary rocks. For these reasons, sedimentary mineral deposits have had a profound effect on the life of every person in the developed world. As technology continues to advance, and as the number of people with access to these resources increases, the demands on Earth's mineral deposits will grow as well.

Robert A. Horton, Jr.

FURTHER READING

Barnes, H. L. *Geochemistry of Hydrothermal Ore Deposits.* 3d ed. New York: John Wiley & Sons, 1997. Contains chapters on the genesis of ore fluids in sedimentary environments, and on the theory behind various geochemical techniques used for the study of sedimentary mineral deposits. Suitable for college-level readers with some knowledge of chemistry.

Blatt, Harvey. *Sedimentary Petrology.* 2d ed. San Francisco: W. H. Freeman, 1992. Offers complete coverage of sedimentary rocks, and the design and conduct of research projects. Appropriate for high school students and general readers.

Blatt, Harvey, Robert J. Tracy, and Brent Owens. *Petrology: Igneous, Sedimentary, and Metamorphic.* 3d ed. New York: W. H. Freeman, 2005. Features a clear and very well illustrated section on sedimentary rocks. Introduces readers to sedimentary rocks in general and classification in particular. Appropriate for general readers, as well as undergraduate and graduate students.

Boggs, Sam, Jr. *Petrology of Sedimentary Rocks.* New York: Cambridge University Press, 2009. Explains the classification of sedimentary rocks. Provides information on different types of sedimentary rocks. Describes siliciclastic rocks and discusses limestones, dolomites, and diagenesis.

Coe, Angela L., ed. *The Sedimentary Record of Sea-Level Change.* New York: Cambridge University Press, 2003. Discusses changes in the sea level throughout time and the factors influencing the sea level, including the ice ages and sedimentation. Includes multiple case studies, chapter summaries, references, and an index. Appropriate for undergraduate and graduate students.

Craig, James R., David J. Vaughan, and Brian J. Skinner. *Earth Resources and the Environment.* 4th ed. Upper Saddle River, N.J.: Prentice Hall, 2010. Contains information on various types of sedimentary mineral deposits. Appropriate for a general audience.

Edwards, Richard, and Keith Atkinson. *Ore Deposit Geology and Its Influence on Mineral Exploration.* London: Chapman and Hall, 1986. An introductory text on the subject of economic geology. Covers various types of sedimentary mineral deposits. Uses British rather than American spellings for many technical terms, particularly mineral names. Suitable for college-level readers with some knowledge of geology.

Hsu, Kenneth J. *Physics of Sedimentology.* 2d ed. New York: Springer, 2010. Discusses physics and mathematics of sedimentary rock dynamics. Contains suggested readings, a subject index, references, and appendices with each chapter. Designed as a university-level text; assumes middle-school-level knowledge.

Kesler, S. E. *Our Finite Mineral Resources.* New York: McGraw-Hill, 1976. Covers many aspects of mineral exploitation, including geology, exploration, refining, and ultimate use. Deals with sedimentary mineral deposits. Treats geological aspects in varying detail; however, geological coverage is generally rudimentary. Written for nonscientists at all levels.

Laznicka, Peter. *Giant Metallic Deposits.* 2d ed. Berlin: Springer-Verlag, 2010. Discusses the location, extraction, and future use of many metals. Organized by topic rather than by metal. Topics include geodynamics that result in large metal ore deposits, the composition of metals within deposits, hydrothermal deposits, metamorphic associations, and sedimentary associations.

Mackenzie, F. T., ed. *Sediments, Diagenesis, and Sedimentary Rocks.* Amsterdam: Elsevier, 2005. Compiles articles on such subjects as diagenesis, chemical composition of sediments, biogenic material recycling, geochemistry of sediments and cherts, and green clay minerals..

Maynard, J. B., E. R. Force, and J. J. Eidel, eds. *Sedimentary and Diagenetic Mineral Deposits: A Basin Analysis Approach to Exploration.* Chelsea, Mich.: Society of Economic Geologists, 1991. Discusses the geochemistry of most major types of metallic sedimentary mineral deposits in a geological context, and the strengths and weaknesses of various models that have been proposed to explain their formation. Suitable for general readers with some knowledge of chemistry and geology.

Middleton, Gerard V., ed. *Encyclopedia of Sediments and Sedimentary Rocks.* Dordrecht: Springer, 2003. Cites a vast number of scientists. Subjects include biogenic sedimentary structures, Milankovitch cycles, deltas and estuaries, and vermiculite. Includes a subject index. Designed to cover a broad scope and a degree of detail useful to students, faculty, and professionals in geology.

Ridge, J. D. *Economic Geology: Seventy-fifth Anniversary Volume.* El Paso, Tex.: Economic Geology Publishing, 1981. Provides detailed reviews of the geology of some of the most important types of sedimentary mineral deposits. Gives brief accounts of some other types that are not themselves the subjects of detailed reviews in this volume. Suitable for college-level readers with some background in geology.

_____. *Ore Deposits of the United States, 1933-1967.* 2 vols. New York: American Institute of Mining, Metallurgical, and Petroleum Engineers, 1968. Describes ore deposits of many types, including major sedimentary mineral deposits in the United States. Contains information on the history of discovery and production as well as geological information. Suitable for college-level readers.

Tucker, M., ed. *Sedimentary Rocks in the Field.* 4th ed. New York: John Wiley & Sons, 2011. An introductory text aimed at British undergraduates. Devoted in large part to classification, in terms that are appropriate for the general reader.

United States Bureau of Mines. *Minerals Yearbook.* Washington, D.C.: Government Printing Office, issued annually. Contains statistics on many aspects of mining and mineral exploitation.

See also: Aluminum Deposits; Building Stone; Cement; Chemical Precipitates; Coal; Diamonds; Dolomite; Earth Resources; Evaporites: Fertilizers; Gold and Silver; Industrial Metals; Industrial Nonmetals; Iron Deposits; Manganese Nodules; Mining Processes; Oil and Gas Exploration; Pegmatites; Platinum Group Metals; Salt Domes and Salt Diapirs; Uranium Deposits.

SEDIMENTARY ROCK CLASSIFICATION

Because sedimentary rocks are formed by several different processes—precipitation, crystallization, and compaction—no single classification scheme is applicable to all of them. Used in various combinations, the main elements of classification are mode of origin, mineralogy, the size of the individual mineral grains that make up the rock, and the origin of these grains.

PRINCIPAL TERMS

- **arkose:** a sandstone in which more than 10 percent of the grains are feldspar or feldspathic rock fragments; also called feldspathic arenite
- **carbonate rock:** a sedimentary rock composed of grains of calcite (calcium carbonate) or dolomite (calcium magnesium carbonate)
- **clastic rock:** a sedimentary rock composed of broken fragments of minerals and rocks; typically a sandstone
- **clay stone:** a clastic sedimentary rock composed of clay-sized mineral fragments
- **greywacke:** a sandstone in which more than 10 percent of the grains are mica or micaceous rock fragments; also called lithic arenite
- **limestone:** a carbonate sedimentary rock composed of calcite, commonly in the form of shell fragments or other aggregates of small calcite grains
- **orthoquartzite:** a sandstone in which more than 90 percent of the grains are quartz
- **sandstone:** a clastic sedimentary rock composed of sand-sized mineral or rock fragments

EVAPORITES

Because there are several very different processes that lead to the formation of sedimentary rocks, no single classification scheme is suitable to all sedimentary rocks. The main elements of classification, however, are mode of origin, mineralogy, the size of the individual mineral grains making up the rock, and the origin of the individual grains. These elements are used in various combinations to categorize several major groups of sedimentary rocks.

One of the major groups is the evaporites. All natural waters contain some dissolved solids that will precipitate when the water evaporates. The crust that forms in a teakettle that has been used for a long time is an example. Seawater contains about 33 parts per thousand dissolved solids and is the major source of the sedimentary rocks classified as evaporites. When a body of seawater in an area with low rainfall is cut off from the sea, as when a sandbar builds up across the mouth of a bay, the trapped seawater tends to evaporate. During this evaporation, several minerals are precipitated in a predictable order. The first to precipitate is the mineral gypsum, hydrated calcium sulfate. If the water is very hot, the precipitating calcium sulfate will not be hydrated, and the mineral anhydrite will form. Further evaporation will cause the precipitation of halite (sodium chloride, or ordinary table salt), and still further evaporation will lead to the precipitation of a complex series of potassium and magnesium salts.

In some environments, most commonly closed depressions in desert areas (the Dead Sea, for example), fresh water evaporates to produce evaporite minerals that are quite different from the minerals produced by the evaporation of seawater. The natron (hydrated sodium carbonate) that was used by the Egyptians in embalming mummies and the borax that originally made Death Valley, California, famous are freshwater evaporite minerals.

From the standpoint of classification, the rocks of evaporative origin—that is, masses of individual crystals of minerals produced by evaporation of seawater or fresh water—are not usually given distinctive names. A fist-sized piece of evaporative sedimentary rock composed of gypsum is normally called gypsum. When it is necessary to indicate clearly that a rock rather than a mineral is being mentioned, the term "rock gypsum" is used. An exception is a rock composed entirely of crystals of halite, which is almost invariably referred to as rock salt.

CLASTIC ROCKS

A second major group of sedimentary rocks is the clastic rocks. "Clastic" comes from the Greek for "broken"; the individual grains of clastic rocks are the product of the mechanical and chemical breakdown, or weathering, of older rocks. Beach sands are composed of such grains. They are composed of residue from weathering of granites and many other kinds of igneous, sedimentary, and metamorphic rocks. Common soil, or mud, also is a residue of the weathering of rocks. Mud differs from beach sand primarily in its content of fine-grained clay minerals. Clay minerals are similar to the mineral mica, but the individual grains are very small, by definition less than 0.004 millimeter in size. When subject to prolonged attack by water and the atmosphere, many minerals that are common in igneous and metamorphic rocks—principally feldspars—are changed into micalike clay minerals. Between beach sands and the clay-sized component of muds are an intermediate size of mineral grains called silts. Silts range in size, again by definition, from 0.0625 millimeter to 0.004 millimeter. Muds are mixtures of silt-sized and clay-sized mineral grains. Grains larger than 2 millimeters in diameter are classified as granules, pebbles, and boulders with increasing size.

The primary basis for classification of clastic sedimentary rocks is grain size. Coarse-grained rocks composed of pebbles and granules are called conglomerates, rocks composed of sand-sized grains are called sandstones, rocks composed of mud-sized materials are mudstones, and rocks composed only of clay-sized grains are called clay stones. Mudstones and clay stones that split readily along flat planes, the bedding planes, are referred to as shales. Collectively, rocks made of silt and clay-sized particles are called mudrocks.

Sandstones are further classified by mineral composition. Sandstones that contain less than 10 percent of fine material are termed arenites, whereas those with more than 10 percent fine material are called wackes. In most sandstones more than 90 percent of the sand grains are quartz, but some sandstones contain appreciable amounts of feldspar grains, volcanic rock fragments, mica, and micaceous rock fragments. These grains are the basis for further classification. Sandstones that are nearly pure quartz sand (more than 90 percent quartz grains, by one common definition) are called orthoquartzites, or quartz arenites ("arenite" is from the Latin for "sand"). Sandstones containing more than 10 percent feldspar grains and volcanic rock fragments are called arkoses or feldspathic arenites, and sandstones with more than 10 percent mica flakes and rock fragments are called greywackes or lithic arenites. Generally, rocks with abundant feldspar and rock fragments have undergone little weathering and are said to be immature, whereas those made only of very stable minerals, such as quartz, have undergone prolonged weathering and are said to be mature.

LIMESTONES

Limestones are another major group of sedimentary rocks. Limestones are composed predominantly of the mineral calcite (calcium carbonate), although some limestones may include some clastic material, typically quartz sand or clay. A similar group of rocks is the dolomites. Dolomites consist predominantly of the mineral dolomite (calcium-magnesium carbonate) and form almost invariably by chemical alteration of preexisting limestone. Therefore, many workers prefer to lump the limestones and dolomites together in a rock group named carbonates.

The greater part of the calcite in limestones is secreted by marine organisms that make their shells from the mineral; clams and oysters are good examples of such organisms. Once these organisms die, their shells are washed about by waves and currents and are broken and abraded into fragments. The fragments may range in size from pebbles to mud, and most limestones are composed of this biogenic detritus.

Two common components of limestones appear to be of inorganic origin. Oölites are round grains, composed of very small crystals of calcite, that have a superficial resemblance to fish eggs. A shell fragment or quartz grain, the nucleus, at the center of the oölite, is surrounded by a coating of fine-grained calcite crystals layered like tree rings. It appears that oölites grow by inorganic deposition of calcite, directly from seawater, on the surface of the nucleus. Field observation of modern oölites suggests that they form only on sea bottoms that are shallow and are periodically agitated by strong waves or currents. It also appears that mud-sized calcite crystals precipitate directly from seawater in some circumstances.

Finally, small organisms, principally marine worms, ingest calcite mud to extract whatever useful organic matter it may contain and excrete it as fecal pellets. The fecal pellets, held together by mucus from the gut of the organism, survive if the bottom currents are not too strong.

Most limestones consist of aggregates of the materials described above. One classification, originally introduced by Robert L. Folk in 1959, and the most widely used classification, is based on the nature of the aggregates and the material that occurs between the grains and cements them together. Mud-sized calcium carbonate tends to accumulate in quiet waters, and a limestone consisting of only mud-sized material is called micrite. Sand-sized calcite grains are deposited in areas with stronger currents, generated by waves, winds, or tides. After final deposition, open spaces between sand-sized grains are often filled by inorganic calcite cement, called spar. In Folk's classification, the abbreviated name of the sand-sized material followed by the name of the material between the sand-sized grains is the rock name, with the traditional rock-name ending "-ite." Typical examples are oösparite, pelmicrite, and biosparite (where "bio-" refers to shell fragments).

CHEMICAL ROCKS

A fourth major group of sedimentary rocks is the chemical rocks. These rocks are divided into two subgroups: chemical precipitates and chemical replacements. Chemical precipitates are sediments that accumulate directly on the sea bottom as a result of chemical reactions that do not involve evaporation. Deposits of iron minerals (most typically the iron oxide mineral hematite) and phosphate minerals (most typically calcium phosphate, or the mineral apatite) are the most common, and economically important, examples. The process of formation of these rocks is not well understood, but research suggests that in most cases bacteria are involved in producing the proper chemical environment for formation of these important, although rather rare, deposits.

The second subgroup is the chemical replacements. In some cases, the original sediment is dissolved and a new mineral takes its place. Typical examples are the solution of calcite and its replacement by dolomite and the solution of calcite and its replacement by fine-grained quartz. The replacement of calcite by fine-grained quartz in limestones is especially common. The replacement product, the very fine-grained quartz replacement, is called chert, but most people are more familiar with the popular term "flint."

IDENTIFYING SEDIMENTARY ROCKS AND COMPONENT MINERALS

Of the more than five thousand minerals identified, only a few dozen are common in rocks at the earth's surface; these are easily identified in the field by observation and simple tests of physical properties. Even most fine-grained rocks can be identified in the field with the aid of a twelve-power hand lens. Gypsum, for example, is easily identified by its satiny

sheen and the fact that it can be scratched by a fingernail. Calcite and dolomite have very similar appearances and physical properties, but a drop of dilute hydrochloric acid will cause calcite, but not dolomite, to effervesce, or fizz. A pocket-sized dropper bottle of dilute hydrochloric acid is standard equipment for the sedimentary geologist.

More detailed studies are done on samples brought to the laboratory. A very common method is the study of thin sections of rocks. To prepare a thin section, a thin slice about a centimeter thick is cut from the sample. The slice is glued to a glass slide and further thinned. When the thickness has been reduced to about 0.03 millimeter, a thin cover glass is glued to the top of the slice, and the thin section is complete.

Thin sections are studied with a microscope, usually a petrographic microscope that has a polarized light source. The effect of passage of the polarized light through individual mineral crystals can be analyzed and much information about the arrangement of the atoms in the crystals obtained. For example, passage of the light through halite does not affect the planar vibration of the polarized light, but when it passes through quartz, it is forced to vibrate in two planes that are perpendicular to each other and parallel to the two crystallographic axis directions of quartz. Therefore, even microscopic-sized grains of halite and quartz are easily distinguished.

Clay minerals, which are too small to be studied effectively by optical methods, are commonly studied by X-ray diffraction. X rays have very short wavelengths and can be diffracted by the regularly arranged planes of atoms in a crystal in the same way that light is diffracted by a diffraction grating. The sample to be analyzed is irradiated with X rays of a single wavelength at steadily varying angles of incidence. The angles at which diffraction occurs represent planes of atoms at different spacings. If the clay mineral kaolin, for example, is present in the sample, a strong diffraction will occur at an angle that corresponds to an atomic plane spacing of 7.2×10^{-10} meters (usually expressed as 7.2 angstroms), and weaker diffractions will occur at angles corresponding to several other spacings that are characteristic of the mineral.

Scanning electron microscopy is a powerful tool for the study of all types of sedimentary rocks but is especially useful for the very fine-grained varieties. Photographic images at 50,000 magnifications are routinely obtained, revealing remarkable details of individual grains and the openings between grains. In addition, semiquantitative chemical analyses of individual grains for sodium and elements heavier than sodium can be made.

INDUSTRIAL APPLICATIONS

Classification of sedimentary rocks has led to the recognition of predictable associations of sedimentary rock types with particular geologic conditions. Sandstones that are derived from and deposited near mountains with granitic cores—the Front Ranges of the Rocky Mountains, for example—most commonly contain more than 10 percent feldspar, mostly of the potassium-rich variety, and are classified as arkoses. Sandstones derived from mountains with metamorphic rock cores, such as the Appalachian Mountains, contain more than 10 percent mica and are classified as greywackes, or lithic arenites. Sandstones deposited far from any mountain chain normally are nearly pure quartz, in some cases more than 99.9 percent quartz. These rocks have generally undergone many cycles of deposition, weathering, erosion, and redeposition, so that only quartz has survived. Such rocks are said to be mature. For quality-control purposes, glass manufacturers prefer sand that is nearly pure quartz, and the sedimentary geologist who specializes in industrial minerals will begin the search for glass sands far from any large mountain chain.

The organic-matter content of muds that lie on the sea bottom for long periods of time tends to be destroyed by scavenging organisms. Sedimentary geologists who specialize in petroleum exploration know that one of the requirements for the generation of petroleum is a mudstone with a high organic-matter content. They look for mudstones that were buried by new mud shortly after deposition—that is, mudstones that had a high rate of sedimentation. As a general rule, such mudstones will be associated with arkoses or greywackes rather than with nearly pure quartz sandstones.

Robert E. Carver

FURTHER READING

Blatt, Harvey. *Sedimentary Petrology.* 2d ed. San Francisco: W. H. Freeman, 1992. Offers complete coverage of the subject of sedimentary rocks. Intended as a college-level textbook but perfectly accessible to interested high school students or laypersons. The final two chapters cover the design and conduct of research projects.

Blatt, Harvey, Robert J. Tracy, and Brent Owens. *Petrology: Igneous, Sedimentary, and Metamorphic.* 3d ed. New York: W. H. Freeman, 2005. Intended for the college sophomore or junior but suited to the interested high school student or general reader. Offers a clear and well-illustrated section on sedimentary rocks. Covers sedimentary rocks in general and classification in particular.

Boggs, Sam, Jr. *Petrology of Sedimentary Rocks.* New York: Cambridge University Press, 2009. Begins with a chapter explaining the classification of sedimentary rocks. The remaining chapters provide information on different types of sedimentary rocks. Describes siliciclastic rocks and discusses limestones, dolomites, and diagenesis.

Carver, Robert E., ed. *Procedures in Sedimentary Petrology.* New York: John Wiley & Sons, 1971. Covers the methods commonly used to study sedimentary rocks, including the methods of grain-size analysis and mineralogical analysis that are the basis of their classification. Where appropriate, a chapter on obtaining the data is followed by a chapter on mathematical or statistical analysis of the data. The mathematics involved is not difficult.

Fairbridge, Rhodes W., and Joanne Burgeois, eds. *The Encyclopedia of Sedimentology.* New York: Springer, 1978. Ideal reference for readers needing information on specific aspects of the classification of sedimentary rocks. Extensively and usefully cross-referenced.

Ham, W. E., ed. *Classification of Carbonate Rocks: A Symposium.* Tulsa, Okla.: American Association of Petroleum Geologists, 1962. A classic work on the title subject by the originators of various schemes of classification of limestones and dolomites, written at a time when there was still much discussion of what was the most appropriate classification. Required reading for all students of carbonate rock classification.

Hatch, F. H., R. H. Rastall, and J. T. Greensmith. *Petrology of the Sedimentary Rocks.* 7th ed. New York: Springer, 1988. An older reference from the time of widespread fascination with classification. Contains details of composition and classification of some rock groups (the carbonaceous group, for example) that cannot be found elsewhere. Illustrations are informative line drawings of thin sections.

Hsu, Kenneth J. *Physics of Sedimentology*. 2d ed. New York: Springer, 2010. Written by a knowledgeable professor. Discusses physics and mathematics of sedimentary rock dynamics, assuming a middle-school level understanding. Contains suggested readings, a subject index, references, and appendices.

Mackenzie, F. T., ed. *Sediments, Diagenesis, and Sedimentary Rocks*. Amsterdam: Elsevier, 2005. Volume 7 of *Treatise on Geochemistry*, edited by H. D. Holland and K. K. Turekian. Compiles articles on such subjects as diagenesis, chemical composition of sediments, biogenic material recycling, geochemistry of sediments and cherts, and green clay minerals.

Middleton, Gerard V., ed. *Encyclopedia of Sediments and Sedimentary Rocks*. Dordrecht: Springer, 2003. Cites a vast number of scientists, who are listed in the author index. Subjects range from biogenic sedimentary structures to Milankovitch cycles. Provides a subject index. Designed to cover a broad scope and a degree of detail useful to students, faculty, and professionals in geology.

Pettijohn, F. J., P. E. Potter, and R. Siever. *Sand and Sandstone*. 2d ed. New York: Springer-Verlag, 1987. A thorough, well-illustrated, and clearly written treatment of the subject. Requires a general familiarity with the geology and mineralogy of sandstones; some preliminary study of the general subject is advised.

Sasowsky, Ian D. and John Mylroie, eds. *Studies of Cave Sediments: Physical and Chemical Records of Paleoclimate*. Rev. ed. Dordrecht: Springer, 2007. Discusses sedimentary rock locations and characteristics found around the United States. Many caves, karsts, and speleothems are examined. Contains scientific articles with technical writing; suited for professional researchers, professors, and advanced students.

Tucker, M., ed. *Sedimentary Rocks in the Field*. 4th ed. New York: Wiley & Sons, 2011. Introductory text aimed at British undergraduates, who learn principally through independent study. Largely devoted to classification in terms that are accessible to the general reader.

Williams, Howel, F. J. Turner, and C. M. Gilbert. *Petrography: An Introduction to the Study of Rocks in Thin Sections*. 2d ed. San Francisco: W. H. Freeman, 1982. An introductory college text accessible to younger students and general readers with some familiarity with the literature. As in the work by Hatch, Rastall, and Greensmith, the sedimentary rock illustrations are primarily line drawings of thin sections, but they make the point very well.

See also: Biogenic Sedimentary Rocks; Building Stone; Carbonates; Chemical Precipitates; Diagenesis; Dolomite; Limestone; Petroleum Reservoirs; Siliciclastic Rocks.

SILICATES

Silicates are chemical compounds comprising silicon, oxygen, and a metal. Most of the earth's crust is made up of silicates, as is about one-third of the world's minerals. A silicate's chemical base is a tetrahedron, a pyramid-shaped molecule with negatively charged ions. Silicates are considered the building blocks of the earth's crust, appearing in a wide range of forms, including sand, quartz, and gemstone.

PRINCIPAL TERMS

- **glass:** crystal formed from superheated quartz silicate (sand)
- **mesodesmic:** molecular bond in which one of the molecule's oxygen ions is capable of bonding with other ions
- **Portland cement:** common form of cement used in industry and construction; comprises hydrated, ground, and heated limestone, clay, and calcium silicates
- **sand:** granular sedimentary silicate
- **tetrahedron:** a molecule in which the central atom forms four bonds

BASIC PRINCIPLES

Silicates are considered the building blocks of the planet, constituting more than 90 percent of the earth's crust. While there are a wide range of types of silicates of varying size and degrees of solidity, each has a basic chemical composition consisting of a silicon and oxygen-based tetrahedron (a molecule in which the central atom forms four bonds). In this case, the basic formula is SiO_4.

The chemical bond forming SiO_4 is said to be mesodesmic: In the case of a silicate, the silicon and oxygen ions share an electrostatic charge, causing them to bond with one another. However, one of the oxygen ions is not shared with a silicon ion, leaving it capable of bonding with other ions. The result of a mesodesmic bond is that the molecule will be able to bond with ions from outside the group.

There are seven basic types of silicates, all of which are classified based on the degree of polymerization (a process by which single molecules bond to form larger compounds known as polymers) that occurs within the SiO_4 tetrahedron. These seven types of silicates are nesosilicates (island silicates), sorosilicates (group silicates), cyclosilicates (ring silicates), two types of inosilicates (single- and double-chain silicates), phyllosilicates (sheet silicates), and tectosilicates (framework silicates).

Because of their mesodesmic nature, silicates appear in a broad spectrum of forms. For example, nearly all igneous rocks (rocks that have been formed from magma thrust from deep within the earth's surface) are silicates, as are most metamorphic rocks (stones that change in structure when subjected to certain pressure, temperature, and chemical conditions). Furthermore, silicates are found in an overwhelming majority of sedimentary rocks—stones that are weathered from other rocks. Sand is one of the best-known of these types of sedimentary silicates.

DIFFERENT TYPES OF SILICATES

There are seven general types of silicates. These categories are determined based on how the silicate tetrahedron's four oxygen atoms have connected to the single silicon atom. Island silicates (nesosilicates), for example, are configured in such a way that the oxygen atoms surround the silicon atom at the corners of the tetrahedron. In this configuration, the remaining oxygen atom is shared with other charged ions, such as magnesium, calcium, and iron.

When two silicate tetrahedra share a single oxygen atom, the dual configuration is known as a double island silicate (sorosilicate). The unusual, hourglass-shaped configuration of the sorosilicate makes it somewhat rare when compared with other silicates. However, one type of double island silicate, the mineral epidote, is found in large volume in certain metamorphic rocks (rocks that, under intense heat and pressure, change in structure to adapt to the environment).

In some cases, two of the oxygen atoms are shared among tetrahedra. The resulting structure appears in the shape of a ring. One of the best-known examples of these cyclosilicates is the mineral beryl. Varieties of beryl include such gemstones as emeralds and aquamarine.

In other cases, two of the oxygen atoms are shared in such a way that the resulting configuration of linked tetrahedra appears in the shape of a single, linked chain. Single-chain silicates (inosilicates) typically make up igneous rocks (rocks that have been pushed from the earth's interior through volcanic eruptions) known as clinopyroxenes. Other examples of single-chain silicates are minerals known as orthopyroxenes that are found in meteorites primarily. Inosilicates also appear in double-chain configurations (amphiboles). The mineral crocidolite, an example of such a silicate, is common in naturally formed asbestos.

The sixth type of silicate is the sheet silicate (phyllosilicate). This group is formed when three of the oxygen atoms in a silicate tetrahedron are shared. The sheet variety of silicate includes such examples as mica, clay, and talc. Phyllosilicates are easily cleaved, which means that they will break along certain planes where chemical bonds are weakest.

The seventh and final manifestation of silicates is formed when all of the corner oxygen atoms in the silicate tetrahedron are shared by another silicate tetrahedron. The resulting framework configuration is three-dimensional rather than linear, creating a complex and diverse mineral that may include other elements. Some examples of these tectosilicates or framework silicates are quartz and feldspar, the latter of which is one of the most prevalent minerals found in the earth's crust.

SILICATES AND WEATHERING

The mesodesmic properties of silicates do not simply help form minerals and other compounds. They also deconstruct rocks and minerals through the process known as weathering.

In weathering, elements contained in the atmosphere and in water bond with the elements on the surface of igneous, metamorphic, and sedimentary rocks. The compounds that are sloughed off these origin rocks are carried to sedimentary basins. These sedimentary rocks vary in size, from granular sand to larger pieces.

The presence of silicates in most of the earth's minerals means that, when weathering occurs, rocks and minerals are more likely to be broken down because of silicates' bonding characteristics. Many sedimentary rocks are again bonded with other igneous, metamorphic, and sedimentary rocks to form clastic rocks while in these sedimentary basins. The ability of many silicates to exchange charges between different external elements plays a key role in the formation of such clastic rocks as well.

SILICATES AND THE CARBON CYCLE

In light of the way charges are exchanged within silicate tetrahedra, scientists believe that silicates play an integral role in many of the earth's major systems. One such system is the carbon cycle.

Under the framework of the carbon cycle, carbon (which is essential to life and to Earth's geochemical processes) is first released from within Earth's interior through carbonate rocks and carbon dioxide. Carbon dioxide is absorbed from the soil by plants, which are in turn eaten by animals (including humans). Animals breathe carbon dioxide, sending it into the atmosphere, where it acts to block harmful radiation from permeating the atmosphere. Precipitation (rain and snow) collects the carbon from the air and returns it to the soil and to bodies of water. The carbon cycle also entails the degradation of dead plant and animal tissue and the weathering process, returning carbon to Earth's crust.

Scientific evidence reveals that silicates play an important role in many facets of the carbon cycle. For example, when metamorphic rocks are formed under extreme conditions beneath glaciers and oceans (where vertical force pushes downward on the earth's tectonic plates, creating intense pressure and temperatures), the silicates in the changing rock bond with the resulting carbon dioxide, bringing this compound back to the surface and into the atmosphere. Additionally, during chemical weathering (whereby water is the key factor in the degradation of rocks), the hydrogen ions that are released bond quickly with silicates, producing clays and other metal compounds in sedimentary basins. Carbon dioxide, freed from the weathered rocks, returns to the atmosphere.

Although scientists understand chemical weathering and its role in the carbon cycle, a missing element of the study of this framework is the extent to which weathering may be predicted. In a recent examination of this issue, silicates provided a great deal of clues. The 2011 study analyzed 338 North American rivers and the chemical weathering that occurs therein. Scientists concluded in part that the manner and volume by which charged particles (ions) are transferred—a key characteristic of silicates—may potentially serve as a predictor for the rate of carbon breakdown in chemical weathering.

SILICATES AND HYDRATION

Just as they differ in configuration, so silicates vary widely in terms of their textures. Some silicates, like phyllosilicates, are easily broken, while others, like quartz, are extremely hard and difficult to cleave. The resulting minerals and compounds are typically different, too: Some are gemstones, while others (like sand) are granular.

Many silicates are used for industrial purposes. For example, superheated sand (which is largely made of the tectosilicate quartz) may be re-formed as the crystal glass. Silicates also play a critical role in the creation of cement. In this latter example, heated and ground limestone is mixed with clay to form Portland cement powder. Silicates (specifically, calcium silicates) are added to form nodules known as clinker.

These silicates are the major contributors to the strength of cement. When water is added (a process called hydration), the silicates in the clinker bond, creating the solid mass with which walls and building foundations are constructed. Recent studies show that repeated hydration (including reheating the cement mixture with the silicate additive) of Portland cement, one of the world's most popular cements, enhances the strength of the cement. This improvement comes from the bonding caused by the silicate tetrahedron configuration.

RESEARCH TOOLS

Researchers employ a number of tools in the identification and study of silicates. One of these tools is X-ray spectroscopy, which radiates X rays onto the subject, which in turn absorbs the X rays, creating a visible profile for researchers to examine.

Spectroscopy can help scientists assess the shape of a given mineral's silicate tetrahedron, enabling scientists to identify the mineral within a larger rock. Spectroscopy has evolved to include other types of radiation, such as ultraviolet light and infrared, all of which can further help profile silicate minerals and their compositions.

To study the processes in which silicates play a role (such as weathering) and the processes by which the silicates themselves form, scientists are increasingly using computer modeling software. Modeling systems help researchers deconstruct a silicate (down to the molecular and even submolecular levels), compiling what is commonly a large amount of data on each component. Additionally, these systems can provide a detailed profile of silicate crystallization as it occurs through time and under certain conditions. Computer modeling also has become useful for researching the development and application of artificial silicates (such as insulation and cement additives), helping to create more efficient products and to reduce environmental waste.

In addition to the use of computer modeling, scientists have another useful tool in studying silicates: the Internet. Because of the global reach provided by the Internet, scientists can compare data on similar silicate deposits and on environmental conditions in different parts of the world. Such collaboration, which can take place in real time, can greatly enhance the pursuit of silicate study.

IMPLICATIONS AND FUTURE RESEARCH

With nearly all of the planet's crust and nearly one-third of all minerals derived from silicates, it is understandable that silicates also play a major role in so many of the planet's processes. Silicates are major contributors, for example, to the carbon cycle, helping to break down rocks and release carbon back into the atmosphere through weathering.

In light of the ongoing scientific pursuit of greenhouse gas reduction, researchers are exploring chemical weathering as an area of interest. In this regard, scientists hope to prevent emissions in the weathering process by sequestering carbon dioxide.

Some studies are examining the silicate-induced dissolution of certain minerals. In one such case, researchers are using their knowledge of weathering to artificially enhance a silicate mineral like olivine. Using such an approach could lead to new methods of carbon sequestration and a reduction in the overall release of carbon dioxide into the atmosphere.

In a similar vein, researchers are examining the use of silicates in fertilizer to reduce greenhouse gases in agriculture. According to one study of rice growth, an increase in silicate-based fertilizers can lead to the suppression of methane, another greenhouse gas. The study concluded that silicate-based fertilizer can help reduce the volume of greenhouse gas emissions. As research continues on the reduction and sequestration of greenhouse gas emissions, scientists will likely continue to explore silicates as an important factor in this pursuit.

Michael P. Auerbach

FURTHER READING

Hillebrand, W. F. *The Analysis of Silicate and Carbonate Rocks.* Ann Arbor: University of Michigan Library, 2009. This book, which is an update of Hillebrand's original 1919 text, provides an extensive review of silicates and their interactions with carbonate rocks.

Icon Group International. *The 2009 World Forecasts of Silicates and Commercial Alkalai Metal Silicates Export Supplies.* San Diego, Calif.: Author, 2009. Provides an assessment of the export market for a specific type of silicate. The 150 key markets outlined in this report underscore the international need for silicates for construction and other commercial applications.

Lauf, Robert. *The Collector's Guide to Silicate Crystal Structures.* Atglen, Pa.: Schiffer, 2010. A comprehensive review of the different types of silicate crystals and the spectrum of silicates (such as chain, sheet, and framework).

McMillan, Nancy J., et al. "Laser-Induced Breakdown Spectroscopy Analysis of Complex Silicate Materials: Beryl." *Analytical and Bioanalytical Chemistry* 385, no. 2 (2006): 263-271. Discusses the use of laser-induced breakdown spectroscopy (LIBS) to study samples of beryl, a cyclosilicate. Presents a case for the use of LIBS in studying the complex and diverse configurations of different types of beryl.

Murillo-Amador, B., et al. "Influence of Calcium Silicate on Growth, Physiological Parameters, and Mineral Nutrition in Two Legume Species Under Salt Stress." *Journal of Agronomy and Crop Science* 193, no. 6 (2007): 413-421. Analyzes the effects of silicates on nutrients used in the growth of cowpea and kidney bean plants in a salty environment. The study found that the addition of silicates greatly improved the plants' growth, physiology, and nutrient absorption.

Strohmeier, Brian R., et al. "What Is Asbestos and Why Is It Important? Challenges of Defining and Characterizing Asbestos." *International Geology Review* 52, nos. 7/8 (2010): 801-872. Describes in detail one of the best-known types of chain silicates: asbestos. Discusses the various forms of asbestos, their composition, and formation and the public health risks of certain types of asbestos.

See also: Carbon-Oxygen Cycle; Cement; Gem Minerals; High-Pressure Minerals; Igneous Rock Bodies; Quartz and Silica; Siliciclastic Rocks; Toxic Minerals; Ultra-High-Pressure Metamorphism.

SILICICLASTIC ROCKS

Siliciclastic rocks, which include siltstone, sandstone, and conglomerate, are second only to mudrocks in abundance among the world's sedimentary rock types. They form major reservoirs for water, oil, and natural gas and are the repositories of much of the world's diamonds, gold, and other precious minerals. In their composition and sedimentary structures, siliciclastic rocks reveal much about paleogeography and, consequently, Earth history.

PRINCIPAL TERMS

- **clastic rock:** a sedimentary rock composed of particles without regard to their composition; this term is sometimes used, incorrectly, as a synonym for "detrital"
- **detrital rock:** a sedimentary rock composed mainly of grains of silicate minerals as opposed to grains of calcite or clays
- **diagenesis:** chemical and mineralogical changes that occur in a sediment after deposition and before metamorphism
- **lithic fragment:** a grain composed of a particle of another rock; in other words, a rock fragment
- **mudrock:** a rock composed of abundant clay minerals and extremely fine siliciclastic material

WEATHERING AND EROSION

Weathering and erosion constantly strip the earth's surface of its rocky exterior. Rocks experience fluctuating temperatures, humidity, and freeze-thaw cycles that physically disaggregate (break up) mineral grains. Simultaneously, these mineral grains are exposed to water, oxygen, carbon dioxide, and dissolved acids or bases that chemically attack them. As a result, a hard, rocky surface is eventually transformed into a collection of mineral grains, clay minerals, and ions dissolved in water.

Weathering and erosion are slow but continuous agents of destruction. Because mountains and other high regions are constantly being uplifted, weathering and erosion provide a more or less steady supply of sediment to streams and rivers and, from there, into the ocean for deposition. Deposited sediment is eventually compacted and cemented into rocks. The rocks that are formed by this process are known collectively as siliciclastic rocks because they are made mostly of particles of silicate minerals.

SILICICLASTIC ROCK COMPOSITION

Siliciclastic rocks differ in the sizes of their grains and their composition. Three main size classes are

recognized: silt, sand, and gravel. Silt includes particles between 0.004 and 0.0625 millimeter (1/256 to 1/16 millimeter) in diameter; sand is composed of particles between 0.0625 (1/16) and 2 millimeters in diameter; and gravel generally includes particles larger than 2 millimeters. Most sedimentologists refer to sand and sandstone when discussing siliciclastic rocks.

Siliciclastic rocks are composed of three broad classes of material: framework grains, matrix, and cement. Framework grains are particles of minerals or small fragments of other rocks that usually make up the bulk of a siliciclastic rock. Matrix is extremely fine-grained material such as clay that is deposited at the same time as framework grains. Cement is any material precipitated within the spaces, or pores, between grains in a sediment or rock. Framework grains and matrix are primary deposits, whereas cement is a secondary deposit because it is precipitated in pores after the primary material.

QUARTZ AND FELDSPAR MINERALS

The main minerals that compose siliciclastic rocks are quartz, various types of feldspar, and micas. The ferromagnesian minerals, such as olivine, pyroxene, and amphibole, are not as common in sedimentary rocks as they are in igneous and metamorphic rocks. Ferromagnesian minerals are more easily dissolved during weathering, and they fracture more readily during erosion, than the more durable quartz and feldspar. Consequently, quartz and feldspar (and, to a lesser extent, the micas) are concentrated in sediments because the other minerals are selectively removed beforehand. Carbonate rocks dissolve very easily and seldom form much residue to contribute to sand.

Quartz is the most chemically and physically stable mineral that forms siliciclastic rocks. Although quartz is uncommon in some igneous and metamorphic rocks, it is sufficiently abundant in granite and gneiss to contribute a large amount of bulk to almost all sands. Subtle differences may provide useful clues as to the origin of some quartz grains. For example, quartz from plastically deformed metamorphic rocks (schist, gneiss) usually occurs as polycrystalline aggregates or displays a distorted crystal structure, revealed by the petrographic microscope. Quartz from volcanic rocks often possesses planar crystal faces or embayments (deep, rounded indentations) that were formed as the crystal grew in magma. Quartz grains eroded from older sedimentary rocks sometimes retain the previous rock's quartz cement as a rind, which is called an inherited overgrowth.

Feldspar minerals are also common in most siliciclastic rocks. Two main groups of feldspar are recognized: plagioclase feldspar and the potassium feldspars (microcline, orthoclase, and sanidine). The plagioclase group is actually a collection of many similar minerals with different chemical compositions. Anorthite is plagioclase feldspar composed of calcium, silicon, and oxygen, whereas albite is plagioclase composed of sodium, silicon, and oxygen. Calcium and sodium substitute rather easily for each other, so most plagioclase feldspars have both calcium and sodium. In general, plagioclase with more calcium is more susceptible to chemical weathering than is sodium-rich plagioclase. Consequently, sodium-rich albite is more common than anorthite in most siliciclastic rocks.

Potassium feldspars all have the same chemical composition of potassium, silicon, and oxygen. They differ from one another in their chemical structure. Microcline is the most highly organized, crystallographically, followed by orthoclase and then sanidine. Microcline is formed in igneous and metamorphic rocks that crystallize very slowly, permitting the greatest amount of crystallographic ordering. At the other extreme, sanidine is formed in volcanic rocks where little time is available for ions to get into their "proper places." During chemical weathering of detrital potassium feldspars, sanidine dissolves much more readily than does microcline or orthoclase. Microcline is usually about equal in abundance to orthoclase in siliciclastic rocks, whereas sanidine is usually rare.

MICAS AND ACCESSORY MINERALS

The mica minerals biotite and muscovite are common in some siliciclastic rocks and rare in others. In general, muscovite (white mica) is more durable and, consequently, more abundant than biotite (brown mica). Chlorite is a mineral with a sheetlike crystal structure similar to the micas. It is common in a few sandstones but generally less abundant than the micas. A large variety of grains is often present in siliciclastic rocks as "accessory minerals." Altogether they seldom constitute more than 1 percent of any siliciclastic rock. Sometimes these accessory minerals are called heavy minerals because most of them have a much greater density than does quartz or feldspar. This group includes zircon, tourmaline, rutile, and garnet; the ferromagnesian minerals pyroxene, amphibole, and olivine; and iron oxides such as hematite, magnetite, chromite, and limonite. If one examines a handful of sand, the accessory minerals are usually the dark grains.

Accessory minerals are important in several ways. First, they may reveal information about the source rock from which the sand was eroded. For example, the accessory mineral chromite is generally formed in basalt, so its presence in a sand indicates that basalt (usually from the sea floor) had previously been uplifted and eroded nearby. Similarly, garnets of particular composition may be representative of particular metamorphic source rocks. Second, accessory minerals may tell geologists how "mature" a sand is—in other words, how great has been its exposure to chemical and physical weathering. Zircon, tourmaline, and rutile are far more durable than the other accessory minerals. If they are the only accessory minerals in a sand, then the sand probably experienced prolonged weathering in the period before its final deposition. Third, accessory minerals may be economically valuable. Diamonds and gold are accessory minerals in a few cases. Much of the United States' titanium comes from unusual concentrations of rutile and ilmenite in beach sands along the coast of South Carolina. Zircon in sands provides an important source of zirconium, used in high-temperature ceramics.

LITHIC FRAGMENTS AND MATRIX

Pieces of rocks (lithic fragments) may form an important part of coarse-grained sandstone and conglomerate. Lithic fragments may be composed of pieces of almost any igneous, metamorphic, or sedimentary rock. Their presence is a powerful indicator of the source rock from which the sand or gravel was weathered and eroded. As one might expect, not all rock fragments are equally durable in the sedimentary environment. Limestone and shale fragments disintegrate very rapidly, whereas granite and gneiss rock fragments are quite robust. Consequently, both the presence and the absence of certain rock fragments may reveal information about the origin of a siliciclastic sediment.

Matrix in siliciclastic rocks is usually clay or fine silt. This substance is the "mud" that often accompanies sand or gravel during deposition. It is commonly deformed between the more rigid framework grains as the sediment is compacted. Some types of matrix were not originally deposited as fine-grained matrix. This "pseudomatrix" is actually squashed fragments of soft grains of, for example, shale or schist. During compaction of the sediment, these fragments are more easily deformed than durable quartz and feldspar, and the fragments superficially resemble original matrix. Considerable skill is required to distinguish between true matrix and pseudomatrix.

SANDSTONES

Sand-sized siliciclastic rocks are often classified based on their mineral composition. Sandstones may be named with the term "arenite" (after the Latin *harena*, "sand"). Relative abundances of quartz, feldspar, and rock fragments determine the name. One of the major types of sandstone is quartz arenite, which is composed almost entirely of quartz, with less than 5 percent of other framework minerals and less than 15 percent clay matrix (usually much less). This rock is usually white when fresh. Another type is feldspathic arenite, which is sometimes called arkose. This rock has abundant feldspar and up to 50 percent lithic fragments. The typical composition of arkose is perhaps 30 percent feldspar, 45 percent quartz, and 25 percent lithic fragments. Up to 15 percent detrital clay matrix may also be present. This rock is often pinkish-gray in color. A third type is lithic arenite, a kind of sand composed largely of lithic fragments and up to 50 percent feldspar. The typical composition of lithic arenite might be 60 percent lithic fragments,

30 percent feldspar, and 10 percent quartz. Up to 15 percent matrix may be present as well. Almost all conglomerates are lithic arenites. Lithic arenites are often dark gray and sometimes feature a salt-and-pepper appearance from the lithic fragments. A fourth type is wacke, sometimes called greywacke, which has between 15 and 75 percent clay matrix at the time of deposition. The term "wacke" may be modified by prefixing with the name of the dominant nonmatrix component; examples might include "quartz wacke," "feldspathic wacke," or "lithic wacke." Some wackes may be "formed" diagenetically from lithic arenites by alteration of shale or schist fragments so that they appear similar to originally detrital clay matrix. This rock is usually gray in color (thus the name "greywacke").

The composition of sandstones holds important clues as to the kinds of rocks from which they were derived. These clues may help petrologists to reconstruct the plate tectonic setting of the sand's source region even though the source region has long since

In this Cambrian sequence at the head of Nordenskiold Fjord, Greenland, a layer of carbonate grainstones (RG2) is overlain by siliciclastic rocks at the top. (Geological Survey of Canada)

disappeared because of erosion or tectonic destruction. For example, quartz arenite often reflects sedimentation on or near a stable craton (a piece of the earth's crust), where physical and chemical weathering are permitted to eliminate unstable minerals for an extended period of time. At the other extreme, a lithic arenite with abundant basaltic rock fragments, calcic plagioclase, and rare quartz may represent detritus shed from an island arc during plate collision. Careful observation of the mineral composition of sandstones has become an important part of reconstructing the earth's history.

DIAGENESIS

The diagenesis of siliciclastic rocks is a complex collection of processes that include cementation, replacement, recrystallization, and dissolution. Diagenesis includes chemical reactions that begin at the time of deposition and may continue until metamorphism takes place. Some sedimentary petrologists consider low-temperature cementation as part of diagenesis, whereas others consider diagenesis to include only those reactions that occur at temperatures exceeding 100 degrees Celsius. The upper range of diagenesis grades into metamorphism; for that reason, sedimentary petrologists and metamorphic petrologists share some common territory in diagenesis.

Siliciclastic sediments are transformed into rocks by compaction and cementation. Common cements in siliciclastic rocks are quartz, calcite, clays, and iron oxides such as limonite or hematite. It is common for a rock to have several cements, each deposited at different times or under different conditions of pore-water chemistry. Cement is introduced into sediment in the form of elements or compounds dissolved in water. For example, silica may be dissolved from quartz grains or given off during low-grade metamorphism and then may saturate water in the pores of a rock. One means of dissolving silica from quartz grains is called pressure solution. Quartz grains exert very high pressures upon one another at their points of contact. Quartz is more soluble under high pressure; therefore it is more easily dissolved where grains touch each other. Well-compacted sands often display flattened grain-to-grain contacts as a consequence of pressure solution. The quartz lost from these grains is carried away, dissolved in pore water, to be precipitated elsewhere.

As the water moves slowly through the pores of the rock, it may encounter different temperatures, pressures, acidity, or gas concentrations. Changes in these conditions may cause the water to lose some of its ability to dissolve silica, and the silica may therefore precipitate as crystals. Similar factors influence the dissolution and precipitation of other cements in siliciclastic rocks. Cement not only holds sediment grains together, forming a rock, but it also fills the pores between grains, which may inhibit further flow of fluids. As a consequence, the porosity and permeability of most rocks are reduced by cementation. Rocks with abundant cement generally make poor reservoirs for petroleum or water.

In some cases, the pore-water chemistry conducive to precipitation of one cement is capable of dissolving another cement. A common example is the generally inverse relationship between quartz and calcite cement. Pore water that precipitates quartz often dissolves calcite at the same time, or the other way around. Sometimes dissolution occurs without immediate precipitation of another mineral type in the void that is formed. Pores formed by dissolution are called secondary porosity; such porosity may be the main cause of porosity in many sandstones. Minerals commonly dissolved include carbonates such as calcite and aragonite, silicates such as feldspars and ferromagnesian minerals, and, though rarely, accessory minerals such as garnet. In some cases, earlier cements may be dissolved to form secondary porosity.

RECRYSTALLIZATION AND REPLACEMENT

Recrystallization is a common feature in diagenesis. In recrystallization, one mineral is transformed into another mineral with the same (or nearly the same) chemical composition or into differently sized (usually larger) crystals of the original mineral. The most common example is the transformation of aragonite into calcite in limestones. In siliciclastic rocks, microcrystalline quartz may recrystallize into more coarsely crystalline quartz, or rutile may recrystallize into anatase.

A more prevalent form of diagenesis among ancient sandstones is replacement. In this process, one mineral is replaced volume-for-volume by another mineral. In some cases, it appears that the replacing mineral has combined material from its host grain with dissolved ions from pore water. The most common replacement examples involve clay minerals. For example, kaolinite may replace potassium feldspar, forming a grain of kaolinite that looks superficially like a potassium feldspar grain.

EXAMINING SILICLASTIC ROCKS

The principal means of studying siliciclastic rocks is with the petrographic microscope. This microscope is similar to other microscopes except that light passes through the specimen rather than illuminates its surface. Rocks are normally not transparent, however, and they must be cut into very thin slices in order for light to pass through them. Magnifications of up to four hundred times are routinely used. In thin section, it is possible to view closely individual grains and their relations to other grains, identify the mineral grains in the rock, and view matrices and cements. Thin sections are routinely used to estimate the amount of pore space in rocks, revealing their potential as reservoirs for fluids such as water, oil, and natural gas.

When illuminated by polarized light, different minerals display characteristic interference colors that reveal the identity of a mineral to the experienced observer. Other clues are the existence and pattern of cleavages in the minerals, alteration products resulting from diagenesis, and overall grain shape. In addition, it is possible to view cements that surround grains and partially or completely fill pores. The sequence of cementation can be interpreted by the relative positions of cements filling pores. Knowing the sequence of cements can help geologists understand how deeply buried the rock was when it was cemented and how the chemical composition of the pore water changed through time.

Another means of examining siliciclastic rocks is with the scanning electron microscope (SEM). The SEM permits examination of clastic grains with much greater magnification than with the conventional petrographic microscope. Magnifications of up to ten thousand times are easily accomplished, with details as small as 0.001 micrometer being easily resolved. The SEM is very useful in examining the surface texture of minerals. The presence of pits, grooves, and cracks can reveal clues as to the environment under which the sand grains were deposited. Were the sand grains blown by wind, carried by water, or transported by glaciers? In some cases, this question may be answered by examination under the extreme magnification of the SEM. The SEM is also very useful for studying the cements that bind sandstones.

ELECTRON MICROPROBE AND X-RAY STUDY

When accurate data concerning chemical composition of minerals are needed, the electron microprobe is used. The microprobe is actually an elaborate version of an SEM. It differs in that its beam does not normally scan but instead remains fixed in one spot, and its detectors are carefully calibrated to give accurate readings of the elemental composition of the material being examined. Microprobes collect X rays and can determine from them the amounts of different elements in almost any mineral. In some instances, the abundance of trace elements in minerals such as zircon, tourmaline, garnet, or rutile may provide clues as to the source region from which the grains were derived, giving an indication of the source for a sedimentary formation. The electron microprobe is a very useful tool in these studies.

There are limitations to the microprobe. It cannot analyze for elements lighter than sodium, so the amount of oxygen or carbon cannot be determined. Furthermore, rock material must be very carefully prepared for readings to be accurate. Also, electron microprobes are rather expensive and require considerable maintenance; however, X rays are also employed in the study of rocks. X-ray diffraction of powdered or whole specimens can identify minerals and their degree of crystal ordering. X-ray fluorescence identifies the chemical composition of minerals. Both techniques are commonly used in the study of clays in the matrix of siliciclastic rocks. Sample preparation is relatively easy, and beginners can operate most modern X-ray machines. X-ray analysis usually requires destruction of the sample by grinding, so the sizes, shapes, and relationships between grains are lost.

ECONOMIC SIGNIFICANCE

Sandstone and its cousins siltstone and conglomerate are common sedimentary deposits. These rocks, collectively known as siliciclastic rocks, are important for a number of reasons. For example, they are porous—that is, they contain small spaces between grains that may be filled with valuable fluids. Siliciclastic rocks are the major type of aquifer in most of the world. Also, they form reservoirs for oil and natural gas. Sand and gravel are essential parts of modern construction because they form the "bulk" in concrete. In this role, they represent the most economically significant mineral resource in the United States, ahead of petroleum, coal, and precious metals.

Michael R. Owen

FURTHER READING

Blatt, Harvey. *Sedimentary Petrology.* 2d ed. San Francisco: W. H. Freeman, 1992. A useful introduction to sedimentary petrology. Suitable for general readers.

Blatt, Harvey, Robert J. Tracy, and Brent Owens. *Petrology: Igneous, Sedimentary, and Metamorphic.* 3d ed. New York: W. H. Freeman, 2005. A standard textbook on sedimentary rocks, covering more than siliciclastics. Includes a complete discussion of the sources of grains, their transportation, deposition, and deformation.

Boggs, Sam, Jr. *Petrology of Sedimentary Rocks.* New York: Cambridge University Press, 2009. Begins with a chapter explaining the classification of sedimentary rocks. Remaining chapters provide information on different types of sedimentary rocks. Describes siliciclastic rocks and discusses limestones, dolomites, and diagenesis.

Houseknecht, David W., and Edward D. Pittman, eds. *Origin, Diagenesis, and Petrophysics of Clay Minerals in Sandstones.* Tulsa, Okla.: Society for Sedimentary Geology, 1992. A collection of essays written by leading experts in their respective fields. Examines the geochemical and geophysical properties of sandstone, clay minerals, and other sedimentary rocks. Filled with illustrations. Includes an index and bibliographical references.

Lauf, Robert. *The Collector's Guide to Silicate Crystal Structures.* Atglen, Pa.: Schiffer Publishing, 2010. Contains many photos and diagrams. Discusses crystal structures, crystallography, and silicate classifications. Written for the mineral collector. Useful for students of mineralogy.

Mackenzie, F. T., ed. *Sediments, Diagenesis, and Sedimentary Rocks.* Amsterdam: Elsevier, 2005. Volume 7 of *Treatise on Geochemistry*, edited by H. D. Holland and K. K. Turekian. Compiles articles on such subjects as diagenesis, chemical composition of sediments, biogenic material recycling, geochemistry of sediments and cherts, and green clay minerals.

McDonald, D. A., and R. C. Surdam, eds. *Clastic Diagenesis.* Memoir 37. Tulsa, Okla.: American Association of Petroleum Geologists, 1984. A thorough discussion of sandstone diagenesis, complemented by well-documented case studies. Superbly illustrated.

Middleton, Gerard V., ed. *Encyclopedia of Sediments and Sedimentary Rocks.* Dordrecht: Springer, 2003. Cites a vast number of scientists, listed in the author index. Subjects range from biogenic sedimentary structures to Milankovitch cycles. Provides a subject index as well. Designed to cover a broad scope and a degree of detail useful to students, faculty, and professionals in geology.

Pettijohn, F. J., P. E. Potter, and R. Siever. *Sand and Sandstone.* 2d ed. New York: Springer-Verlag, 1987. Perhaps the most comprehensive, authoritative, and readable book on sandstone in English. Very well illustrated, with copious references. The basic reference for all sandstone studies.

Prothero, Donald R., and Fred Schwab. *Sedimentary Geology: An Introduction to Sedimentary Rocks and Stratigraphy.* 2d ed. New York: W. H. Freeman, 2003. A thorough treatment of most aspects of sediments and sedimentary rocks. Well illustrated with line drawings and black-and-white photographs. Contains a comprehensive bibliography. Focuses on carbonate rocks and limestone depositional processes and environments. Suitable for college-level readers.

Scholle, Peter A. *Color Illustrated Guide to Constituents, Textures, Cements, and Porosities of Sandstones and Associated Rocks.* Memoir 28. Tulsa, Okla.: American Association of Petroleum Geologists, 1979. An excellent guide to the major types of grains, matrices, and cements common in sandstones. Includes microphotographs, complete with short description of the features they show.

Scholle, Peter A., and D. R. Spearing, eds. *Sandstone Depositional Environments.* Memoir 31. Tulsa, Okla.: American Association of Petroleum Geologists, 1982. Well-illustrated compendium of the environments in which siliciclastic rocks may be found.

Tucker, M., ed. *Sedimentary Rocks in the Field.* 4th ed. New York: Wiley & Sons, 2011. Covers the major techniques, practices, and policies of sedimentary petrology. Well referenced.

See also: Aluminum Deposits; Biogenic Sedimentary Rocks; Building Stone; Cement; Chemical Precipitates; Clays and Clay Minerals; Coal; Diagenesis; Earth Resources; Evaporites; Fertilizers; Limestone; Oil and Gas Exploration; Oil Shale and Tar Sands; Petroleum Reservoirs; Sedimentary Mineral Deposits; Sedimentary Rock Classification.

SOIL CHEMISTRY

Soils are complex chemical factories. Regardless of the type of soil, chemical processes such as plant growth, organic decay, mineral weathering, and water purification are ongoing processes.

PRINCIPAL TERMS

- **cation:** an element that has lost one or more electrons such that the atom carries a positive charge
- **clay mineral:** a group of minerals, commonly the result of weathering reactions, composed of sheets of silicon and aluminum atoms
- **hydroxl:** a combination of an oxygen and hydrogen atom, which behaves as a single unit with a negative charge
- **lysimeter:** a simple pan or porous cup that is inserted into the soil to collect soil water for analysis
- **stable isotopes:** atoms of the same element that differ by the number of neutrons in their nuclei yet are not radioactive
- **weathering:** reactions between water and rock minerals at or near the earth's surface that result in the rock minerals being altered to a new form that is more stable

WEATHERING PROCESS

Soil chemistry has been studied as long as there has been sustainable agriculture. Although they did not recognize it as such, those first successful farmers who plowed under plant stalks, cover crops, or animal wastes were actively managing the soil chemistry of their fields. These early farmers knew that to have productive farms in one location season after season, they had to return something to the soil. It is now understood that soil chemistry is a complex of chemical and biochemical reactions. The most obvious result of this complex of reactions is that some soils are very fertile whereas other soils are not. Soil itself is a unique environment because all of the "spheres"—the atmosphere, hydrosphere, geosphere, and biosphere—are intimately mixed there. For this reason, soil and soil chemistry are extremely important.

Soil chemistry begins with rock weathering. The minerals making up a rock exposed at the earth's surface are continually bathed in a shower of mildly acidic rain—not polluted rainwater but naturally occurring acid rain. Each rain droplet forming in the atmosphere absorbs a small amount of carbon dioxide gas. Some of the dissolved carbon dioxide reacts with the water to form a dilute solution of carbonic acid. A more concentrated solution of carbonic acid is found in any bottle of sparkling water.

Most of the common rock-forming minerals, such as feldspar, will react slowly with rainwater. Some of the chemical elements of the mineral, such as sodium, potassium, calcium, and magnesium, are very soluble in rainwater and are carried away with the water as it moves over the rock surface. Other chemical elements of the mineral, such as aluminum, silicon, and iron, are much less soluble. Some of these elements are dissolved in the water and carried away; most, however, remain near the original weathering, where they recombine into new, more resistant minerals. Many of the new minerals are of a type called clays.

Clay minerals tend to be very small crystals composed of layers of aluminum and silicon combined with oxygen or hydroxyl ions. Between the layers of aluminum and silicon atoms are positively charged ions (cations) of sodium, potassium, calcium, and magnesium. The cations hold the layers of some clays together by electrostatic attraction. In most cases, the interlayer cations are not held very tightly. They can migrate out of the clay and into the water surrounding the clay mineral, to be replaced by another cation from the soil solution. This phenomenon is called cation exchange.

The weathering reactions between rainwater and rock minerals produce a thin mantle of clay mineral soil. The depth to fresh, unweathered rock is not great at first, but rainwater continues to fall, percolating through the thin soil and reacting with fresh rock minerals. In this way, the weathering front (the line between weathered minerals and fresh rock) penetrates farther into the rock, and the overlying soil gets thicker.

BIOLOGICAL PROCESSES

Throughout the weathering process, biological processes contribute to the pace of soil formation. In the very early stages, lichens and fungi are attached to what appear to be bare rock surfaces. In reality, they are using their own acids to "digest" the rock minerals. They absorb the elements of the mineral they need, simultaneously extending fine filaments into the rock for attachment, and the remainder is left to form soil minerals. As the soil gets thicker, larger plants and animals begin to colonize it. Large plants send roots down into the soil looking for water and nutrients. Some of the necessary nutrients, such as potassium, are available as exchangeable cations on soil clays or in the form of deeper, unweathered minerals. In either case, the plant obtains the nutrients by using its own weathering reaction carried on through its roots. The nutrient elements are removed from minerals and become part of the growing plant's tissue.

Without a way to replenish the nutrients in the soil, the uptake of nutrients by plants will eventually deplete the fertility of the soil. Nutrients are returned to the soil through the death and decay of plants. Microorganisms in the soil, such as bacteria and fungi, speed up the decay. Since the bulk of the decaying plant material is found at the surface (the dead plant's roots also decay), most of the nutrients are released to the surface layer of the soil. Some of the nutrients are carried down to roots deep in the soil by infiltrating rainwater. Most of the nutrients, however, are removed from the water by the shallow root systems of smaller plants. The deeper roots of typically large plants can mine the untapped nutrients at the deep, relatively unweathered soil-rock boundary.

The soil and its soil chemistry are now well established, with plants growing on the surface and their roots reaching toward mineral nutrients at depth. Water is flowing through the soil, carrying dissolved nutrients and the soluble by-products of weathering reactions. Yet the soil continues to evolve even after it is well established. Eventually, when the downward migration of the weathering front matches the rate at which soil is eroded from the surface, the soil reaches its maximum evolution. In tropical and subtropical climates, the end stage of soil evolution is very deep soil in which most of the nutrients are gone. Virtually all nutrients are contained in the vegetation, and people often assume, falsely, that such soils are extremely productive.

STUDY OF SOIL WATER

The study of soil chemistry is concerned with the composition of soil water and how that composition changes as the water interacts with soil atmosphere, minerals, plants, and animals. Soil and its chemistry can be studied in its natural environment, or samples can be brought into the laboratory for testing. Some tests have been standardized and are best conducted in the laboratory so that they can be compared with the results of other researchers. Most standardized tests, such as measurements of the soil's acidity and cation-exchange capacity, are related to measurements of the soil's fertility and its overall suitability for plant growth. These tests measure average values for a soil sample because large original samples are dried and thoroughly mixed before smaller samples are taken for the specific test.

Increasingly, soil chemists are looking for ways to study the fine details of soil chemical processes. They know, for example, that soil water chemistry changes as the water percolates through succeeding layers of the soil. The water flowing through the soil during a rainstorm has a different chemical composition from that of water clinging to soil particles, at the same depth, several days later. Finally, during a rainstorm, the water flowing through large cracks in the soil has a chemical composition different from that of the same rainwater flowing through the tiny spaces between soil particles.

SAMPLING TECHNIQUES

Soil chemists use several sampling techniques to collect the different types of soil water. During a rainstorm, water flows under the influence of gravity. After digging a trench in the area of interest, researchers push several sheets of metal or plastic, called pan lysimeters, into the wall of the trench at specified depths below the surface. The pans have a very shallow V shape. Soil water flowing through the soil collects in the pan, flows toward the bottom of the V, and flows out of the pan into a collection bottle. Comparing the chemical compositions of rainwater that has passed through different thicknesses of soils (marked by the depth of each pan) allows the soil chemist to identify specific soil reactions with specific depths.

After the soil water stops flowing, water is still trapped in the soil. The soil water clings to soil particles and is said to be held by tension. Tension water can spend a long time in the soil between rainstorms. During that time, it reacts with soil mineral grains and soil microorganisms. Tension water is sampled by placing another type of lysimeter, a tension lysimeter, into the soil at a known depth. A tension lysimeter is like the nozzle of a vacuum cleaner with a filter over the opening. Soil chemists actually vacuum the tension water out of the soil and to the surface for analysis.

DETERMINATION OF ISOTOPIC COMPOSITION

Nonradioactive, stable isotopes of common elements are being used more often by soil chemists to trace both the movement of water through the soil and the chemical reactions that change the composition of the water. Trace stable isotopes behave chemically just the way their more common counterparts do. For example, deuterium, an isotope of hydrogen, substitutes for hydrogen in the water molecule and allows the soil chemist to follow the water's movements. Similarly, carbon-13 and nitrogen-15 are relatively uncommon isotopes of biologically important common elements. Using these isotopes, soil chemists can study the influences of soil organisms on the composition of soil water. Depending on what the soil chemist is studying, the isotope may be added, or spiked, to the soil in the laboratory or in the field. Spiking allows the movement of the isotope to be tracked by following the unusual concentration of the isotope. Alternatively, naturally occurring concentrations of the isotope in rain or snowmelt may be used. Regardless, soil water samples are collected by one or more of the lysimeter methods, and their isotopic composition is determined.

THE SOIL CHEMICAL FACTORY

The wonderful interactions of complex chemical and biochemical reactions that are soil chemistry are one indication of the uniqueness of planet Earth. Without the interaction of liquid water and the gases in the atmosphere, many of the nutrients necessary for life would have remained locked up in rock minerals. Thanks to weathering reactions, the soil chemical factory started to produce nutrients, which resulted in the exploitation of the soil environment by millions of organisms. The processes involved in soil chemistry—from weathering reactions that turn rock into new soil to the recycling of plant nutrients through microbial decay—are vital to every human

being. Without fertile soil, plants will not grow. Without plants as a source of oxygen and food, there would be no animal life.

Because of the complex chemical interrelationships that have developed in the soil environment, it may seem that nothing can disrupt the "factory" operation. As more is understood about soil chemistry and the ways in which human activities stress soil chemistry, it is apparent that the factory is fragile. Not only do humans rely on soil fertility for their very existence, but they also are taking advantage of soil chemical processes to help them survive their own past mistakes. Soil has been and continues to be used as a garbage filter. Garbage, whether solid or liquid, has been dumped on or buried in soil for ages. Natural chemical processes broke down the garbage into simpler, less toxic or unsanitary forms and recycled the nutrients. When garbage began to contain toxic chemicals, those chemicals, when in small quantities, were either destroyed by soil bacteria or firmly attached to soil particles. The result is that water—percolating through garbage, on its way to the local groundwater, stream, or lake—does not carry with it as much contamination as one might expect. Soil chemistry has, so far, kept contaminated garbage from ruining drinking water in many cases. There are well-known cases, however, where the volume and composition of waste buried or spilled were such that the local soil chemistry was overwhelmed. In cases of large industrial spills, or when artificial chemicals are spilled or buried, the soil needs help to recover. The recovery efforts are usually very expensive but, faced with the possible long-term loss of large parts of the soil chemical factory, humankind cannot afford to neglect this aspect of the environment.

Richard W. Arnseth

FURTHER READING

Albarede, Francis. *Geochemistry: An Introduction.* 2d ed. Cambridge, Mass.: Cambridge University Press, 2009. An introduction for students looking to gain knowledge in geochemistry. Covers basic topics in physics and chemistry, including isotopes, fractionation, geochemical cycles, and the geochemistry of select elements.

Berner, Elizabeth K., and Robert A. Berner. *The Global Water Cycle: Geochemistry and Environment.* Englewood Cliffs, N.J.: Prentice-Hall, 1987. Designed to teach environmental geochemistry to college students. Covers the role of water in soil chemistry. Emphasizes the hydrologic cycle. Discusses the unique properties of water and the water cycle. Features chapters on rainwater and soil water. Includes an extensive bibliography of more technical references.

Bohn, Heinrich L., Rick A. Myer, and George A. O'Connor. *Soil Chemistry*. 3d ed. New York: John Wiley & Sons, 2001. Fairly widely used college text on soil chemistry. Introductory chapter and chapter on weathering are accessible to any reader with a college background. Provides an annotated bibliography at the end of each chapter for further reading.

Brill, Winston. "Agricultural Microbiology." *Scientific American* 245 (September, 1981): 198. Discusses the link between soil chemistry and microbiological processes. Emphasizes nitrogen biochemistry and genetic engineering. Several figures nicely illustrate the complexity and cyclic nature of soil nitrogen chemistry. Appropriate for those with a background in the terminology of microbiology.

Evangelou, V. P. *Environmental Soil and Water Chemistry: Principles and Applications*. New York: Wiley, 1998. Examines soil and water chemistry, the pollution levels of water and soils, and the chemical processes used to determine those levels.

Lloyd, G. B. *Don't Call It Dirt*. Ontario, Calif.: Bookworm Publishing, 1976. Provides a backyard gardener's point of view on soil fertility, microorganisms, and the interrelationships between soil moisture and soil processes. Information is pertinent and accessible to the general reader.

McBride, Murray B. *Environmental Chemistry of Soils*. New York: Oxford University Press, 1994. Focuses on the chemical make-up and properties of soils in different regions. Written for the college-level reader.

Millot, Georges. "Clay." *Scientific American* 240 (April, 1979): 108. Emphasizes the chemistry and industrial uses of clay minerals. Offers a brief discussion of weathering as the source of clays and as part of the geologic cycling of clays. Discusses how weathering reactions vary in different climates and yield different clay minerals in the local soils. Provides several spectacular scanning electron microphotographs of representative clay types. Accessible to the general reader.

Randolph, John. *Environmental Land Use Planning and Management*. Washington, D.C.: Island Press, 2004. Describes basic principles and strategies of land-use planning and management. Discusses various land features, types, and environmental issues, such as soils, wetlands, forests, groundwater, biodiversity, and runoff pollution. Provides case studies and specific examples.

Sparks, Donald S., ed. *Soil Physical Chemistry*. 2d ed. Boca Raton, Fla.: CRC Press, 1999. Introduction to soil chemistry. Offers a clear look into the practices, procedures, and applications of the field. Appropriate for the careful reader.

Sposito, Garrison. *The Chemistry of Soils*. 2d ed. New York: Oxford University Press, 2008. Provides a comprehensive discussion of soil composition. Describes chemical makeup and basic soil components. Discusses physical and chemical characteristics and processes within soil types. Includes a list of references for further reading.

Storer, Donald A. *The Chemistry of Soil Analysis*. Middletown, Ohio: Terrific Science Press, 2005. Contains valuable information on common tests conducted in soil chemistry. Designed for practical use by students, environmental laboratory technicians, agronomists, archaeologists, and engineers. Describes specific analyses and techniques in a manner accessible to those with a limited background in geology.

Tan, Kim Howard. *Principles of Soil Chemistry*. 4th ed. Boca Raton, Fla.: CRC Press, 2011. Expanded in this third edition. Written as an introduction to soil chemistry; explains the processes and application of the science in an easy-to-understand manner, accompanied by plenty of illustrations. Includes a twenty-five-page bibliography, as well as an index.

Walther, John Victor. *Essentials of Geochemistry*. 2d ed. Burlington, Mass.: Jones & Bartlett Publishers, 2008. Contains chapters on radioisotope and stable isotope dating and radioactive decay. Geared more toward geology and geophysics than chemistry; provides content on thermodynamics, soil formation, and chemical kinetics.

See also: Desertification; Expansive Soils; Land Management; Landslides and Slope Stability; Land-Use Planning; Land-Use Planning in Coastal Zones; Mining Wastes; Soil Erosion; Soil Formation; Soil Profiles; Soil Types.

SOIL EROSION

Soil erosion, which plays an important role in shaping the landscape, is a process that takes extended periods, although people have greatly accelerated it by removing vegetation. Controlling soil erosion is essential to maintaining the world's food supply.

PRINCIPAL TERMS

- **granules:** small grains or pellets
- **porosity:** the ability to admit the passage of gas or liquid through pores
- **rills:** small rivulets in channels
- **suspension:** a condition in which particles are dispersed through a supporting medium
- **understrata:** material lying beneath the surface of the soil

SHAPING THE LANDSCAPE

Soil erosion is a natural process whereby rock and soil are broken loose from the earth's surface at one location and moved to another. Erosion creates and transforms land by filling in valleys, wearing down mountains, and making rivers appear and disappear. Generally, this process takes thousands or even millions of years.

The active forces of erosion are water and wind. Water erosion involves the movement of soil by the action of rainwater and melted snow moving rapidly over exposed land surfaces. The principal types of water erosion are splash erosion and gully erosion.

SPLASH EROSION

Splash erosion is the removal of thin layers, one drop at a time, from land surfaces. Fine-grained soil such as silt loams, fragile sandy soil, and all soils deficient in organic matter are especially vulnerable. Most troublesome are lands that tend to slope, lands subjected to heavy rainfall, and lands composed of shallow surface soils overlying dense clay subsoil.

When rainwater is absorbed into the ground, small sievelike openings are made in the soil. Eventually, fine particles carried by the water plug the openings, thereby causing the rainwater to flow off the land. As the soil is softened by the rainfall, clods, lumps, and granules break down and form a pasty mass, which resists penetration by rainwater. As a result, the runoff increases and forms a relatively impervious, skinlike film sometimes referred to as puddled soil. As rain continues, the abrasive force of the rain results in cutting, which penetrates the skinlike layer and starts trenching.

When splash erosion reaches an advanced stage, it is referred to as rill erosion. Rill erosion can be defined as localized small washes in channels that can usually be eliminated by ordinary plowing. Rilling is the most common form of erosion on soft, freshly plowed soils that are high in silt content where the slopes are deeper than 4 or 5 percent.

There are several specialized forms of splash erosion. Pedestal erosion occurs when a stone or tree root protects easily eroded soil from splash erosion. Consequently, isolated pedestals containing resistant material are left standing. This type of erosion occurs after several years, primarily on bare patches of grazing land, and often can be seen on a small scale on pebbly ground after a rain. The erosion patterns formed in highly erodible soils are called pinnacle erosion. This erosion is usually found in deep vertical rills in the sides of gullies. When these gullies cut back and join, they form pinnacles. When pinnacle erosion is present, reclamation is difficult. Another specialized form of splash erosion is piping, which is associated with the formation of continuous pipes or channels underground. Piping usually occurs in soil types that are subject to pinnacle erosion. It involves water penetrating the soil surfaces and moving downward until it reaches a less permeable layer. The fine particles of the more porous soil may be washed out if the water flows over the less permeable layer through an outlet. The more rapid lateral flow increases the sideways erosion, causing the entire surface flow to disappear down a vertical pipe. The water then flows underground until it reappears in the side of a gully. Once pipe erosion starts, it cannot be controlled; this has been a factor in some dam collapses. Fortunately, pipe erosion is restricted to the "bad lands." The last type of splash erosion is slump erosion. It is prominent in areas of high rainfall with deep soils. Slumping can become the chief agent in the development of gullies, probably as a result of flood flow in channels. Riverbank collapse and coastal erosion are the other main cases of slumping.

GULLY EROSION

The second principal type of water erosion is gully erosion. This type of erosion occurs in places where runoff from a slope is sufficient in volume and velocity to cut deep trenches. It also takes place in areas where concentrated water continues cutting in the same groove long enough to form deep rills. Gullies usually begin in slight depressions in or below fields where water concentrates, in ruts left by farm machinery, in livestock trails, or along furrows between crop rows. Gullies ordinarily carry water only during or following rains or melting snows. Most gullies cannot be removed by normal plowing because of their size; in fact, some of them take the form of huge chasms 15-30 meters deep. A field gully is a channel at least 45 centimeters wide and between 25 and 30 centimeters deep; a woodland channel, by comparison, is considered to be a channel deep enough to expose the main lateral roots of trees. Woodland gullies develop in wooded areas that received water from cultivated slopes immediately above or on closely cut, severely burned-over woodland.

There are two main types of gullies. V-shaped gullies, which have sloping sides with narrow bottoms, are the normal type of gully. Gullies that have straight, more or less vertical sides with broad bottoms are U-shaped gullies. They are less winding than the V-shaped gullies because the soft materials at the base of the sides tend to give way to the impact of currents, and irregularities in the channel walls are planed away. Usually, the presence of gullies means that the land has been overused or abused. U-shaped gullies are usually the more serious because the soft, unstable materials commonly found in their lower depths, such as sand, loose gravel, or soft rocks, are easily cut out by floodwaters. This cutting near the bottom causes the banks to split off from above in great vertical blocks. When the caved-in material is washed out of the gully, the trench is left box-shaped. Most gullies tend to branch out as they grow, but the tendency is more serious in the U-shaped gullies. They are the most difficult to control because of the instability of the understrata.

Waterfall erosion is an important form of gully erosion because it does so much damage. It is caused by water cascading over the heads and sides of gullies, over dams, and over terraces whose channels have been filled with the debris of erosion. This type of erosion is most commonly associated with flooding.

WIND EROSION

The second active force of soil erosion is wind erosion. Wind erosion can occur in places where water erosion is also active, but it reaches its most serious proportions on both level and sloping areas during dry times. Like water erosion, wind erosion usually proceeds very slowly; however, wind erosion is likely to increase rapidly in relatively flat and gently undulating treeless regions, like the Great Plains. When grass is plowed up, the cultivated soil becomes much less cohesive. Organic material that normally collects under a cover of grass and that serves as a bonding agent when grass is present disappears by decay and oxidation when the grass is gone. After periods of drought, the soil turns into a dry powdery mass, which is easily swept up by the wind and lifted into the pathways of high air currents, which carry it hundreds and, at times, thousands of kilometers. Coarser, heavier particles, known as ground drift, are blown along near the surface of the ground and pile up in drifts about houses, fences, farm implements, and clumps of vegetation. The fine materials that are blown away are dust; the coarser materials that are left behind are sand. The susceptibility of soils to wind erosion depends on the size of the particles and on the content of the organic matter. Coarse sands are more likely than heavy clays to blow away immediately after plowing. Ironically, the finer-textured soils, especially those of granular structure, show the greatest resistance. In fact, they sometimes remain undisturbed through years of cultivation until their organic material is disrupted.

The massive removal of soil particles through the action of the wind takes the form of dust storms. Early in 1934, a dust storm originating in the Texas-Oklahoma Panhandle covered a vast territory extending eastward from the Rocky Mountains to several hundred kilometers over the Atlantic. This one storm deprived the Great Plains states of 200 to 300 million tons of soil.

There are essentially five different types of wind erosion, although there is some overlapping, and several of the processes occur simultaneously. Detrusion is the wearing away of rocks and soil formations by fine particles carried away in suspension. This process often carves large rocks in deserts into grotesque shapes or streamlined hills, called yardangs. Abrasion occurs close to the ground where the moving particles are larger and bound over the surface. The removal of very fine particles, carried off in suspension, is efflation, and the rolling away of large particles is extrusion. The removal of particles of intermediate size bouncing downwind is known as effluxion, and the bouncing process is called saltation. The latter has nothing to do with salt but is derived from the Latin word for "leap."

SEVERITY OF EROSION

The severity of erosion is determined by a great many factors; one of the least controllable factors is climate. The United States has more erosion problems than England because it lacks England's gentle rains and mists. Many areas in the United States have rainstorms of sufficient strength and intensity to erode many centimeters of soil from a field not protected by vegetation within minutes. For example, rain-induced erosion may be severe in the Corn Belt (roughly, the states of Indiana, Illinois, Missouri, Nebraska, and Kansas) because rains frequently occur in these states in June when the plants have not matured enough to protect the soil; in arid areas of the western United States, serious water erosion may exist because there is not enough rain to establish a protective ground cover for the infrequent rains.

Soil type and topography also influence the severity of erosion. Soils in the United States range from poorly drained and saturated to very dry, from sandy to clayish, from acid to alkaline, and from shallow to deep. They also vary in slope characteristics, porosity, organic content, temperature, and the capacity to supply nutrients. Land is better for crops if it is nearly level, with just enough slope for good drainage. About 45 percent of the cropland in the United States falls into this category.

A farmer's cropping practices can also greatly influence the severity of erosion. Erosion is likely to occur when fields are plowed in the conventional way: in straight rows regardless of the topography, with all plant cover and crop residue removed. Erosion can be retarded if the farmer uses the best land for crops and puts any other land to different uses. After that, any number of conservation practices, such as strip cropping, contour planting, crop rotation, terracing, and various conservation tillage practices, can be introduced.

MEASURING EROSION

Soil losses per acre are measured by either the Universal Soil Loss Equation (USLE) or the Wind Erosion Equation (WEE). These formulas have been developed from field experiments in various parts of the country. Both equations measure the tons of each soil type that are lost annually through the action of climate, cropping systems, management practices, and topography.

The use of these equations involves a number of limitations. Although both the USLE and the WEE measure the movement of the soil, they do not reveal the distance the soil has traveled or where the soil was deposited. Thus, the USLE could overestimate the severity of the erosion. Another drawback to the USLE is its failure to measure losses resulting from snowmelt. A revised version of the USLE, the RUSLE, has been developed to overcome these limitations.

The soil losses measured by the RUSLE or WEE are average figures taken from measurements over an extended period. They are usually reported in tons per acre. The computed losses are often connected with soil-loss tolerances, or T-values. These figures are the maximum soil losses that can be sustained without adversely affecting productivity. Some erosion is natural even without human disturbance. As long as soil evolution keeps pace with erosion, the soil loss is sustainable. Excess soil erosion is frequently defined as amounts greater than T-values. The U.S. Department of Agriculture has assigned T-values that usually range from 1 to 5 tons per acre, depending on the properties of the soil. Unfortunately, the validity of these numbers in representing maximum sustainable soil losses is doubtful. A soil loss of 5 tons per acre per year translates into a net loss of 2.54 centimeters (1 inch) of soil every thirty years. Even in fairly steep topography, natural long-term erosion rates are only a few millimeters in thirty years. T-values have been set too high on some soils to assure long-term maintenance of the soil. T-values are also limited in their value because they do not reflect the impact of technology on crop yields.

Erosion can also be studied according to how it will be affected by different types of rain and how it will vary for different types of soil. Therefore, the amount of erosion that occurs depends on a combination of the ability of the soil to withstand rain and the power of the rain to cause erosion. This relationship between factors can be expressed in mathematical terms. Erosion is a function of the erosivity (of the rain) and the erodibility (of the soil); that is, erosion equals erosivity multiplied by erodibility.

For given soil conditions, one rainstorm can be compared quantitatively with another, and a numerical scale of values of erosivity can be created. For given rainfall conditions, one soil condition can be compared quantitatively with another, and a numerical scale of values of erodibility can be created. Erodibility of the soil can be subdivided into two parts: the inherent characteristics of the soil (that is, mechanical, chemical, and physical composition) and the way the soil is managed. Management may, in turn, be subdivided into land management and crop management.

COSTS AND BENEFITS OF EROSION

Soil erosion can be both beneficial and harmful. Erosion benefits people by contributing to the formation of soil by breaking up rocks. Erosion also creates rich, fertile areas as it deposits soil at the mouths of rivers and on the floors of valleys. The Nile River valley in Egypt has been sustained for thousands of years by sediments eroded from farther upstream. Erosion is also important from the standpoint of aesthetics. The Grand Canyon, for example, was created over millions of years through the eroding action of the Colorado River.

Yet soil erosion is one of the leading threats to the food supply, as it robs farmland of productive topsoil. Soil scientists estimate that it takes nature between three hundred and one thousand years to produce about 2.5 centimeters of topsoil, although humans can replace topsoil at a much faster rate. Still, considering that about 70 percent of the United States is subject to erosion, the prospect of building productive soil becomes overwhelming. Once the topsoil has been removed, a heavy layer of clay often remains, which may not contain enough porosity to support a good crop.

Crops can also be damaged by the loss of nutrients to erosion. Most serious are losses of nitrogen, sulfur, and phosphorus. In addition to the natural nutrients taken from the soil, synthetic fertilizers and chemicals are often washed from fields and into lakes and rivers, thereby contributing to pollution.

Finally, erosion that forms deep gullies presents problems to farmers. Gullies caused by flowing water are often too deep for farm machinery to cross. Unable to accommodate tractors and other farm equipment, fields riddled with gullies face ruin.

Alan Brown

FURTHER READING

Agassi, Menachem, ed. *Soil Erosion, Conservation, and Rehabilitation.* New York: Marcel Dekker, 1996. Assesses soil erosion processes through an examination of rainfall and water runoff. Deals with methods used in soil conservation and rehabilitation programs. Illustrations, index, and bibliography.

Batie, Sandra S. *Soil Erosion: Crisis in America's Croplands?* Washington, D.C.: Conservation Foundation, 1983. Attempts to structure an improved public policy for soil conservation. Assesses the extent and effects of erosion and explains the techniques for reducing soil erosion in language that can be easily understood by the nonscientist. Contains photographs and diagrams.

Bennett, Hugh Hammond. *Elements of Soil Conservation.* 2d ed. New York: McGraw-Hill, 1955. Although some of the information is dated, the chapter describing the processes and effects of erosion is up-to-date. Contains many photographs. Also includes a bibliography of books, pamphlets, and films concerning soil conservation.

Boardman, John, and David Favis-Mortlock, eds. *Modelling Soil Erosion by Water.* New York: Springer-Verlag, 1998. Covers the effects of climatic factors and climatic changes on soil erosion. Discusses possible climatic causes and effects and uses computer-simulated erosion models to illustrate complex processes. Technical at times.

Cornforth, Derek H. *Landslides in Practice.* Hoboken, N.J.: John Wiley & Sons, 2005. Explains how and why landslides occur. Describes the resulting changes to a landscape and prevention methods. Compares various soil types and slope stability, and examines a number of case studies. Provides the tools and techniques needed to evaluate landslide potential and determine preventative strategies.

Hudson, Norman. *Soil Conservation.* 3d ed. Ames: Iowa State University Press, 1995. Covers the history of soil conservation and demonstrates how the methods of soil conservation practiced in the United States can be applied in developing countries. Adopts an engineering approach to explain soil conservation. Includes many charts, illustrations, and photographs.

Lal, Rattan, ed. *Soil Quality and Soil Erosion.* Boca Raton, Fla.: CRC Press, 1999. Based on papers presented at a 1996 symposium on soil and water conservation. Offers an examination of the quality, management, and erosion of soils and water. Deals with methods of determining those levels.

Morgan, Royston Philip Charles. *Soil Erosion and Conservation.* 3d ed. Malden, Mass.: Blackwell Science, 2005. Explains the procedures and protocol used in determining the rates and effects of soil erosion. Describes how the application of these techniques is used in soil conservation programs. Illustrations, maps, thirty-page bibliography, and index.

Newson, Malcolm. *Land, Water and Development: Sustainable and Adaptive Management of Rivers.* 3d ed. London: Routledge, 2008. Presents land-water interactions. Discusses recent research, study tools and methods, and technical issues, such as soil erosion and damming. Appropriate for undergraduates and professionals. Covers concepts in managing land and water resources in the developed world.

Norris, Joanne E., et al., eds. *Slope Stability and Erosion Control: Ecotechnological Solutions.* New York: Springer, 2010. Discusses erosion processes, mass wasting, and landslides as they occur on varying slopes. Evaluates vegetation as a solution to slope instability. Cites many studies but is not overly technical. Appropriate for engineers, landscape architects, ecologists, foresters, and undergraduate students.

Simms, O. Harper. *The Soil Conservation Service.* New York: Praeger, 1970. Primarily a record of the activities of the U.S. Soil Conservation Service. Discusses the methods that the government recommends for preventing soil erosion. Appropriate for the general reader.

Stallings, J. H. *Soil: Use and Improvement.* Englewood Cliffs, N.J.: Prentice-Hall, 1957. Focuses on proper land management. Explains past and present methods of soil conservation in terms appropriate for the general reader.

Troeh, Frederick R., and Louis M. Thompson. *Soils and Soil Fertility.* 5th ed. New York: Oxford University Press, 1993. Introductory textbook in soils for students of agriculture. Covers every aspect of soil, including the physical properties of soil and soil conservation. Highly readable and well supplemented with charts, illustrations, and diagrams, although the reference sections are dated.

U.S. Soil Conservation Series. *Our American Land: Use the Land, Save the Soil.* Washington, D.C.: Government Printing Office, 1967. Promotes the need for soil conservation by providing statistical evidence of damage that has been done in the United States by erosion. Also recommends proper land management techniques. Appropriate for the general reader.

See also: Clays and Clay Minerals; Desertification; Expansive Soils; Land Management; Landslides and Slope Stability; Land-Use Planning; Land-Use Planning in Coastal Zones; Soil Chemistry; Soil Formation; Soil Profiles; Soil Types.

SOIL FORMATION

A soil is formed when decomposed organic material is encompassed into weathered mineral material at the earth's surface. The climate, the organisms living in the soil, the type of parent material, the local topography, and the amount of time the soil has been developing all influence the resulting soil characteristics. Changes in the soil arise from the main processes of soil formation: addition, removal, transfer, and transformation of parts of the soil.

PRINCIPAL TERMS

- **horizon:** a layer of soil material approximately parallel to the surface of the land that differs from adjacent related layers in physical, chemical, and biological properties
- **leaching:** the dissolving out or removal of soluble materials from a soil horizon by percolating water
- **sediment:** rock fragments such as clay, silt, sand, gravel, and cobbles
- **soil profile:** a vertical section of a soil, extending through its horizons into the unweathered parent material
- **weathering:** the mechanical disintegration and chemical decomposition of rocks and sediments

STAGES OF FORMATION

Soil formation and development are continuing processes. A soil is a layer of unconsolidated weathered mineral matter on the earth's surface that can support life. It is initially formed when the remains of plants are mixed into weathered mineral fragments. Soil characteristics can change, however, for soil develops over time as the environment dictates.

Many processes contribute to the two stages of soil formation, which may grade indistinctly into each other. The first stage is the accumulation of rock fragments known as parent material. These sediments may have formed in place from the physical and chemical weathering of hard rocks below or by the accumulation of sediments moved by wind, water, gravity, or glaciers. The second stage is the formation of soil horizons, which may either follow or occur simultaneously with the first stage. The near-surface horizon, where materials are removed, is called the A horizon. Some of the removed material accumulates in the subsurface, in a zone called the B horizon. If the soil has an organic-rich top layer, that horizon is called the O horizon. In some cases, the A horizon has so much iron removed that it looks bleached. A very light zone at the base of the A horizon is called the E

horizon. Unmodified parent material is termed the C horizon.

The characteristics of the horizons and their degree of differentiation depend upon the relative strengths of four major processes in the soil: weathering, transfers, gains, and losses of the soil constituents. These processes have been observed directly in laboratory experiments but also are implied by relating physical, chemical, and morphological properties in different parts of the soil profile.

TRANSFORMATIONS AND TRANSFERS

Weathering is the fundamental process of soil formation, as it transforms soil constituents into new chemicals. The primary minerals found in the initial sediments and rocks are decomposed by physical and chemical agents into salts that are soluble in water and new minerals. Clay is a very important new mineral formed from the breakdown of primary minerals by the agents of water and air. Calcite, or calcium carbonate, is another mineral formed through weathering, especially of rocks such as basalt. Iron and aluminum hydroxides form a third major group of weathering-produced minerals. Organic matter is added to the soil when roots die and vegetation residues fall onto the surface. These are transformed by microorganisms into a mixture of chemically stable, organic substances called humus.

The transfer of soil constituents is generally in a downward direction. Rainwater is the major transporting agent. As it moves through the soil, it picks up salts and humus in solution and clays and humus in suspension. These materials are usually deposited in another horizon when the water is withdrawn by roots or evaporation, or when the materials are precipitated as a result of differences in pH (degree of acidity) or salt concentration. The clays normally coat the exteriors of sediment grains and line the interiors of pores. This transfer of materials through solution is commonly called leaching and requires good drainage. The soil profile often becomes vertically zoned according to the solubility of the dissolved substances. In forest soils, leaching generally removes iron and aluminum, leaving behind a bleached-looking zone. This process is called podzolization. In arid soils, calcium minerals precipitate in the subsoil to form calcite or gypsum, a deposit called caliche.

GAINS AND LOSSES

Gains consist of additions of new materials to the soil. Wind can deposit fine sands and silts if there is a source nearby. Chemical reactions during weathering add oxygen and water to the soil minerals. Groundwater brings new minerals into the soil pores. Floods can add new sediment to the soils in

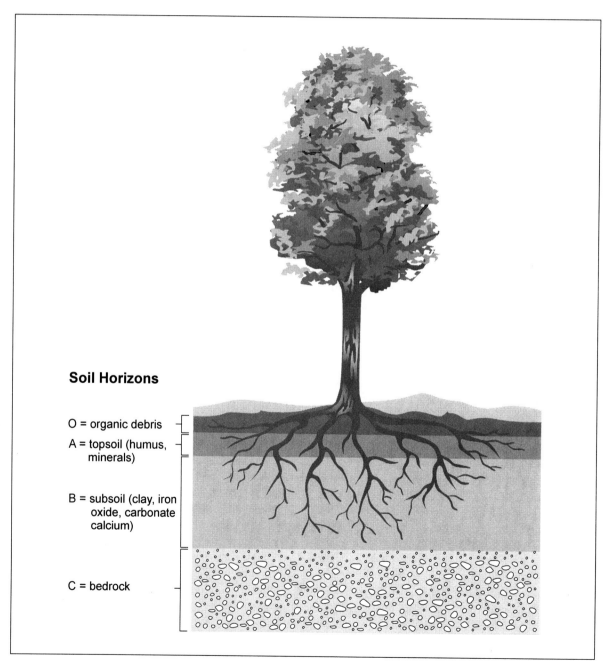

Soil Horizons

O = organic debris

A = topsoil (humus, minerals)

B = subsoil (clay, iron oxide, carbonate calcium)

C = bedrock

a floodplain. The largest addition to soils, however, comes from organic matter from the decomposition of plant material. Annual additions of new organic material between 255 and 22,900 kilograms per hectare have been estimated for soils in different environments ranging from deserts to tropical rain forests.

Losses from the soil profile generally occur when there is so much water moving through the system that the dissolved and suspended materials exit at the bottom of the profile and move downslope and eventually enter a stream. Soil erosion also occurs on the surface by water runoff, especially where there is minimal vegetation to control the sediment. Soil erosion is one of the major environmental problems facing the world today. Oxidation of plant matter balances some of the additions of new organic material.

CLIMATE

After describing many soil profiles around the world and classifying them, scientists concluded that this weathered material at the earth's surface was quite variable and differed from environment to environment. In 1980, Hans Jenny summarized the work of early researchers and listed the five main soil-forming factors that created this variation: climate, organisms, parent material, topography, and time. The rates and degrees of these processes in different soils depend upon these five factors. These factors are interdependent, and, therefore, very different soils may be formed from the same parent material.

Of the five factors, climate has probably the greatest impact on soil development. This dominance is best shown on a large-scale geographic basis, in which the distribution of soils on soil maps is related to vegetation type, which is in turn related to climatic zones. Precipitation provides the water for the transfer processes in the soils. In dry climates, where little water is moving through the system, calcium carbonate and salts build up in the soils. In humid climates, the abundance of water moving through the systems allows leaching and clay movement processes to dominate in the soils. Temperature is the second important element of climate, as it controls the rates of chemical weathering reactions in the soil. In colder environments, soil weathering and development are slower because weathering reactions are not active all year round. Some extreme climatic conditions also can control soil development, such

as permafrost, which limits soil depth and drainage. Overall, climate controls vegetation, which also controls soil development.

ORGANISMS AND PARENT MATERIALS

Living organisms, primarily vegetation, control the distribution and types of humus that are formed in soils. In turn, humus controls some of the important soil-forming processes. Grassland vegetation produces humus that is highest in mineral matter and not very acid, so resulting soils have thick A horizons that are high in organics. Conifer forests produce humus that is low in mineral constituents and highly acidic. These forest soils are dominated by leaching and the production of O, E, and B horizons (B horizons are rich in iron and aluminum). Deciduous forest soils are located somewhere between the conifer forest and the grassland soils in characteristics. Therefore, it is rare to find white, bleached E horizons under grassland vegetation, and vice versa—it is rare to find thick A horizons under conifer forest vegetation. Through its leaves and root systems, vegetation also protects the soil from erosion and, therefore, promotes soil development.

The parent material is the rock or sediment deposit from which the soil has developed. The composition of the original minerals controls the complexity of the soil processes by determining the weathering by-products, and controls the rate of these processes by porosity. A parent material of pure quartz sand would undergo minimal weathering, producing a simple soil profile. A parent material of wind-deposited silt, or loess, produces a more complex soil, as it has many minerals initially that can weather to by-products such as clays, salts, carbonates, and oxides. A soil produced on granite would yield a fairly deep, sandy soil, whereas a soil produced on basalt would yield a fairly thin, clay-rich soil. A soil developed in shale would be clay-rich, whereas one developed on sandstone would be sandy. Soils formed on rocks rich in iron tend to produce very red soils. Soils formed on limestone are difficult to acidify and podzolize because the parent material neutralizes those chemical processes. The porosity, or the amount of pore space, of the parent material gives ready access of gases and liquids to weathering. In porous, or permeable, soils, weathering and transfers of soil constituents are more rapid than in soils with low permeability.

TOPOGRAPHY AND TIME

Topography is the geometric configuration of the soil surface and is defined by the slope gradient, orientation, and elevation. This factor essentially controls the microclimate. A slope modifies rainfall by allowing rapid runoff on a steep slope or ponding in low places. Slope orientation modifies soil temperatures by the differences at which the sun's rays strike the soil. In the northern latitudes, southfacing slopes are warmer and drier than those on north-facing slopes. In mountain ranges, rainfall increases and temperature decreases with increases in elevation.

Time is the factor of the duration of the processes working in the soil. All the characteristics in the soil take time to form. The A horizon forms most rapidly, taking as little as 10 years with the aid of humankind to more than 1,000 years in a cold and dry arctic or alpine environment. B horizons take longer to form because minerals weather slowly. In the dry western parts of the United States, discoloration from the weathering of iron-rich minerals appears at about 1,000 years, but the strongest red colors require 100,000 years of development in the same environment. In the same part of the United States, initial detection of clay movement into the B horizon in a sandy parent material occurs after about 10,000 years of development, with maximum development taking more than 100,000 years. Calcium carbonate deposition in arid regions of the world takes between 100,000 and 500,000 years to produce a strongly cemented horizon. It is estimated that it takes more than 1 million years to produce a highly weathered lateritic soil (decayed iron-rich rock that is red in color) in the tropics.

Soils change with time just as humans change as they grow. Young or poorly developed soils in a grassland environment have only A and C horizons. With time and the movement of iron oxides and clay into the lower parts of the profile, a faint B horizon develops between the A and C horizons. An old or well-developed soil in this environment would have a thick A horizon and a thick B- horizon with abundant clay moved into it, all overlying a C horizon.

LAND-USE PLANNING

Scientists are finding that soils can be useful tools in the interpretation of past environments of an area and also in the formulation of land-use plans for that region. First, the soil scientists must determine the exact soils that are in the area through a mapping program. During this project, they determine the regional developmental sequence of soils—that is, the poorly developed and the well-developed soils of the different places on the landscape. They assign ages to each soil in the sequence using radiocarbon-dated samples from the soils. As a result, when a particular soil is located in that region and compared with the sequence, an age can be assigned to the land surface upon which the soil developed. Geologists commonly use soils to estimate the ages of deposits, especially on glacial landscapes. The idea of multiple glaciations has been developed through the use of soils. Poorly developed soil on glacial sediments represents a recent glaciation compared with a site that features well-developed soil lying on sediment deposited by an older glacial advance.

The frequency of geological hazard events such as landslides, rockfalls, and flooding can be estimated by the study of buried soils in sediment deposits. An ancient, buried A horizon within a soil profile means that some geological event covered an old land surface with sediment. If the organic matter of that buried A horizon is dated using radiocarbon dating, an approximate time of burial can be determined. If a valley bottom has many buried soils, it probably means that the nearby stream floods quite frequently. These decisions are important for land-use planning, for one does not want to build in an area where hazards occur frequently. For example, nuclear power plants must not be located in areas where there are active faults (ground breaks). Soils are commonly used to determine if faults are active in an area and, if so, the frequency of movement on the fault.

The U.S. Department of Agriculture produces maps of soils across the country. The maps describe the productivity of crop growth and manageability related to land use of each soil. These data help farmers grow crops that are best suited for their soils; they tell the farmers which soils are most susceptible to erosion so that extra precautions can be taken with the fragile soils. The data also help land-use planners keep the most productive soils in agriculture instead of paving them. In addition, the maps help to predict how much fertilizer to put on the fields. Natural soil fertility is related to geological activity and weathering, as the oldest soils are leached of most of their nutrients and are least fertile. Young soils have had little weathering and leaching and are most fertile, especially those on floodplains, young glacial deposits, and volcanic deposits. Soil maps are also used to recognize where shrink-swell clays occur in the soils. The presence of shrink-swell clays in the soil is important for home owners. If present, these clays can crack the foundation of a house. To prevent these damages, special foundations must be constructed and the clays stabilized with the use of lime.

Scott F. Burns

FURTHER READING

Birkeland, P. W. *Soils and Geomorphology*. 3d ed. New York: Oxford University Press, 1999. One of the best books on the subject written from the perspective of a geologist. Offers many examples from around the world. Covers soil formation and development. Discusses the applications of soil formation to geology in the final chapter. Suitable for advanced high school and university-level readers.

Buol, S. W., F. D. Hole, and R. J. McCracken. *Soil Genesis and Classification*. 5th ed. Ames: Iowa State University Press, 2003. An update of a classic soil-science text. Features an update on the new soil horizon symbol nomenclature. Includes an in-depth discussion on soil development. Suitable for university-level readers.

Gerard, A. J. *Soils and Landforms*. London: Allen & Unwin, 1981. Covers all factors of soil formation, emphasizing the topographic factor. Excellent section on catenas (distribution of soils on a slope). Best suited to university-level readers.

Jenny, H. *Factors of Soil Formation: A System of Quantitative Pedology*. New York: Dover, 1994. Expands on the five factors of soil formation and development. Aimed at university-level readers. Illustrations, bibliography, and index.

_____. *The Soil Resource: Origin and Behavior*. New York: Springer-Verlag, 1980. Rewrites Jenny's classic book, written in 1941. Expands on the five factors of soil formation and development. Aimed at university-level readers.

Paton, T. R., G. S. Humphreys, and P. B. Mitchell. *Soils: A New Global View*. New Haven, Conn.: Yale University Press, 1995. Aimed at readers at the university level. Covers good examples of soils and soil formation from around the world. Color illustrations, maps, bibliography, and index.

Retallack, G. J. *Soils of the Past*. 2d ed. New York: Blackwell Science, 2001. Offers information on paleosols. Discusses soil classification and formation, identifying paleosols and their characteristics, and factors involved in paleosol formation. Cover the soil-fossil records with discussion of ancient landscapes and ecology.

Schaetzl, Randall J., and Sharon Anderson. *Soils: Genesis and Geomorphology*. New York: Cambridge University Press, 2005. Provides a large amount of fundamental content. Discusses soil morphology and classification, the formation of soil, other soil dynamics, paleosol characteristics, and the reconstruction of environments through the study of paleosols.

Simonson, R. W. "Outline of a Generalized Theory of Soil Genesis." *Soil Science Society of America Proceedings* 23 (1959): 152-156. A classic paper in which the author outlines the four types of processes involved in soil formation.

Singer, M. J., and D. N. Munns. *Soils*. 6th ed. Upper Saddle River, N.J.: Prentice Hall, 2005. One of the most readable elementary soil science texts available. Includes a good basic section on soil formation. Appropriate for high school-level readers.

Soil Science Society of America. *Glossary of Soil Science Terms*. Madison, Wis.: Author, 2001. Defines soil science terms; written by soil science professionals.

U. S. Department of Agriculture, Soil Survey Staff. *Keys to Soil Taxonomy*. Honolulu: University Press of the Pacific, 2005. Contains the most used soil classification system in the world. A challenging read but an abundant source of information.

Van Breemen, Nico, and Peter Buurman. *Soil Formation*. 2d ed. Boston: Kluwer Academic Publishers, 2002. Examines all stages and processes of soil formation. Illustrations and diagrams help to clarify difficult concepts. Intended for college students, but useful for the layperson.

Walther, John Victor. *Essentials of Geochemistry*. 2d ed. Burlington, Mass.: Jones & Bartlett Publishers, 2008. Contains chapters on radioisotope and stable isotope dating and radioactive decay. Geared toward geology and geophysics rather than chemistry. Covers thermodynamics, soil formation, and chemical kinetics.

See also: Desertification; Expansive Soils; Land Management; Landslides and Slope Stability; Land-Use Planning; Land-Use Planning in Coastal Zones; Soil Chemistry; Soil Erosion; Soil Profiles; Soil Types.

SOIL PROFILES

A soil profile is the vertical section of a soil extending through its horizons into the unweathered parent material. Scientists use the different arrangements of horizons to give classification names to the soil profiles, which can be used to interpret landscape history and determine future land use.

PRINCIPAL TERMS

- **horizon:** a layer of soil material approximately parallel to the surface of the land that differs from adjacent related layers in physical, chemical, and biological properties
- **leaching:** the dissolving or removal of soluble materials from a soil horizon by percolating water
- **sediment:** rock fragments such as clay, silt, sand, gravel, and cobbles
- **structure:** the arrangement of primary soil particles into secondary units called peds
- **texture:** the relative proportions of varying sediment sizes in a soil
- **weathering:** the mechanical disintegration and chemical decomposition of rocks and sediments

HORIZON NOMENCLATURE

The term "soil" has many definitions. To the soil scientist, it is anything that will support plant growth. To the civil engineer, it is an unconsolidated surficial material that can be penetrated by a shovel. To the geologist, it is an unconsolidated layer of weathered mineral matter arranged in layers at the earth's surface. A soil profile is a vertical arrangement of soil horizons down to and including the parent material in which the soil has developed. Two sets of horizon nomenclature are used to describe soil profiles: symbols for field descriptions and diagnostic horizon names for classification. The field description symbols have been in use since the nineteenth century, whereas the diagnostic horizon names came into use only in the 1960's.

The field description symbols used to identify various kinds of horizons were developed in Russia in the 1880's. There, soil scientists such as V. V. Dokuchayev applied the letters A, B, and C to the main horizons of the black soils of the steppes of Russia. The A horizon was designated the zone of maximum organic material accumulation, the C horizon was the unaltered parent material, and the B horizon was the layer in between the A and the C. These concepts spread to the rest of the world, with the B horizon concept being modified to be a zone of the accumulation of iron, aluminum, and clays that had moved down the profile. Horizons may be prominent or so weak that they can be detected only in the laboratory; they can be thick or thin. The use of the A, B, and C master horizons has remained the backbone of horizon symbols. Three additional master horizons, however, have also been included: O, for organic-rich layers; E, for layers that have been intensively leached; and R, for rock. Soil terms like "podzol" (from *pod*, meaning "underneath," and *zola*, meaning "ash") and "chernozem" (from *cherny*, meaning "black," and *zemlya*, meaning "earth") are still used and reflect the Russian roots of soil classification.

Since the 1930's, many sets of horizon symbols have been developed around the world. With the United States and many other countries adopting a horizon nomenclature similar to that produced by the Food and Agricultural Organization (FAO) of the United Nations in the early 1980's, the world has come closer to a universally accepted set of horizon symbols.

REFINING HORIZON NOMENCLATURE

There are three kinds of symbols used in soil horizon descriptions: capital letters, lowercase letters, and Arabic numerals. The capital letters describe the master horizons, the lowercase letters depict some specific characteristic of a master horizon, and the numeral characterizes a further subdivision or parent material layering. An example of a horizon symbol might be Bg1.

The "g" describes a "gleyed" horizon, or one where the iron has been removed during soil-forming processes, or saturation with stagnant water in a high water table has preserved the reduced state of the iron, giving the layer a neutral or gray color. This letter also can describe a soil with red and gray specks from the oxidized and reduced forms of iron. A horizon that is more than 90 percent cemented is given the "m" designation. Horizons with abundant silica are described with a "q" and abundant sodium with an "n." If the horizon is very densely packed, which commonly occurs in silt-rich soils, the "x" designation is used. A "b" denotes a buried soil horizon. A "c" signifies that weathering has formed concretions, or nodules, in the soil.

Transitions between master horizons are commonly found in nature. Where the properties of both horizons are mixed in the same layer, both capital letters are used, with the first letter denoting the horizon whose properties dominate. The term BA horizon would be a transition between the A and B horizons, where the B horizon characteristics prevail. Where the horizon has distinct parts of both horizons, the two capital letters are separated by a slash mark (A/B, E/B, B/C).

Organic Soil Horizons

The O horizon is a master horizon in which there is an accumulation of mainly organic matter that overlies a mineral soil. It is dominated by fresh or partly decomposed organic litter such as twigs and needles, and many times the original form of most of the vegetative matter is visible to the naked eye. This horizon contains between 20 and 30 percent organic material, sometimes more depending upon the clay content of the underlying mineral horizons. Three subdivisions of this master horizon are based on the amount of decomposition in the organic material and range from the Oi (least decomposition) to the Oe (intermediate decomposition) and the Oa (most decomposition).

The A horizon is a master horizon in which decomposed organic matter, humus, is mixed with mineral sediments. The organic matter content is not great enough to be classified as an O horizon. This horizon is generally a surface horizon, except when located below an O horizon. Because of the presence of organic material, this layer can be darker than underlying horizons. The organic material is assumed to be derived from the decomposition of plant and animal remains deposited on the surface. This layer is the zone of maximum biological activity. A subdivision is the Ap, or "plow horizon," where the A horizon has been disturbed by cultivation.

Mineral Soil Horizons

The E horizon is the master horizon commonly located below the O or A horizons in forest environments where leaching is dominant in the soil profile. Abundant water passing through the O horizon may become very acidic; this water leaches iron, aluminum, and organics from the A horizon as it passes to the lower part of the profile. The remaining mineral matrix is light-colored because of the color of the primary mineral sand grains that remain. Sometimes the horizon is almost pure white. For many years, this horizon was called an A2 horizon but has been removed from the category, as it is not a zone of organic material accumulation. The E horizon sometimes is light enough to look like ash, explaining the Russian term "podzol"; the process that forms such soils is still called "podzolization."

The B horizon is the master zone of accumulation of materials that have been moved down by water from the O, A, and/or E horizons. These suspended materials include clay, iron, aluminum, and organic matter. The soil horizon shows little or no evidence of the original sediment or rock structure. Several kinds of B horizons are recognized depending upon the materials moved into them and, of course, not all horizons are present in any given soil. A Bh horizon is an accumulation of abundant organic material and therefore is very black in color. In a Bs horizon, the deep red color depicts the accumulation of abundant iron and aluminum (the latter of which is colorless). A Bk horizon has an abundant amount of calcium carbonate and a white color throughout. A Bt horizon includes a large amount of clay that has moved down from the horizons above. Clay films are noted on the sediments, and the soil samples are very sticky. A By horizon is an accumulation of gypsum. A Bz horizon contains salts more soluble than gypsum. A Bw horizon is a weakly developed B horizon in which a reddish color has developed through weathering, but few or no apparent materials have been moved into it from above.

The C master horizon is the subsurface layer of partially weathered parent material. Soil-forming processes, in which particles and chemicals move downward from the O, A, E, and B horizons, have not affected this horizon, yet weathering of the sediments has slightly changed the color of the parent material. If the horizon material has the structure of the parent rock but is weathered enough to get a shovel through it, the horizon is called a Cr horizon; geologists also refer to this horizon as a saprolite. The R master horizon is the unweathered, consolidated bedrock underlying the soil.

SOIL TAXONOMY

A second set of diagnostic horizon nomenclature was developed in the latter half of the twentieth century in the United States, using a new system called Soil Taxonomy. Information from diagnostic surface horizons (epipedons), subsurface horizons, soil-moisture, and temperature regimes together are used to determine the soil classification. Soil Taxonomy is based mainly on measurable soil properties based in seventeen epipedons and subsurface horizons rather than on theories of soil formation, which form the basis for older classification systems. The classification system is very exact and offers exotic combinations of Latin and Greek syllables, such as Pachic Cryumbrepts and Natraqualfic Mazaquerts. Each syllable is a descriptive term for some aspect of the soil. For example, "natraqualfic" means sodium-rich (natr-), wet (aq-), and having a B horizon rich in iron and aluminum (alfic). Field observations must be supplemented by laboratory analyses of properties plus climatic data. Similar approaches using this type of horizon nomenclature have also been developed in Canada and by the FAO (the UN Food and Agriculture Organization). In the United States, soils are classified into twelve soil orders on the basis of their horizon materials and structures. Most soil types are defined by the distinctness of their horizons and degree of evolution, but some, like soils developed from peat, volcanic ash, swelling clays, and on permafrost, are defined by those special characteristics.

Soil Taxonomy describes four different epipedons. The Mollic epipedon is a thick, dark-colored surface horizon rich in mineral nutrients. The Umbric epipedon is similar to the Mollic visually but is low in mineral nutrients. The Ochric epipedon is a thin surface horizon that does not meet the requirements of the Mollic or Umbric. The Histic epipedon is basically an O horizon.

The additional Soil Taxonomy diagnostic horizons are subsurface horizons. The Albic horizon is basically an E horizon, as it has been highly leached. An Argillic horizon is a B horizon that has been enriched significantly with clay from higher in the profile. The Natric horizon contains a high abundance of salt. The Spodic horizon is a B horizon with abundant iron, aluminum, and/or organics that have been moved from above. The Cambic horizon is a B horizon that is slightly enriched in red color but, unlike the Spodic and Argillic horizons, lacks an accumulation of products from above. The Oxic horizon is a B horizon with an abundance of highly weathered minerals. A Calcic horizon has an abundance of calcium carbonate in it. A Petrocalcic horizon is a cemented Calcic horizon, often called caliche. The Gypsic horizon contains an abundance of gypsum. A Petrogypsic horizon is a cemented Gypsic horizon. A Duripan is cemented with silica. A Fragipan horizon is very dense and is many times formed in silt particles.

STUDYING A SOIL

The first step in the study of a soil is producing a description of the soil profile. A scientist studying a particular soil exposes a soil profile by either digging a soil pit or finding a road cut. First, the soil horizons are delineated (A, B, C, and so on). Second, each horizon is characterized by thickness, color, presence of films, and boundaries. The horizon texture (the relative amount of sediment sizes) is recorded as well. For example, a sandy loam soil has an abundance of sand with some silt in it. The structure of that horizon, or how the sediments are put together, is described as well. Third, samples from each horizon are gathered to be analyzed in the laboratory. Finally, an approximate classification is made based on the minimal information available in the soil pit.

The soil samples are returned to the laboratory for a chemical and physical characterization. The pH (degree of acidity) and amounts of chemical nutrients are generally determined using pH meters and titrations. The exact texture is determined using sieves and water-settling columns. The percentage of organic matter is calculated using titration. Additional information can be obtained using an X-ray machine to determine the types of clays in the soil.

Once the field and laboratory data have been assembled, diagnostic horizons can be selected and the exact classification is determined using Soil Taxonomy. Twelve major classification groupings, or orders, are used in this system. Entisols are young soils that have no B horizon, whereas Inceptisols have a weak B horizon. In humid climates the major mature soil types are Mollisols, Alfisols, and Spodosols. The Mollisols are soils with a Mollic epipedon and are often grassland soils. Alfisols have an Argillic horizon and abundant nutrients; they are often deciduous forest soils. Spodosols are highly leached soils that have Spodic (E) horizons and are common in conifer forests. Ultisols are highly weathered soils of warm climates, and Oxisols are extremely weathered soils of the tropics. Aridisols are arid climate soils. Histosols are peaty soils that have thick Histic epipedons. Vertisols are high in clays that shrink and swell with seasonal moisture variation and exhibit extensive cracking during dry seasons. The Andisols are volcanically produced soils. Gelisols are soils formed above permafrost. Using the proper classification, scientists can interpret the soil's age, future land use, the crops it can best support, and its past history.

A FUNDAMENTAL TOOL

For years, the soil profile has been the basic descriptive tool of scientists. This cross-section constitutes the format from which scientists make their soil classifications. Without a proper soil profile description, one cannot perform the soil classification required before environmental interpretations and soil maps can be produced. The symbols that are used to describe the horizons of the soil profile may have changed, but the basic A-B-C concept that was developed during the nineteenth century has not changed much. Different systems of description of the soil profile and the resulting soil classification systems are present around the world, but with time they are becoming more similar.

In the United States, the Soil Taxonomy system has been officially used since 1975. This new classification offers a very exact hierarchical system similar to what biologists use to classify plants and animals. The twelve major orders are divided into suborders, great groups, subgroups, and finally soil series. The subgroup name, such as "Typic Paleudult," is similar to the genus/species designation of a plant or animal. Soil Taxonomy is the most comprehensive soil classification system in the world and therefore is used more than any other approach. It is such an exact system that one can give a scientist familiar with the taxonomy a subgroup name and he or she can construct a complete soil profile description very close to the actual one without even seeing the soil. Soil classification may be the most difficult classification problem in science, as a large number of factors must be considered and a wide variety of client groups want to apply the results to very different problems.

Scott F. Burns

FURTHER READING

Al-Rawas, Amer Ali, and Mattheus F. A. Goosen, eds. *Expansive Soils: Recent Advances in Characterization and Treatment.* New York: Taylor & Francis, 2006. Discusses the classification of expansive soils. Provides information on swelling and consolidation, and the effects these processes have on buildings. Discusses how site characteristics and stabilizing agents can alter expansion processes.

Birkeland, P. W. *Soils and Geomorphology.* 3d ed. New York: Oxford University Press, 1999. One of the best books written on soils from the perspective of a geologist. Offers many examples from around the world. Covers the newly adopted soil profile description symbols. Discusses Soil Taxonomy in the early chapters. Suitable for university-level readers; early chapters are appropriate for high school-level readers.

Buol, S. W., F. D. Hole, and R. J. McCracken. *Soil Genesis and Classification.* 5th ed. Ames: Iowa State University Press, 2003. An update of a classic text on soil science. New to the edition is an update on the new soil horizon symbol nomenclature. Includes an in-depth discussion on soil development. Suitable for university-level readers.

Carr, Margaret, and Paul Zwick. *Smart Land-Use Analysis: The LUCIS Model.* Redlands, Calif.: ESRI Press, 2007. Discusses the use of ArcGIS software to analyze land suitability. Provides a step-by-step guide to ArcGIS software. The LUCIS model is explained with case studies to provide examples of potential conflicts of land use.

Idriss, I. M., and R. W. Boulanger. *Soil Liquefaction During Earthquakes.* Oakland, Calif.: Earthquake Engineering Research Institute, 2008. Covers the basic principles of the liquefaction behavior of soils and analysis methods.

Jenny, Hans. *Factors of Soil Formation: A System of Quantitative Pedology.* New York: Dover, 1994. Describes the factors involved with variations in soils. Suitable for university-level readers. Illustrations, index, and bibliography.

_____. *The Soil Resource: Origin and Behavior.* New York: Springer-Verlag, 1980. A rewrite of the classic book from 1941. Describes variations in soils. Suitable for university-level readers.

Kapur, Selim, and George Stoops, eds. *New Trends in Soil Micromorphology.* New York: Springer, 2010. Compiles articles suited for graduate students and researchers. Discusses the identification and analysis of paleosols, and soil micromorphology of the Quaternary period. Addresses vertisols, andosols, and other paleosols.

Landa, Edward R., and Christian Feller, eds. *Soil and Culture.* New York: Springer, 2010. Provides a full history of the relationship between humans and the earth. Organized into parts discussing soil's function in art, philosophy, culture, and health. Provides interesting information from an anthropological perspective.

Paton, T. R., G. S. Humphreys, and P. B. Mitchell. *Soils: A New Global View.* New Haven, Conn.: Yale University Press, 1995. Aimed at readers at the university level. Offers many good examples of soils and soil formation from around the world. Color illustrations, maps, bibliography, and index.

Singer, M. J., and D. N. Munns. *Soils.* 6th ed. Upper Saddle River, N.J.: Prentice Hall, 2005. Includes well-explained soil profile descriptions and classification. Suitable for high school-level readers.

Soil Science Society of America. *Glossary of Soil Science Terms.* Madison, Wis.: Author, 2001. Defines soil science terms; written by soil science professionals.

Sparks, Donald S., ed. *Soil Physical Chemistry.* 2d ed. Boca Raton, Fla.: CRC Press, 1999. Offers a clear look into the practices, procedures, and applications of the field.

Sposito, Garrison. *The Chemistry of Soils.* 2d ed. New York: Oxford University Press, 2008. Provides a comprehensive discussion of soil composition. Describes chemical makeup and basic soil components. Discusses physical and chemical characteristics and processes within soil types. Concludes with a list of references for further reading.

Tan, Kim Howard. *Principles of Soil Chemistry.* 4th ed. Boca Raton, Fla.: CRC Press, 2011. Written as an introduction to soil chemistry. Explains the processes and application of the science in an easy-to-understand manner accompanied by plenty of illustrations. Includes a twenty-five-page bibliography, as well as an index.

U.S. Department of Agriculture, Soil Survey Staff. *Keys to Soil Taxonomy.* Honolulu: University Press of the Pacific, 2005. A difficult but important text on the official classification system of the United States.

See also: Desertification; Expansive Soils; Land Management; Landslides and Slope Stability; Land-Use Planning; Land-Use Planning in Coastal Zones; Soil Chemistry; Soil Erosion; Soil Formation; Soil Types.

SOIL TYPES

The classification of soils is of crucial importance, the taking soil surveys a primary task of the U.S. Department of Agriculture. Knowledge of soil types indicates the crops that are best suited for a particular soil. Soil classification also provides farmers with the most appropriate methods for preventing soil erosion.

PRINCIPAL TERMS

- **A horizon:** the surface soil layer; also known as top-soil
- **belt:** a geographic region that in some way is distinctive
- **B horizon:** the soil layer just beneath the topsoil
- **leach:** to remove, or be removed from, by the action of a percolating liquid
- **monolith:** a large soil profile created by dovetailing monolith containers together and combining small, individual profiles
- **pedologist:** a soil scientist
- **polypedons:** bodies of individual kinds of soil in a geographic area
- **profile:** a representative soil sample containing different layers of soil
- **sesquioxide:** an oxide consisting of three oxygen atoms for every two atoms of other elements, the most common being aluminum (Al_2O_3) and iron (Fe_2O_3) sesquioxides
- **zone:** an individual soil group within a horizon

THREE BASIC TYPES

Because soils consist of particles varying greatly in size and shape, specific terms are required to give some indication of their physical properties. Soils can be classified according to different characteristics. The simplest way to classify soils is by their texture. Three broad yet fundamental groups of soils are now recognized by pedologists: sands, clays, and loams. All class names that have been devised over the years are based on these three groups. The sand group includes all soils, of which sand makes up more than 70 percent of the material by weight. In contrast to the heavier groups of soils, which are stickier and more clayey by nature, this group is characteristically sandy in texture. Two specific classes within the sand group are recognized: sand and loamy soil. Soils included in the clay group must consist of at least 35 percent clay and in most cases not less than 40 percent. The names are "sandy clay," "silty clay," or simply "clay," which is the most common of all. Sandy clays often contain more sand than clay; similarly, the silt content of silty clays usually exceeds that of the clay fraction itself. The loam group, which is the most important for agriculture, is more difficult to explain. An ideal loam is a combination of sand, silt, and clay particles. It also exhibits light and heavy properties in roughly equal proportions. In most cases, the quantities of sand, silt, or clay require a modified class name. For example, a loam in which sand is present is classified as a sandy loam.

Qualifying attributes such as stone, gravel, and various grades of sand must be considered when placing some soils in classes. For example, one refers to stony-clay loams, gravelly sands, or fine-sandy loams. Pedologists also classify soils according to their composition. A great soil group consists of many soil types whose profiles have major features in common. Every soil type has the same numbers and kinds of definitive horizons, but they need not be expressed in the same profile to the same degree. For example, the Fayette, Dubuque, Downs, and Quandahl soils are all members of a single great soil group.

Geographic belts marked by certain combinations of great soil groups can be shown on maps of small scale. A schematic soil map of the world consists of six broad belts. One belt consists of rough landscapes, such as mountains, in which many of the soils are stony, shallow, or both. The patterns of soil types in these areas are especially complex. The other five broad belts have simpler patterns, but each includes a number of great soil groups. The soil types within a single farm, for example, commonly represent two or more great soil groups. The thousands of soil types in the United States are classified in twelve major soil orders. Collectively, the groups have wide ranges in their characteristics. They also vary greatly in such qualities as fertility, ability to hold available moisture, and susceptibility to erosion.

PODZOLIC AND LATOSOLIC SOILS

Although terms like "podzol" and "chernozem" have long been superseded by more precise classifications, they still are useful as broad labels to describe major groups of soils.

491

Podzolic soils dominate a broad belt in the higher latitudes of the Northern Hemisphere and some smaller areas in the southern half of the world. They include the brown podzolic soils, gray-brown podzolic soils, and gray wooded soils. These groups were formed under forest vegetation in humid, temperate climates. The B horizons of these soils vary. Some podzolic soils have B horizons that are composed primarily of humus, sesquioxides, or both; others have B horizons that are mainly accumulations of clay. Podzolic soils are more strongly weathered and leached than many of the other soil groups. Because they are usually acidic, low in bases such as calcium, and low in organic matter, levels of fertility are moderate to low. As a group, however, these soils are responsive to scientific management. According to Soil Taxonomy, the system used in the United States, these soils are mostly termed spodosols if they show strong leaching, or alfisols if they show B horizons rich in iron and aluminum.

The equatorial belts of Africa and South America are dominated by latosolic soils, which are termed oxisols. These soils are also dominant in the southeastern parts of Asia and North America, in northeastern Australia, and in the larger islands of the western Pacific Ocean. Latosolic soils include the great soil groups known as laterites, reddish-brown lateritic soils, yellowish-brown lateritic soils, red-yellow podzolic soils, and several kinds of latosols. Red-yellow podzolic soils are so named because they share features with both the latosolic and podzolic groups, but they are more closely related to the latosols. Latosolic soils have been formed under forest and savanna vegetation in tropical and subtropical to fairly dry climates. Although these soils do not extend into arid regions, they may be found in alternately wet and dry areas with low rainfall. The latosols are the most strongly weathered soils in the world. As a result of the large amounts of iron oxides formed through intense weathering, the profiles commonly display red and yellow colors. Except for a darkened surface layer, most of these soils lack distinct horizons; therefore, the profile may remain unchanged for many feet. Although supplies of plant nutrients are usually low, these soils contain a high capacity to fix phosphorus in unavailable forms. Most of these soils are resistant to erosion and easily penetrated by plant roots. In fact, some tree roots in southeastern Brazil have extended more than 60 feet below the surface layer. The moisture capacities in latosolic soils are mostly moderate to high, although they are low in some. Without the benefit of modern science and industry, productivity is normally low.

CHERNOZEMIC AND DESERTIC SOILS

Chernozemic soils, termed mollisols in the United States, have formed under prairie or grass vegetation in climates that are humid to semiarid, temperate to tropical. Although these soils are most extensive in temperate zones, some large areas exist in the Tropics. Chernozemic soils include the great soil groups known as chernozems, prairie soils (or brunizems), reddish prairie soils, chestnut soils, and reddish chestnut soils in temperate regions. In tropical and subtropical regions, these soils are often known as black cotton soils, grumusols, regurs, and dark clays. The A horizons of chernozemic soils are normally very thick, fertile, and slightly weathered. The B horizons are usually much less distinct. The A horizons of chernozemic soils are high in organic matter and nitrogen in temperate zones but not in tropical and subtropical zones. These soils are less acidic and higher in bases and plant nutrients than podzolic or latosolic soils; therefore, they are among the most fertile soils in the world, producing about 90 percent of the world's grain. It is no exaggeration to say that temperate chernozemic soils (mollisols) feed the human race. Within the United States, the chernozemic soils form the heart of the Corn Belt and wheat-producing regions. Because the soils extend from the margins of humid zones into semiarid zones, production varies with seasonal weather. Chernozemic soils in tropical and subtropical zones are not suitable for cultivation because they are high in clay, are plastic, and subject to great shrinking and swelling. Handling these soils is difficult.

Desertic soils, or aridosols, have formed alternately under shrub or grass vegetation in arid climates ranging from hot to cold. These soils are commonly found in the great deserts of Africa, Asia, and Australia, and in the smaller deserts of North and South America. They include the great soil groups known as desert soils, red desert soils, sierozems, brown soils, and reddish-brown soils. Desertic soils have been slightly weathered and leached because of the shortage of moisture. The lack of adequate rainfall also limits plant growth; consequently, these soils are low in organic matter and nitrogen. Limited rainfall is also reflected in the shallow profiles. Most of these soils feature faint horizons. The A horizon is lighter in color than the B horizon because it has commonly lost carbohydrates and perhaps some bases and clay. The slightly darker B horizon has some accumulation of clay but is very low in organic matter. Levels of nutrients other than nitrogen are usually moderate to high in these soils. Available moisture capacities vary, depending on the thickness of the profiles and textures of horizons. Land management can have a tremendous impact on these soils. Often the B horizon is cemented by calcite in a deposit known as caliche.

SOIL SAMPLES

The most simple and common method for determining the class name of a soil is by feel. In fact, rubbing the soil between the thumb and fingers can probably tell a scientist as much about a soil as any other superficial means. To estimate plasticity more accurately, it is helpful to wet the sample; this technique causes the soil to "slick out," thereby giving the observer a good idea of the amount of clay present. There is ample information available in the way the soil particles feel: sand particles are gritty; silt has a floury or talcum-powder feel when dry and is only moderately plastic or sticky when wet; and silt and clay often take the form of clods. Wetting the soil and rolling it into a rope or flattening it into a ribbon are simple but effective ways to gauge relative amounts of silt and clay.

A surveyor can learn more about the nature of soils by using simple tools. The only tool on which the fieldworker can absolutely rely is the spade. Although many spades tend to compact the soil layers somewhat, the gouge or cheese-sampler types of spade obtain samples of near perfection. Good samples that reveal the differentiation of horizons can also be taken with a butcher's knife. A horizontal cut is made between the horizons and two vertical cuts are made in the face. Individual elements can then be picked out and used as representative zone samples.

A permanent record of the soil character can be taken by means of a monolith. A monolith container is made of steel and is 66 centimeters long. Since these containers are made to dovetail into each other, soil profiles can be taken at any depth. After

the sample has been taken, a fixing solution is applied to preserve it. As soon as the sample has been "set," it can be erected in a vertical position so that the soil layers can be examined in their natural order. Individual soil fragments taken by hand can also be preserved. The specimen is first air-dried and then oven-dried at 50 degrees Celsius until it reaches the same temperature throughout. While it is still warm, the specimen is immersed in a bath containing a solution of 3 percent cellulose acetate in acetone until bubbles cease to appear. It is then allowed to drain and dry on a sieve. Specimens fixed in this manner may be handled without breaking down.

A field map illustrating the soil types present in a large area can also be made by taking samples. A surveyor bores a hole with an auger until sure that a representative sample of the site has been acquired. After describing the profile in a notebook and then marking the site with a triangle and number, the surveyor moves to a neighboring site and continues taking samples at regular intervals.

AERIAL PHOTOGRAPHS AND DIAGRAMS

Much can also be learned about soil characteristics from the "layout" of fields in aerial photographs. These are especially useful for examining soils in more inaccessible areas, such as vast forests. Aerial photographs show only the external or environmental characteristics of soil. With experience, however, scientists can deduce from the study of these features a good idea of what lies underneath. Aerial photographs can generally be studied for soil recognition without the aid of sophisticated photographic apparatus. Quality prints at a scale of about 15 centimeters to 1.5 kilometers usually give enough definition for general purposes. Later, simple stereoscopic studies of "stereo pairs" of contact prints provide three-dimensional detail.

Probably the most accurate method of determining the class of soils has been devised by the U.S. Department of Agriculture. This method involves the creation of a diagram that can be obtained through mechanical analysis. The diagram reveals that a soil is a mixture of different sizes of particles and that a close correlation exists between particle size distribution and the properties of soils. A surveyor can readily check the accuracy of class designations once mechanical analyses of the field soils under scrutiny are available.

The U.S. Conservation Service groups soils into eight land-capability classes. Soils in classes I to III are generally suitable for cultivation. Soils in classes IV to VIII are severely limited in their utility for farming.

FARMING APPLICATIONS

Knowledge of soil types is essential for effective farming. Farming methods have to be tailored according to the type of soil that is being cultivated. As people from the eastern United States moved to the prairies and the plains, for example, the farming practices that they brought with them had to change according to the different soils types that they encountered: chernozem, prairie, and chestnut soils. Even today, farming practices are still undergoing change because no one knows just what type of farming is best suited to the chestnut and brown soils.

Failure to adapt farming practices to the soil type can have a devastating impact on a community. A decline of productivity often occurs because of the inability of people in some regions to establish an appropriate type of farming. This is an especially severe problem with latisols (oxisols), which look superficially rich but cannot sustain intensive agriculture. People may attempt to force a pattern of use on soils that are not suited to it, as on the reddish-chestnut soils of the southern plains in the United States. Soil erosion and soil depletion, which often occur because of inappropriate practices, are symptoms of bad relationships between people and soil. It follows that not only do people live differently on different soils, but they also must live differently to live at all. This fact may be illustrated by civilizations that largely developed on one particular type of soil: the Egyptians on the alluvial soils along the Nile River; the ancient Greeks and Romans on the terra rossa ("red earth") soils around the Mediterranean Sea; the Arabians on the soils of the deserts; and the Westerns on the gray-brown podzolic of western Europe, the northeastern United States, eastern Canada, and parts of Australia, New Zealand, and South Africa. Not only were the types of crops upon which these cultures relied determined in great part by the soil type, but important artifacts were as well, such as the kinds of pottery that they used.

Alan Brown

FURTHER READING

Al-Rawas, Amer Ali, and Mattheus F. A. Goosen, eds. *Expansive Soils: Recent Advances in Characterization and Treatment.* New York: Taylor & Francis, 2006. Discusses classification of expansive soils. Provides information on swelling and consolidation and the effects these processes have on buildings. Discusses how site characteristics and stabilizing agents can alter expansion processes.

Bear, Firman E. *Earth: The Stuff of Life.* 2d ed. Norman: University of Oklahoma Press, 1990. Explains how soil came into being; primarily a treatise on soils in relation to the growth of plants, animals, and humans, as well as a plea for the sensible conservation of soil. Recommended for the general reader.

Birkeland, P. W. *Soils and Geomorphology.* 3d ed. New York: Oxford University Press, 1999. One of the best books written on soils from the perspective of a geologist. Offers many examples from around the world. Covers the newly adopted soil profile description symbols in the early chapters. Includes a very elementary discussion of soil taxonomy in the early chapters. Suitable for university-level readers, but the early chapters are easily understood by high-school-level readers.

Brady, Nyle C., and Ray R. Weil.. *The Natures and Properties of Soils.* 14th ed. Upper Saddle River, N.J.: Prentice Hall, 2007. The most complete book on soil science available. Covers soil composition, formation, and reactions, and soil's effect on the world's food supply. Contains ample maps, diagrams, and photographs, as well as a complete glossary. For college-level students.

Buol, S. W., F. D. Hole, and R. J. McCracken. *Soil Genesis and Classification.* 5th ed. Ames: Iowa State University Press, 2003. An update of a classic soil-science text. Includes an update on the new soil-horizon symbol nomenclature. Includes an in-depth discussion on soil development. Suitable for university-level readers.

Carr, Margaret, and Paul Zwick. *Smart Land-Use Analysis: The LUCIS Model.* Redlands, Calif.: ESRI Press, 2007. Discusses the use of ArcGIS software to analyze land suitability. Provides a step-by-step guide to using ArcGIS software. Explains the LUCIS model with case studies to provide examples of potential conflicts of land use.

Clarke, G. R. *The Study of Soil in the Field.* 5th ed. Oxford, England: Clarendon Press, 1971. A comprehensive introduction to soil science. Includes such areas as soil site characteristics, sampling, surveys, and evaluation. Charts and diagrams clarify some of the esoteric terminology. Includes a short section of references. For college-level students.

Kellogg, Charles E. *Our Garden Soils*. New York: Macmillan, 1952. Intended as a manual for gardeners. Explains the composition and characteristics of various types of soil. Introduces the general reader to the terminology that scientists use to describe soils.

_____. *The Soils That Support Us*. New York: Macmillan, 1956. A comprehensive and easy-to-read introduction to soil science, with an emphasis on the relationship between people and soil. Includes photographs, charts, maps, and several useful appendices. For high school and college students.

Landa, Edward R., and Christian Feller, eds. *Soil and Culture*. New York: Springer, 2010. Provides a full history of the relationship between humans and the earth. Organized into parts discussing soil's function in art, philosophy, culture, and health, among others. Provides interesting information from an anthropological perspective.

Paton, T. R., G. S. Humphreys, and P. B. Mitchell. *Soils: A New Global View*. New Haven, Conn.: Yale University Press, 1995. Aimed at university-level readers. Easily comprehensible and offers many good examples of soils and soil formation from around the world. Color illustrations, maps, bibliography, and index.

Retallack, G. J. *Soils of the Past*. 2d ed. New York: Blackwell Science, 2001. An excellent source of information on paleosols. Discusses soil classification and formation, identifying paleosols and their characteristics, and the factors involved in paleosol formation. Covers soil fossil records with discussion of ancient landscapes and ecology.

Simonson, Roy W. "What Soils Are." In *Soil*, edited by the U.S. Department of Agriculture. Washington, D.C.: Government Printing Office, 1957. Explains soil formation and composition in a somewhat technical but very comprehensive way. Includes a chart and diagram. For the college-level student.

Sparks, Donald S., ed. *Soil Physical Chemistry*. 2d ed. Boca Raton, Fla.: CRC Press, 1999. An excellent introduction to soil chemistry. Offers a clear look into the practices, procedures, and applications of the field and is easily understood by the careful reader.

Troeh, Frederick R., and Louis M. Thompson. *Soils and Soil Fertility*. 5th ed. New York: Oxford University Press, 1993. Appropriate for an introductory college course in soils for students of agriculture. Covers every aspect of soil, including the physical properties of soil and soil conservation. Highly readable and well supplemented with charts, illustrations, and diagrams.

See also: Desertification; Expansive Soils; Land Management; Landslides and Slope Stability; Land-Use Planning; Land-Use Planning in Coastal Zones; Soil Chemistry; Soil Erosion; Soil Formation; Soil Profiles.

SOLAR POWER

Solar power is seemingly boundless, but it has yet to be established as a major energy resource for modern civilization. A variety of technologies exist or are under development to harness that power. Many of these technologies work very well; the problem, however, is cost.

PRINCIPAL TERMS

- **barrel:** a measure of energy consumption, 1 barrel is equal to the energy in an average barrel (42 gallons) of crude oil; about 5.8 million Btu
- **Btu (British thermal unit):** the amount of heat necessary to raise the temperature of 1 pound of water 1 degree Fahrenheit; equivalent to about 0.25 calorie
- **insolation:** the radiation from the sun received by a surface; generally expressed in terms of power per unit area, such as watts per square meter
- **kilojoule:** a unit of electrical energy equivalent to the work done to raise a current of electricity flowing at 1,000 amperes for 1 second (1,000 coulombs) by 1 volt; equivalent to 4.184 calories; approximately the energy needed to raise 100 kilograms one meter
- **kilowatt:** 1,000 watts, or about 1.34 horsepower
- **megawatt:** 1 million watts, or about 1,340 horsepower
- **photolytics:** the technology that makes use of sunlight's ability to alter chemical compounds in ways that can produce energy, fuels, or both
- **photovoltaics:** the technology employed to convert radiant solar energy directly into an electric current, using devices called solar cells
- **quad:** 1 quadrillion Btu; equivalent to 8 billion barrels of gasoline

INSOLATION

The sun's hydrogen fusion reaction (the same as humankind's hydrogen bomb) produces a relatively steady and uninterrupted 380 billion kilowatts of energy. This energy is released into space at the solar surface at a rate of over 56 million watts per square meter. Spreading out radially as it travels outward, this power is greatly reduced by distance. By the time it reaches Earth, 1 square meter of the surface receives about 1.1 kilowatts. Still, the total of the sun's energy received on the planet is about 85 trillion kilowatts. The portion that falls on the United States amounts to five hundred times more energy than the nation's total need. If the United States could tap into this energy with only 10 percent efficiency, it could meet all of its energy requirements from what falls on merely 2 percent of its surface. The opportunity is similarly great for most countries.

Although the human inhabitants of Earth are only beginning to learn how to harness this power, other life-forms have been using it for at least 2 billion years. The solar energy consumed by plants through photosynthesis is enormous. Within the United States, the amount of solar energy used by vegetation equals at least 20 million barrels of oil per day. Solar energy is the power that drives Earth's weather and stirs ocean currents; therefore, any utilization of wind power, ocean thermal power, or the energy stored in plants (whether burning firewood or fuels distilled from biomass) is indirectly an application of solar power. For that matter, fossil fuels represent solar energy stored in plants and animals that lived hundreds of millions of years ago, but such broad interpretations of solar power have little operational meaning. In the context of this discussion, solar power will mean insolation—the radiant energy of the sun as it falls on Earth's surface.

CHALLENGES TO UTILIZATION

The practical application of solar power to modern civilization's energy requirements presents some difficult challenges. One problem is that solar energy, although abundant and widely distributed, is not concentrated, whereas humankind's energy demands tend to require a considerable amount of power at very confined locations. Virtually all of North America south of the 48th parallel receives more than 12,000 kilojoules of solar energy per square meter of its surface every day, with large areas of the West and South enjoying more than 16,000 kilojoules. Yet it takes about 55,000 kilojoules to heat the hot water required by a family of four; this power application alone theoretically requires 3 to 4 square meters of collection device to satisfy one small family's demand for water for cooking, washing, and bathing. In actuality, it requires closer to 10 square meters, because the collection devices operate at far below 100 percent efficiency.

Another serious problem in utilizing solar power is the pattern of its availability. Solar power at a given site peaks every day as the sun passes overhead and disappears entirely during the hours of darkness. Seasonal variations in the location of the sun affect the number of hours of energy received, and as distance from the tropics increases, the incoming sunlight strikes at ever more acute angles, further diminishing the power available per square meter.

Finally, cloud cover seriously interferes with solar energy reception, making overcast days almost as unproductive as nighttime. The average number of cloudy days per year at a given location is a factor as significant as latitude. For example, sunny, equatorial Kenya enjoys almost twice the insolation as cloudy Nigeria, only 6 degrees of latitude away. Relatively sunny Washington, D.C., at 38 degrees north latitude, has more insolation to work with than Nigeria.

These problems notwithstanding, it has been demonstrated that solar power can be successfully harnessed over a wide area on Earth's surface. Various schemes to collect and concentrate the energy and manage its utilization and storage to match availability to demand are already at a practical stage of development. Moreover, public opinion is strongly favorable toward solar power as an alternative to the continued use of fossil fuels or the large-scale development of nuclear power. Experts continue to caution that widespread utilization of solar power is still in the future.

Thermodynamic and Photovoltaic Potentials

The essence of the energy problem facing the world is not an energy shortage but the cost of deriving and distributing "high-quality" energy from abundant, renewable, but "low-quality" sources. Quality, as traditionally applied to energy, means temperature first and foremost. The value of a unit of energy depends greatly on the temperature at which it can be delivered because, under the laws of thermodynamics, a unit at high temperature can do more useful work than a low-temperature unit. As it falls upon Earth's surface, solar energy is considered a low-quality thermal resource: It is fine for making "sun tea" and heating swimming pools, but it will neither spin the wheels of industry nor light a modern,

industrialized society. The fact that it is pure, clean energy creating no harmful by-products for the environment is of little consequence if it cannot do the work required of it.

Fortunately for the future, solar power can be collected and concentrated to raise its thermal quality. In fact, solar energy's thermodynamic potential (the highest heat at which it can be realistically supplied) is about 5,000 degrees Celsius, which is 3,000 degrees higher than the thermodynamic potential of conventional nuclear power and equal to the theoretical maximum heat available from the complete combustion of pure carbon in pure oxygen. Moreover, solar radiation is able to generate energy in other forms than heat. The most important of these to modern civilization is its photovoltaic potential—that is, its ability to mobilize electrons in various semiconductor materials, thereby creating an electrical current. Sunlight is also capable of splitting water molecules into hydrogen and oxygen, a fact of enormous significance to more advanced power technologies.

Thermal Systems

Solar power technologies already available are quite varied but can be grouped generally as thermal, photovoltaic, and photolytic. Thermal schemes are those best known and understood by the average consumer, because they apply solar power to energy needs in traditional ways. Photolytic technology is still largely in the future.

Thermal systems that require no mechanisms are termed passive designs. They rely on the choice of materials and the size and careful placement of building elements to control the flow of heat within the living space. Typically, large, south-facing glass walls admit solar heat in winter but are shadowed by deep roof overhangs in summer. Interior design facilitates the natural convective flow of warm and cold air and the storage of energy. In favorable locations, good passive thermal architecture can provide as much heating and cooling as an efficient building needs.

"Flat plate" collectors used to heat water are an excellent example of active solar thermal technology. Water is pumped through piping inside a collector designed to gather and hold the solar heat that strikes it. As the water circulates through the collector, its temperature may be raised to as much as 100 degrees Celsius. Such systems most commonly supply domestic hot water but may also be used for space heating and cooling of residences and larger buildings. (Refrigerated cooling with solar-heated water involves a process different from mechanical refrigeration, but is not a new idea. Absorption refrigeration, as it is called, was used in early refrigerators.) Other active solar thermal systems for space heating and cooling use the flow of water over building surfaces to add or subtract heat as needed.

Another application of active solar thermal power is in so-called heat engines, which use the expansion, contraction, or evaporation of a fluid or gas to obtain mechanical motion. A number of designs exist, dating back to the early eighteenth century, but all have suffered from low efficiency until now, when they are being interfaced with advanced thermal collectors whose much higher temperatures provide the efficiency needed. These designs promise to be increasingly important in both terrestrial and space applications.

Also requiring much higher temperatures (from 250 to more than 500 degrees Celsius) are schemes that use the sun to create superheated steam to operate turbine-powered generators. "Solar farms," such as a 194-megawatt plant in the Mojave Desert 225 kilometers northeast of Los Angeles, use parabolically shaped troughs lined with mirror strips to produce temperatures of 375 degrees Celsius. A working fluid is pumped through the collectors and then to a central heat exchanger to create steam for a generating plant. Another approach, first introduced in Italy, involves reflecting the insolation captured by an array of mirrors to a central "power tower," within which the sum of all the reflected insolation produces operating temperatures in excess of 1,000 degrees Celsius.

PHOTOVOLTAICS

Photovoltaics—the direct generation of electricity from sunlight—was pioneered in the 1950's as an offshoot of semiconductor research. As the space age dawned, devices called solar cells became familiar as the power sources for many satellites. Their very high cost and low efficiency were of little consequence in these applications, as alternatives were either even more expensive or nonexistent. Their high cost stemmed partly from the fact that the silicon disks had to be obtained from ultrathin slices of pure silicon crystal, laboriously grown in laboratories

to avoid contamination. Even the slicing operation added significantly to the cost, because half of every crystal produced was consumed by the saw cuts. Low efficiencies compounded the problem, as a great many of the expensive cells were needed to provide even minimal current. Yet persistent research into new materials and techniques has been changing this gloomy picture. In fact, photovoltaic power generation, though not yet economically competitive with traditional energy sources, has dropped so much in cost that a few major utility companies have put photovoltaic generating stations online to supplement power from other sources and pave the way for more widespread use in years to come. Photovoltaic panels are now commonly used to store power for outdoor ornamental lighting and are common sights on highways as power sources for remote monitors and emergency telephones. In 2009, more than 100,000 photovoltaic systems were installed in the United States.

A more visionary application for photovoltaics is the Satellite Solar Power Station (SSPS), for which designs but no serious implementation plans exist. Such a satellite would be a gigantic platform in geosynchronous orbit, giving large arrays of solar cells twenty-four-hour-a-day exposure to solar energy undiluted by passage through the atmosphere. The power would be beamed back to receiving sites on Earth as microwaves and there reconverted to electricity. Obviously, such concepts pose severe hazards, such as the loss of proper alignment of the satellite power transmitter.

PHOTOLYTIC TECHNOLOGY

The most challenging of all solar energy technologies may be ultimately the most rewarding. Research has shown that a small percentage of photons from the sun are energetic enough to split water molecules, from which three atoms—two hydrogen and one oxygen—are liberated. If this process becomes feasible on a large scale, oxygen and hydrogen could be readily stored and either burned as fuels or recombined as water in a reaction that releases electricity. Again, the space age has shown the way through development of "fuel cells," which combine hydrogen and oxygen with almost 100 percent efficiency, yielding significant amounts of fresh water and electricity in the process. If photolysis, as it is called, could draw upon the abundant salt water of the oceans to produce hydrogen, oxygen, and electricity—with fresh water as an added benefit—humanity's quest for acceptable energy sources would move beyond "renewable" to involve environmentally enhancing power.

COMPETITIVENESS WITH OTHER ENERGY SOURCES

The energy crisis precipitated by the Arab oil embargo of 1973 forced industrialized nations to make major adjustments in energy use and policy. Through the Energy Research and Development Administration (ERDA) and the Solar Energy Research Institute (SERI), both created in 1974, the United States began a program of government-sponsored research designed to help meet an announced goal of 10 quads of energy from solar power by the year 2000—enough to replace about 5 million barrels of oil daily. The oil embargo, however, also stimulated a sharp, but unfortunately temporary, reduction in the demand for energy. After years of steady 6 percent annual increases, new energy construction almost came to a standstill. On top of this, oil prices fell sharply again in the 1980's, and environmental constraints on the use of fossil fuels were eased. The net effect was a loss of urgency in the quest for solar power, and federal assistance for research and application was cut to a small fraction of previous levels. Meanwhile, serious problems developed in the fledgling solar power industry, as complaints about inflated pricing and exaggerated savings undermined public confidence and legislative support. A large number of the young solar energy ventures that were launched in the 1980's failed.

Solar energy has yet to make a significant impact on energy consumption for several reasons, including the high initial costs of solar heating systems (a large active space heating system costs up to 20 percent of the total cost of a residence), the tendency of building contractors to be conservative, and the fact that fossil fuels remained relatively inexpensive during the 1990's. Since approximately one-quarter of the United States' annual energy consumption is for heating and cooling, solar energy could have a significant impact. In most regions of the United States, solar energy can best be employed for passive space heating and domestic hot water systems. Active space heating systems are feasible only in areas such as Colorado and Utah, where the winters are cold and sunny. As fossil fuels become depleted and their prices continue to rise

in the twenty-first century, however, solar power is expected to become an increasingly viable energy resource.

Richard S. Knapp

FURTHER READING

Andrews, John, and Nick Jelley. *Energy Science: Principles, Technologies, and Impact.* New York: Oxford University Press, 2007. Discusses various forms of energy, their environmental and socioeconomic impacts, and the principles of energy consumption, along with information on the generation, storage, and transmission of energy. Many mathematical examples require a strong mathematics or engineering background. Contains useful information for undergraduates and professionals with interest in energy resources.

Behrman, Daniel. *Solar Energy: The Awakening Science.* London: Routledge & Kegan Paul, 1979. Written in a popular and anecdotal style that includes attention to the personalities of leading solar proponents, as well as their ideas. A more detailed discussion of solar power. Illustrated with photographs and contains a good index.

Boyle, Godfrey, ed. *Renewable Energy.* 2d ed. New York: Oxford University Press, 2004. Provides a complete overview of renewable energy resources. Discusses various energy resources, including solar energy, as well as the basic physics principles, technology, and environmental impact of renewable energies. Includes references and a further reading list for each chapter. Advanced technical details of power supplies are limited.

Brinkworth, B. J. *Solar Energy for Man.* New York: John Wiley & Sons, 1973. Gives a clear explanation of the principles of solar power from a standpoint that precedes the "energy crisis." Well illustrated and suitable for readers with some grounding in general science.

Camacho, E. F., Manuel Berenguel, and Francisco R. Rubio. *Advanced Control of Solar Plants.* New York: Springer, 1997. A thorough examination of solar power plants, their operation, and the safety systems that govern them. Illustrations, index, and bibliographical references.

D'Alessandro, Bill. "Dark Days for Solar." *Sierra Magazine* (July/August, 1987): 34. Details the profound damage done to America's emerging solar power industry in the 1980's by the combination of low-cost oil, elimination of tax credits, loss of government funding for research, and the industry's own bad business practices.

Glaser, Peter E., Frank P. Davidson, Katinka I. Csigi, et al., eds. *Solar Power Satellites: The Emerging Energy Option.* New York: E. Howard, 1993. One volume in a space science and technology series. Discusses the solar energy option, its possibilities, and future uses. A good introduction to alternative energy sources.

Halacy, D. S. *Earth, Water, Wind, and Sun: Our Energy Alternatives.* New York: Harper & Row, 1977. Appropriate for readers seeking a nontechnical overview of the possibilities and problems of solar power. Also recommended for high school readers and for those wanting a brief survey of the field of renewable energy sources.

Landsberg, Hans H., et al. *Energy: The Next Twenty Years.* Cambridge, Mass.: Ballinger, 1979. Provides a historical perspective of renewable energy sources, and the potential that scientists of that time recognized. Devotes one chapter to solar energy, with emphasis on policy recommendations intended to advance its development in the short and long term. More of a "why to" than a "how to" examination of solar power.

Letcher, Trevor M., ed. *Future Energy: Improved, Sustainable and Clean.* Amsterdam: Elsevier, 2008. Begins with a discussion of the future of fossil fuels and nuclear power. Discusses such renewable energy supplies as solar, wind, hydroelectric, and geothermal power. Focuses on potential energy sources and currently underutilized energy sources. Includes a section covering new concepts in energy consumption and technology.

Mathews, Jay. "Solar Energy Complex Hailed as Beacon for Utility Innovation." *The Washington Post* (March 2, 1989): A25. Details the stunningly successful introduction of a large commercial solar plant in California, which sprang up just as most private ventures in solar power were on the brink of disaster.

Patel, Mukund R. *Wind and Solar Power Systems.* Boca Raton, Fla.: CRC Press, 1999. Provides a thorough examination and clear explanation of wind and solar energy plants, discussing their operations, protocol, and applications. Suitable for the reader without prior knowledge of alternative energy sources. Illustrations, maps, index, and bibliography.

See also: Earth Resources; Geothermal Power; Hydroelectric Power; Nuclear Power; Ocean Power; Tidal Forces; Unconventional Energy Resources; Wind Power.

STRATEGIC RESOURCES

A nation's strategic resources are those resources that are essential for its major industries, military defense, and energy programs. For the United States, these resources include manganese, chromium, cobalt, nickel, platinum, titanium, aluminum, and oil.

PRINCIPAL TERMS

- **alloy:** a substance composed of two or more metals, or a metal and certain nonmetals
- **catalyst:** a chemical substance that speeds up a chemical reaction without being permanently affected by that reaction
- **manganese nodules:** rounded, concentrically laminated masses of iron and manganese oxide found on the deep-sea floor
- **oil rights:** the ownership of the oil and natural gas on another party's land, with the right to drill for and remove them
- **ore deposit:** a natural accumulation of mineral matter from which the owner expects to extract a metal at a profit
- **proven reserve:** a reserve supply of a valuable mineral substance that can be exploited at a future time
- **salt dome:** an underground structure in the shape of a circular plug resulting from the upward movement of salt

WHAT MAKES RESOURCES "STRATEGIC"

Reference is frequently made to strategic resources but, unfortunately, there is no general agreement on what makes resources "strategic." Because the word "strategy" has a military connotation, strategic resources are often considered to be those resources that would be of critical importance in wartime. A somewhat broader definition of strategic resources is that they are those resources that a nation considers essential for its major industries, military defense, and energy programs. Similarly, there is no general agreement as to which of the many earth resources are strategic ones. Several authors have restricted the definition to metals that are in short supply, with some even limiting the definition to the six metals alloyed with iron in the making of steel. Others, taking a broader view, have included nonmetals among the strategic resources, including fertilizers and energy sources such as petroleum. One possible definition is that a strategic resource is any resource whose unavailability would adversely affect a nation's ability to function.

In listing those Earth resources that are strategically significant, it is important to realize that each nation's list will be different. In other words, a resource that is in critically short supply for one nation may be possessed in abundance by a second. Furthermore, what is considered to be a strategic resource today may not be considered one tomorrow or a hundred years from now. Before the 1970's, the opening of large Middle Eastern oil fields had driven the price of oil down to $1.30 per barrel and the price of gasoline at the pump as low as 20 cents per gallon. Excess production capacity continued until the early 1970's, when increasing political tensions in the Middle East finally resulted in a united front on the part of the members of the Organization of Petroleum Exporting Countries (OPEC). First came voluntary production cutbacks, then the Arab oil embargo of 1973 to 1974. Eventually, the price of oil reached $35.00 per barrel, and gasoline was selling for more than $1.50 per gallon, when it could be obtained at all. A resource that everyone had taken for granted suddenly became a strategic resource. During World War II, the Battle of Stalingrad and the attack on Pearl Harbor were both driven by the need of Germany and Japan to secure sources of oil.

PROTECTIONS AGAINST SHORTAGES

To ensure adequate oil supplies for the United States in future emergencies, the U.S. government authorized the establishment of the Strategic Petroleum Reserve. The purpose of this legislation was to purchase 1 billion barrels of oil and to store it in large caverns hollowed out of underground salt domes in coastal Texas and Louisiana. The creation of the Strategic Petroleum Reserve illustrates an important principle. Nations can protect themselves against a possible interruption in supplies of imported strategic resources by stockpiling them during peacetime. Shortages during World War I, for example, caused the U.S. Congress to pass the Strategic Materials Act of 1939 and to begin stockpiling tin, quartz crystals, and chromite—none of which are produced in significant amounts in the United States—in anticipation of the outbreak of another war.

Another way that a major industrial nation protects itself against possible wartime shortages of strategic materials is by arranging for access to supplies of these materials in the event of war. Frequently, this preparation will involve trade agreements with neutral nations or with nations to which a country is bound by political alliances. In extreme cases, it may be necessary to invade a neighboring nation in order to obtain access to strategic resources. The history of Alsace-Lorraine in Europe exemplifies such a situation. This important iron-mining district is traditionally French, but it shares a border with Germany. In both world wars, Germany occupied Alsace-Lorraine in order to assure access to these iron ore deposits.

A third way that a nation can ensure adequate supplies of strategic resources during wartime is to develop substitutes for scarce materials using cheap materials already available. Certain mineral resources, such as mercury and uranium, have unique properties, and for these resources, no satisfactory substitutes can be found. Some of the other metals can be synthesized from related elements but only by prohibitively expensive methods. A number of strategic resources, however, can be synthesized from cheap, readily available materials at comparatively low cost. One good example is quartz crystals, used in electronics and optics, which are now grown synthetically for industrial use. Abrasive diamonds are another material that is now made synthetically.

U.S. STRATEGIC RESOURCES IN SHORTEST SUPPLY

The four strategic resources that would be in shortest supply in the United States were it to go to war are manganese, chromium, cobalt, and nickel. Manganese is a soft, silver-gray metal that is essential in the making of steel. Up to 7 kilograms of manganese is necessary for the production of each ton of iron or steel, and no satisfactory substitute has ever been found. Manganese removes undesired quantities of oxygen and sulfur from the iron during the steel-making process, yielding a hard, tough product suitable for bridge steel, projectiles, and armor plating. U.S. manganese deposits are small, low in grade, and expensive to work; 95 percent of the country's manganese is imported.

Chromium is a hard, silver-gray metal that is also essential in the making of steel and for which no satisfactory substitute has been found. Chromium makes steel resistant to corrosion and appears on the shiny, chromium-plated surfaces on automobiles. The chromium content of stainless steel varies from 12 to 30 percent. Chromium-steel alloys are also used in aircraft engines, military vehicles, and weapons. The United States has large, low-grade deposits of chromium ore in Montana, but they are expensive to process. Consequently, 90 percent of its chromium is imported, primarily from South Africa.

Cobalt is a silver-white metal that is also needed in steel making. The addition of small quantities of cobalt makes steel harder and heat-resistant. Consequently, cobalt steel has important applications in the manufacture of metal-cutting tools, jet engines, and rockets. Cobalt alloys are magnetic, and the magnetism is retained permanently; therefore, these alloys are used for the manufacture of magnets. The United States has a small cobalt production from low-grade ores in Missouri, but 90 percent of its cobalt is imported, primarily from the country of Zaire, in Africa.

Nickel is a nearly white metal that is another important alloying agent in the making of steel. Nickel steels do not corrode or rust, so large quantities of nickel are used in the manufacture of stainless steel. Nickel is also used in plating because of its shine, in coinage (as in the familiar nickel coin), and for alloys that have important applications in the defense industry. The United States has very few high-grade nickel ore deposits, but there is a large, low-grade deposit in Minnesota. About 80 percent of its nickel is imported from Canada, the world's largest nickel producer.

Additional mineral resources that might be in short supply in the United States during wartime include platinum, titanium, and aluminum. Platinum has valuable chemical properties, because it resists corrosion and acts as a catalyst to speed chemical reactions. Titanium is a silver-gray metal used as an alloy. Because it imparts great strength, heat resistance, and resistance to corrosion, it is used in the construction of supersonic aircraft, jet engines, and space capsules. The major use for aluminum in the United States is in beverage cans, 102 billion of which were produced in 1998. Future resources that may become critically short include rare-earth elements, which are important in electronics. Many nations produce rare-earths, but virtually all of the world's refining capacity is in China. The United States is 100 percent dependent on imports for rare-earth elements. Another resource that might become critical in the future is lithium, which is essential for rechargeable batteries.

ESTIMATING RESERVES

Strategic resources can be studied in a variety of ways. The first is to identify which of the various Earth resources are in shortest supply. That is done by analyzing production and usage figures for each of a nation's industrially important minerals in order to determine how much the nation relies on imports for each of these materials. When usage exceeds production, the nation is relying on imports. The greater the reliance on imports, the more strategic the resource becomes.

A second way in which one can study strategic resources is to identify resources that are adequate at present but that may become scarce in the future; this is done by estimating reserves. A reserve is a supply of a mineral substance that still remains in the ground and is available to be extracted at some future time. Two types of reserve can be distinguished: proven reserves and undiscovered reserves. Proven reserves are reserves that have already been outlined by drilling or some other means; there is practically no risk of the desired substance's not being there. Undiscovered reserves, in contrast, are reserves that are believed to be present on the basis of geologic studies but are still inadequately explored.

SEARCH FOR EXPANDED SUPPLIES

A third way of studying strategic resources is to begin a search for expanded supplies before current reserves are exhausted or before imports have become unavailable because of wartime conditions. Various methods have been employed for increasing the supply of a strategic resource. They include stockpiling, trade agreements (or territorial annexation, in extreme cases), manufacture of synthetics or the substitution of other substances, and the development of conservation or recycling programs.

Two additional avenues have also become available in the search for expanded supplies of strategic resources. One is the use of the plate tectonics theory, which proposes that Earth's surface is divided into a few large plates that are slowly moving with respect to one another. Intense geologic activity occurs at plate boundaries, and many mineral deposits are believed to have been formed by such activity. Much exploration for new deposits of strategic resources is being concentrated at plate boundaries. A second new avenue is the exploration of the deep-sea floor. Although manganese nodules were discovered on the deep-sea floor by the *Challenger* expedition in the late 1800's, it was not until the advent of manned submersibles and remote-controlled television cameras that it was realized that 15 percent or more of the ocean floor may be covered by such nodules. They have valuable amounts of iron, nickel, manganese, cobalt, and other strategically important substances. As exploration of the deep-sea floor continues, other supplies of strategic resources may be found there as well.

CONSERVATION MEASURES

Because a nation's strategic resources would be of critical importance in wartime, one would expect not to hear much about them in peacetime. They are substances that must be imported, however, and excessive imports lead to trade deficits. As a result, a government may resort to conservation measures in order to reduce such imports. A good example is the American program for recycling aluminum beverage cans. This program has two objectives: to reduce U.S. imports of costly aluminum ore, and to conserve the large amounts of electricity needed to process aluminum ore.

In troubled times, such as the 1973-1974 Arab oil embargo and the ensuing energy crisis, measures aimed at conserving strategic resources become very noticeable to the general public. To reduce oil consumption, Americans were asked to turn down their thermostats, the nationwide speed limit was reduced to 55 miles per hour, and automobile companies were told to improve the gas mileage of their cars, which led to a whole new generation of midsized cars. Furthermore, ripple effects caused by the oil shortage spread through the economy, triggering a recession, costing people their jobs, and setting off a stock market decline. Overseas, the value of the dollar declined against other currencies, and soon American tourists found themselves paying more for a hotel room or a meal.

In wartime, the need to conserve strategic resources may even result in rationing. This happened in the United States during World War II. Each driver was given a ration card entitling him or her to purchase a certain number of gallons of gasoline each week; drivers were placed in different categories depending on whether they drove for pleasure or to work.

Donald W. Lovejoy

FURTHER READING

Boyle, Godfrey, ed. *Renewable Energy.* 2d ed. New York: Oxford University Press, 2004. Provides a complete overview of renewable energy resources. Discusses specific energy resources, as well as their basic physics principles, technology, and environmental impact. Includes references and a further reading list with each chapter.

Carlson, Diane H., Charles C. Plummer, and Lisa Hammersley. *Physical Geology: Earth Revealed.* 9th ed. New York: McGraw-Hill Science/Engineering/Math, 2010. A thorough and well-illustrated introduction to physical geology. Bibliography and index.

Craig, James R., David J. Vaughan, and Brian J. Skinner. *Earth Resources and the Environment.* 4th ed. Upper Saddle River, N.J.: Prentice Hall, 2010. Provides an excellent overview of Earth's metallic, nonmetallic, and energy resources. Contains helpful line drawings, maps, tables, and charts; few photographs. Includes suggestions for further reading and a list of principal ore minerals and their production figures for 1982. Suitable for general readers.

_____. *Resources of the Earth.* 3d ed. Upper Saddle River, N.J.: Prentice Hall, 2001. Includes numerous black-and-white photographs, color plates, tables, charts, maps, and line drawings. Contains excellent sections on strategic resources, international conflicts, and the reliance of the United States on imports. Suitable for college-level readers.

Davidson, Jon P., Walter E. Reed, and Paul M. Davis. *Exploring Earth: An Introduction to Physical Geology.* 2d ed. Upper Saddle River, N.J.: Prentice Hall, 2001. An introduction to physical geology and the systems of Earth. Explains the composition, history, and constant changes of Earth. Includes colorful illustrations and maps. Appropriate for general readers.

Fridleifsson, Gudmundur O., and Wilfred A. Elders. "The Iceland Deep Drilling Project: A Search for Deep Unconventional Geothermal Resources." *Geothermics* 34 (2005): 269-285. Written for readers with a background in engineering or geophysics. Provides a good overview of common issues in deep-drilling and focuses on the benefits of harnessing geothermal power.

Grotzinger, John, et al. *Understanding Earth.* 5th ed. New York: W. H. Freeman, 2006. A general text on all aspects of geology. Suitable for advanced high school and college students.

Gupta, Chiranjib, and Harvinderpal Singh. *Uranium Resource Processing: Secondary Resources.* Berlin: Springer-Verlag, 2010. Discusses the topics of metallurgy, mineralogy, and leaching as applied to uranium deposits, as well as the recovery of uranium from phosphates and other resources. Includes a further reading list with each chapter, and a complete subject index.

Illinois State Geological Survey. *How to Read Illinois Topographic Maps.* Champaigne, Ill.: Illinois Department of Natural Resources. 2005. Written specifically for Illinois but applicable to any topographic map. Provides an excellent overview of features common to topographical maps. Numerous images, complete with labels, enhance the utility of this reference material.

Jensen, John R. *Remote Sensing of the Environment: An Earth Resource Perspective.* 2d ed. Upper Saddle River, N.J.: Prentice Hall, 2006. Discusses the principles, instruments, and methodology of remote sensing and its use in geology. Analyzes specific resources, such as vegetation, water, soil and rocks, and geomorphic features, through remote sensing. Provides practical information for use by geologists and graduate students.

Jensen, M. L., and A. M. Bateman. *Economic Mineral Deposits.* 3d ed. New York: John Wiley & Sons, 1981. Provides detailed information on metallic and nonmetallic mineral deposits and their modes of formation. Discusses the history of mineral use, the exploration and development of mineral properties, and the role of strategic minerals in international relations and war. For a college-level audience.

Klein, Cornelis, and C. S. Hurlbut, Jr. *Manual of Mineralogy.* 23d ed. New York: John Wiley & Sons, 2008. Provides detailed descriptions of the various metallic and nonmetallic minerals, including their crystallography, physical properties, and chemical characteristics. Includes helpful information on the mode of formation of the various minerals, where they are found, and why they are useful. Includes summary tables.

Lennox, Jethro, ed. *The Times Comprehensive Atlas of the World*. 13th ed. New York: Times Books, 2001. Includes an overview of world resources, followed by large, eight-color maps. The world mineral map shows the global distribution of metals and nonmetals. The world energy map shows the distribution of energy sources and their comparative consumption by nations. Suitable for high school readers.

Newson, Malcolm. *Land, Water and Development: Sustainable and Adaptive Management of Rivers*. 3d ed. London: Routledge, 2008. Presents land-water interactions. Discusses recent research, study tools and methods, and technical issues, such as soil erosion and damming. Suited for undergraduate students and professionals. Covers concepts in managing land and water resources in the developed world.

Pernetta, John, ed. *The Rand McNally Atlas of the Oceans*. Skokie, Ill.: Rand McNally, 1994. A beautifully illustrated atlas with subsections on various energy sources in the ocean, the continental shelves, manganese nodule deposits on the deep-sea floor, and offshore oil. Includes color photographs, maps, and line drawings. Suitable for high school readers.

Tarbuck, Edward J., Frederick K. Lutgens, and Dennis Tasa. *Earth: An Introduction to Physical Geology*. 10th ed. Upper Saddle River, N.J.: Prentice Hall, 2010. Provides a clear picture of Earth's systems and processes that is suitable for the high school or college reader. Includes illustrations, graphics, and an accompanying computer disc. Bibliography and index.

Tennissen, Anthony C. *The Nature of Earth Materials*. 2d ed. Englewood Cliffs, N.J.: Prentice-Hall, 1983. Contains detailed descriptions of 110 common minerals, with a black-and-white photograph of each. Includes helpful sections on the distribution of mineral deposits and the utilization of various Earth materials. Suitable for college students.

Thompson, Graham R., and Jonathan Turk. *Introduction to Physical Geology*. Belmont, Calif.: Brooks/Cole, Cengage Learning, 1997. Provides an easy-to-follow look at what physical geology is and the phases involved. Shows the reader each phase of Earth's geochemical processes. Excellent for high school and college readers. Illustrations, diagrams, and bibliography included.

See also: Clays and Clay Minerals; Earth Resources; Earth Science and the Environment; Evaporites; Hazardous Wastes; Landfills; Mining Processes; Mining Wastes; Nuclear Power; Nuclear Waste Disposal; Pyroclastic Rocks.

SUB-SEAFLOOR METAMORPHISM

Sub-seafloor metamorphism of oceanic ridge basalts by magma-driven, convecting seawater induces significant changes in the chemical composition of both rock and circulating fluid. These changes are a major factor in the exchange of elements between the lithosphere and the hydrosphere and play a critical role in the origin of certain ore deposits.

PRINCIPAL TERMS

- **black smokers:** active hydrothermal vents along seafloor ridges, which discharge acidic solutions of high temperature, volume, and velocity, charged with tiny black particles of metallic sulfide minerals
- **convection:** fluid circulation produced by gravity acting on density differences arising from unequal temperatures within a fluid; the principal means of heat transfer involving fluids of low thermal conductivity
- **facies:** a part of a rock or a rock group that differs from the whole formation in one or more properties—for example, composition
- **metasomatism:** chemical changes in rock composition that accompany metamorphism
- **ophiolite complex:** an assemblage of metamorphosed basaltic and ultramafic igneous rocks that originate at marine ridges and are subsequently emplaced in mobile belts by plate-collision tectonics
- **pillow basalt:** a submarine basaltic lava flow in which small cylindrical tongues of lava break through the surface, separate into pods, and accumulate downslope in a formation resembling a pile of sandbags
- **regional metamorphic facies:** the particular pressure and temperature conditions prevailing during metamorphism as recorded by the appearance of a new mineral assemblage
- **thermal gradient/geothermal gradient:** the rate of temperature increase with depth below the earth's surface
- **water-to-rock ratio:** the mass of free water in a given volume of rock divided by the rock mass of the same volume; processes occurring at water-to-rock ratios less than 50 are "rock-dominated," while those greater than 50 are "fluid-dominated"

IDENTIFICATION OF THE PROCESS

Seafloor spreading generates an estimated 3×10^{13} kilograms of new basaltic crust per year along the world's ocean ridge system. Magma rising from the mantle produces this young ocean-floor crust (termed "juvenile"), which consists of upper layers of permeable lavas and lower layers of dike networks cutting gabbroic intrusive bodies. The thermal gradients and permeabilities of rocks in the vicinity of ocean ridges are both high, resulting in an environment favorable for the rapid circulation of seawater and convective cooling of rocks. Intensive studies of heat-flow patterns across ocean ridge systems in the early 1970's confirmed that large-scale seawater circulation through hot basaltic crust is actually occurring. During the same period, advances in structural and chemical studies of ophiolite complexes on land led geologists to conclude that these peculiar rock sequences are fragments of oceanic crust and upper mantle that were tectonically emplaced by plate collisions.

A principal line of evidence used to support this conclusion is the fact that ophiolitic basaltic rocks had been metamorphosed and intensely veined prior to emplacement. The ophiolitic metabasalts were, in fact, identical in all respects to samples obtained in dredge hauls along the axial valleys of ocean ridges. As a result, a new form of rock metamorphism, regional in extent but restricted to the marine ridge environment, was recognized. The first scientific paper describing this phenomenon in a comprehensive and unified fashion appeared in 1973; this landmark paper referred to the newly recognized metamorphic process as sub-seafloor metamorphism. For brevity, the acronym SFM will be used.

It is important to recognize that SFM is a process involving both heat and mass transfer. Geochemists were quick to realize that large-scale, continuous alteration of ridge basalts by convecting seawater could have profound effects on the exchange of chemical elements between the earth's lithosphere and hydrosphere. This hypothesis was tested by direct sampling of discharge fluids from submarine hot springs in the Galápagos Rift and the East Pacific in the late 1970's. Compared with normal seawater, these hot springs vent a fluid that is distinctly acidic; strongly depleted in magnesium (Mg) and sulfate (SO_4); enriched in silica (SiO_2), calcium (Ca), and hydrogen sulfide (H_2S); and enriched in a wide range of metallic elements. These compositional differences are the result of rock-seawater interaction and prove that SFM involves large-scale metasomatism. This is in marked contrast to regional metamorphism in continental mobile belts, which is a process that involves much less change in chemical composition.

REGIONAL METAMORPHIC FACIES

Samples recovered by dredging the steep escarpments of submarine ridge valleys represent the common regional metamorphic facies (zeolite, prehnite-pumpellyite, greenschist, and amphibolite), but the vast majority record the conditions of the greenschist facies. Most samples are metabasalts with mineralogy indicative of a temperature range of 100 to 450 degrees Celsius and very low pressures (400 to 500 bars) relative to continental metamorphism. The original mineralogy of ridge basalts, which consists of tiny crystals of calcic plagioclase, clinopyroxene, and olivine set in a glassy matrix, is converted by SFM to complex mixtures of actinolite, tremolite, hornblende, albite, chlorite, epidote, talc, clay, quartz, sphene, and pyrite that tend to preserve the original igneous texture of the rock. Such assemblages are well known in basaltic rocks of continental mobile belts and ophiolite complexes, and they record greenschist facies conditions. For the majority of samples, greenschist facies metamorphism has produced an alteration assemblage that consists of albite, actinolite, chlorite, and epidote and is accompanied by a small amount of quartz and pyrite.

The intense hydrothermal veining observed in some dredge samples and especially in ophiolite complexes appears to record former high fracture density and permeability, which is necessary for extensive basalt-seawater interaction. The major vein minerals are chlorite, actinolite, epidote, quartz, pyrite, and, less commonly, sulfides of iron, copper, and zinc. On the basis of mineralogy, greenschist facies alteration is described as chlorite-rich (chlorite greater than 15 percent of rock, epidote less than 15 percent) or epidote-rich (epidote greater than 15 percent, chlorite less than 15 percent). Chlorite-rich alteration is dominant in the dredge-haul samples recovered so far, and this preponderance is taken to mean that the chlorite-forming reactions occur at lower temperatures and closer to the seawater-basalt interface, where seawater influx of the basaltic crust begins. In other words, the transition from chlorite-rich to epidote-rich alteration records prograde metamorphism that, in the case of ridge basalts, correlates with increasing depth and temperature.

MAGNESIUM METASOMATISM

The mechanics of ridge-basalt emplacement ensure that all such rocks have an opportunity to react with seawater and undergo metasomatism. Although the chemical composition of average midocean-ridge basalt (MORB) is known, it is unlikely that it reflects exactly the pristine composition of the initial magma. The metasomatic nature of SFM has mainly been determined on the basis of dredge samples of pillow basalt. Individual pillows typically exhibit intensely altered rims and cores of relatively fresh basalt. On this basis, it is known that chlorite-rich alteration shows greater departures from parent-rock composition than does the higher-grade epidote-rich alteration. Chlorite-rich samples exhibit significant gains of magnesium and corresponding losses of calcium, but their state of oxidation is, for the most part, unaffected by the alteration process. Epidote-rich samples generally undergo minor losses of magnesium and gains of calcium but become somewhat oxidized by the alteration reactions.

Evidently, magnesium is removed from seawater and held in magnesium-rich minerals (especially chlorite) at the water-rock interface. The extent of this reaction depends on both temperature and water-to-rock mass ratio. At water-to-rock ratios less than 50 (in other words, rock-dominated conditions), magnesium is completely removed from seawater, the reaction rate being proportional to temperature. The mineral reactions create acidic conditions as long as magnesium remains in solution to drive the reaction. For rock-dominated conditions, however, magnesium is soon depleted and the acidic conditions associated with the reaction are short-lived. However brief, the acidic stage is important because it permits the circulating seawater to leach metallic elements from the enclosing basalt. At water-to-rock ratios in excess of 50 (that is, fluid-dominated conditions), magnesium cannot be completely removed from solution, so that the acidic, metal-leaching conditions of the reaction are maintained indefinitely. Calcium, sodium, and potassium are dissolved from basalt because these elements are not utilized in the formation of chlorite. In contrast, iron and aluminum are retained in the rock because they participate in the formation of chlorite. The rate and magnitude of the reaction both increase with temperature and, because thermal gradients near ridges are very high, it would be expected that magnesium metasomatism would be restricted to relatively shallow crustal depths. This limitation exists because fluids downwelling from the crust-seawater interface would become quickly heated.

SODIUM METASOMATISM

For temperatures above 350 degrees Celsius, experiments predict that magnesium metasomatism will be replaced by sodium metasomatism of a slightly more complex nature. In this case, seawater sodium (Na) replaces calcium in plagioclase crystals of basalt, which in turn permits the liberated calcium to form epidote and maintain acidic conditions independent of the fluid-to-rock ratio. This reaction shows that the more albite forms, the more calcium is recycled from plagioclase to epidote, and the more acidic the aqueous solution becomes. This reaction depends on the availability of silica, which must be supplied to the solution by the rock. Because basalts with glassy matrices are more susceptible to silica leaching than comparable rocks with crystalline matrices, sodium metasomatism (albitization or spilitization) is expected to be most pronounced in basalts that were formerly glassy.

The complex fluids currently venting from submarine ridge hot springs are viewed as the end product of a sequence of chemical reactions that begin with seawater infiltration of hot, fractured basalt some distance from the ridge axis. Convective circulation, supported by magmatic heat, drives the downwelling fluids to depths of perhaps 1 or 2 kilometers, during which time they are heated to temperatures in excess of 350 degrees Celsius and react extensively with the surrounding rocks. As hot fluids are expelled from the ridge crest, fluids nearby are drawn downward into the crust and heated. The fluids ultimately are returned to the sea floor by upwelling through narrow, focused zones along ridge axes and discharged into the sea as highly evolved hot-spring solutions. Many chemical and physical details of this complex process are yet to be explained, but it seems clear that a calcium-fixing reaction, such as the second reaction

described, must play a major role in producing the high-temperature, acidic, metal-charged solutions that vent from ridge-crest hot springs.

OBSERVATION AND EXPERIMENTATION

Prior to the late 1970's, scientific data were largely gathered along ocean ridges from specially designed research ships operating on the surface. The methods utilized included bathymetry, heat-flow and magnetic measurements, dredge sampling, and core drilling of bottom sediments and ridge basalts. It is impossible to overemphasize the success of these techniques, which provided the confirming evidence for seafloor spreading and the foundation for modern plate tectonics theory. A new era began in 1977, however, when the manned submersible *Alvin* was utilized to make direct observations of submarine hydrothermal activity along the Galápagos Rift. That was soon followed by the immensely successful 1979 RISE expedition, which used *Alvin* to photograph and sample the "black smokers" on the East Pacific Rise. Manned submersibles, supported by conventional research vessels, offer many advantages. Chief among them are that small-scale geological phenomena may be directly observed and revisited to study temporal effects; a wide range of geophysical data may be measured directly on the sea floor; samples of sediment, fluid, and rock may be collected and re-collected directly from precisely known sample sites; the physical characteristics of each data-measurement site can be observed and described; and interactions between biological and geological systems can be observed.

Direct observations, coupled with detailed laboratory experimentation on basalt-seawater reactions, are valuable as a means of investigating processes operating at the rock-seawater interface along ocean ridges where crustal growth occurs. Additional data derive from studies of hot brines in deep boreholes of active geothermal areas such as those in Salton Sea, California; Iceland; and New Zealand. Useful as these approaches are, their value is limited because they cannot tell how metamorphic conditions vary with depth beneath the rock-seawater interface; neither do they predict the effects of prolonged (for example, over millions of years) rock-seawater interaction. Full characterization of SFM as a petrologic process requires traditional geological study of rocks on land that record the entire range of SFM effects. Fortunately, such rocks, known as ophiolite complexes, are relatively common in major orogenic belts such as the Alps, Appalachian-Caledonian system, and numerous circum-Pacific mountain ranges.

VOLCANIC-EXHALATIVE ORE BODIES

From an economic point of view, studies of sub-seafloor metamorphism have greatly advanced knowledge of the important "volcanic-exhalative" class of ore deposits. Volcanic-exhalative ore bodies are major strata-bound metallic deposits that precipitate directly onto the sea floor from metal-laden solutions discharged by submarine hot springs. The resulting ores are mineralogically variable but generally sulfide-rich and, for this reason, are often called massive sulfide. Of the several types of exhalative ore deposits now recognized, the Cyprus type is most closely identified with igneous and metamorphic processes operating along submarine ridges. On the island of Cyprus, massive sulfide copper-zinc ores are hosted by ancient, metamorphosed, seafloor rocks called ophiolites and blanketed by a thin layer of iron manganese-rich marine chert. The Cyprus ores are lensoidal to podlike in shape, 200 to 300 meters across and up to 250 meters thick. The largest of these massive sulfide lenses contain 15 to 20 million tons of ore that typically assays about 4 percent copper, 0.5 percent zinc, 8 parts per million each of silver and gold, 48 percent sulfur, 43 percent iron, and 5 percent silica.

Cyprus-type massive sulfides are of worldwide occurrence, and hundreds of these deposits are now known. As a class, they constitute a major mineral resource produced as a by-product of sub-seafloor metamorphism. They are considered, by most geologists, to form directly on the sea floor from the discharge of black smokers such as those discovered along the East Pacific Rise. The smokers are chimneys ranging from 2 to 10-meters high that pump dense, black plumes of hot, acidic, metal-laden brine onto the sea floor. The turbulent black plumes, for which this special class of hot spring is named, are composed of fine particles of pyrite, pyrrhotite, and sphalerite. Calculations indicate that at their present high flow rates ten black smokers such as those operating on the East Pacific Rise could produce the largest known Cyprus-type massive sulfide body in only two thousand years.

Gary R. Lowell

FURTHER READING

Best, M. G. *Igneous and Metamorphic Petrology*. 2d ed. Malden, Mass.: Blackwell Science Ltd., 2003. A popular university text for undergraduate majors in geology. Well-illustrated and fairly detailed treatment of the origin, distribution, and characteristics of igneous and metamorphic rocks. Covers sub-seafloor metamorphism.

Borradaile, G. J., and D. Gauthier. "Magnetic Studies of Magma-Supply and Sea Floor Metamorphism: Troodos Ophiotile Dikes." *Tectonophysics* 418 (2006): 75-92. Discusses ophiolites, seafloor metamorphism, magnetic fabrics, and chemical remagnetization. Examines seafloor spreading and spatial patterns of magma feeders. Best suited for researchers studying geomagnetism and metamorphic rock.

Bucher, Kurt, and Martin Frey. *Petrogenesis of Metamorphic Rocks*. 7th ed. New York: Springer-Verlag Berlin Heidelberg, 2002. Provides an excellent overview of the principles of metamorphism. Contains a section on geothermometry and geobarometry.

Cann, J. R., H. Elderfield, A. Laughton, et al., eds. *Mid-Ocean Ridges: Dynamics of Processes Associated with Creation of New Ocean Crust*. Cambridge, England: Cambridge University Press, 1999. Describes in depth the processes of ocean ridges and ocean crusts. Offers a clear explanation of the earth's crust and its metamorphism. Illustrations, bibliography, and index.

Condie, Kent C. *Plate Tectonics and Crustal Evolution*. 4th ed. Oxford: Butterworth-Heinemann, 1997. Provides a detailed description of plate tectonics and the ocean crust. Offers a variety of useful illustrations and maps. Includes bibliographical references.

Fyfe, W. S., N. J. Price, and A. B. Thompson. *Fluids in the Earth's Crust*. New York: Elsevier, 1978. Best overview available of the theoretical side of fluid-rock interaction. Argues that subduction and seafloor spreading may be viewed as large-scale metasomatic processes, the end product of which is the crust. Written for advanced students. Extensive bibliography.

Guilbert, J. M., and C. F. Park, Jr. *The Geology of Ore Deposits*. Long Grove, Ill.: Waveland Press, 1986. A splendid edition of a traditional text for undergraduate majors in geology. Chapter 13 treats ore deposits formed by submarine volcanism in comprehensive fashion. Cyprus-type ore bodies are specifically described. Excellent index and illustrations.

Haymon, R. M. "Growth History of Hydrothermal Black Smoker Chimneys." *Nature* 301 (1983): 695-698. Describes the chemistry and mineralogy of the black smokers discovered by the 1979 RISE expedition. Can be found in most university libraries.

Humphris, S. E., and G. Thompson. "Hydrothermal Alteration of Oceanic Basalts by Seawater." *Geochimica et Cosmothimica Acta* 42 (January, 1978): 107-125. A major contribution to the growing field of sub-seafloor metamorphism. Requires some background in geochemistry. Examine the long-term implications of chemical exchange between seawater and the lithosphere.

MacLeod, P. A. Tyler, and C. L. Walker, eds. *Tectonic, Magmatic, Hydrothermal, and Biological Segmentation of Mid-Ocean Ridges.* London: Geological Society, 1996. Looks at the processes of midocean ridges and sub-seafloor spreading. Includes an examination of tectonics and hydrothermal vents, as well as a look at submarine geology and magmatism.

Mottl, M. J. "Metabasalts, Axial Hot Springs, and the Structure of Hydrothermal Systems at Mid-Ocean Ridges." *Geological Society of America Bulletin* 94 (1983): 161-180. Summarizes the status of sub-seafloor metamorphism. Appropriate for college-level readers who have some background in the subject area.

Seyfried, W. E., Jr. "Experimental and Theoretical Constraints on Hydrothermal Alteration Processes at Mid-Ocean Ridges." *Annual Review of Earth and Planetary Sciences* 15 (1987): 317-335. Covers the status of sub-seafloor metamorphism. Emphasis is on experimental aspects of basalt-seawater interaction. Technical, with some background in geochemistry needed.

Spiess, F. N., and RISE Project Group. "East Pacific Rise: Hot Springs and Geophysical Experiments." *Science* 207 (March 28, 1980): 1421-1432. Report on the 1979 RISE expedition that discovered the now-famous black smokers and explored the rift system in the vicinity of 21 degrees north latitude by manned submersibles. Recommended for general readers.

Spooner, E. T. C., and W. S. Fyfe. "Sub-Sea-Floor Metamorphism, Heat, and Mass Transfer." *Contributions to Mineralogy and Petrology* 42, no. 4 (1973): 287-304. Classical work in the ophiolite terrane of the Ligurian Apennines that established a unified model for sub-seafloor metamorphism. Essential reading at a modest technical level.

Vernon, Ron H., and Geoffrey Clarke. *Principles of Metamorphic Petrology.* New York: Cambridge University Press, 2008. Discusses metamorphic processes, anatectic and nonanatectic migmatites, and formation of granite bodies. Contains a glossary, appendices, indexing, references and color plates.

Yardley, B. W. D. *An Introduction to Metamorphic Petrology.* New York: John Wiley & Sons, 1989. An excellent undergraduate text with a good index and extensive current references. For college-level readers, a good place to begin studies of any type of metamorphism.

See also: Blueschists; Contact Metamorphism; Metamorphic Rock Classification; Metamorphic Textures; Metasomatism; Pelitic Schists; Regional Metamorphism.

SULFUR

Sulfur is one of the most widely distributed elements on Earth and is a key to the planet's ability to support life. It is believed that sulfur is one of the few elements at the earth's core, and it is possible that sulfur played a role in the core's formation. Sulfur is one of the key indicators of volcanic activity, surfacing during eruptions in liquid and gaseous forms. Sulfur has a number of useful industrial applications, most notably in agriculture and metallurgy. The element also helps scientists find clues about the earth's continuing development and its geochemical processes.

PRINCIPAL TERMS

- **allotrope:** form of an element with different physical and chemical properties
- **chalcogenides:** nonmetallic elements that are known to form ores with certain metal elements; also called ore formers
- **monoclinic:** a sulfuric type that is less dense than rhombic sulfur and that appears in the form of long needles
- **rhombic:** a more common form of sulfur, with a dense and more stable composition
- **weathering:** a geochemical process whereby rocks and minerals are broken down by water and wind

PHYSICAL AND CHEMICAL PROPERTIES

Sulfur has been an important element throughout human history. The Greeks attributed it to the leader of the gods, Zeus, in light of his perceived power of lightning. Christians later considered sulfur as the fuel of Hell itself. Sulfur also has been known as *brimstone*, a Middle English term meaning "burning stone."

Sulfur (whose chemical symbol on the periodic table is S) is a light, odorless, yellow, and solid element found predominantly in and around areas that experience volcanic activity (including hot springs). It is not soluble in water, but it may be dissolved in other liquids.

Sulfur has a number of allotropes (forms of an element with different physical and chemical properties). The alpha form is more yellow than the paler beta form. The most common allotrope of sulfur is the rhombic form, which is denser and more stable. Monoclinic sulfur, however, is less dense and appears in long needles.

Odorless and tasteless when it is inert, sulfur burns easily, emitting a blue flame and releasing sulfur dioxide gas, a strong and offensive odor. This element is highly reactive, bonding easily with other elements. Samples of pure sulfur are found underground and near volcanoes, but sulfur samples are not readily found above ground outside volcanic areas. However, sulfur is often found in other minerals, such as barium sulfate and iron pyrite (also known as fool's gold).

Ten known sulfur isotopes exist. (Isotopes are forms of an element that have the same atomic number but different atomic weight.) Four of these isotopes are naturally occurring (S-32, -33, -34, and -36). The other six occur when exposed to radiation, as is the case for S-35 (a sulfur isotope commonly used as a tracer in medicine).

Along with oxygen, selenium, and tellurium, sulfur is considered one of the chalcogenides (or ore formers) on the periodic table. It is a nonmetal, but its atoms strongly retain electrons. Sulfur atoms can form long aggregates, sometimes in a stable, circular chain. Its polysulfide (more than one sulfur atom) ions are easily formed as a result, making bonding with other elements and compounds a relatively simple process.

USES OF SULFUR

Sulfur is one of the most prevalent elements on Earth. In addition to its benefits to the natural environment, sulfur and the compounds it forms have a wide range of applications.

For example, sulfur dioxide is often used in winemaking, as it reduces the amount of color lost during the aging process. There is concern, however, that too much sulfur dioxide in the wine barrels could cause illness in consumers. An ongoing effort is trying to establish agreeable standards for the use of sulfur dioxide additives. Nevertheless, this additive remains common in this major industry.

Sulfur also has long been a key ingredient in gunpowder and other explosives and in the vulcanization of rubber, preventing the rubber from melting as the mixture is heated. Furthermore, sulfur is used as an insecticide and in the petroleum refining process. Additionally, it is a key ingredient in fertilizer, providing a basic nutrient to plants and crops.

The compounds in which sulfur is found also have a wide range of uses. For example, sulfuric acid is used in iron and steel production. Sulfur dioxide, meanwhile, is used as a disinfectant and bleaching agent. Magnesium sulfate (more commonly known as Epsom salt) is commonly used as a bath additive and laxative. Calcium sulfate is a common ingredient in alabaster and gypsum, which are used in the creation of such decorative items as vases and other artwork.

Because of its ability to form stable rings that in effect become electron traps (which attract and retain electrons), sulfur is one of the most effective electrical insulators on Earth. Researchers at one point used balls of pure sulfur in the generation of static electricity. One of the most effective high-voltage electrical insulator compounds is sulfur hexafluoride.

Sulfur and sulfuric compounds remain in high demand in light of their many uses. The petroleum industry is one source, as sulfuric acid is a by-product of the refining process. However, because this process only produces a set amount of sulfuric acid, the market becomes inelastic. Businesses are therefore always searching for new sulfur deposits and resources.

THE SULFUR CYCLE

Many different processes are ever-occurring on Earth to sustain life and environmental balance. One such process is the sulfur cycle.

Nearly every form of life on the planet requires sulfur-based compounds. Although it is one of the most abundant elements on Earth, sulfur is not formed on the surface. Rather, it is believed to originate in the planet's core (many scientists theorize that it played a role in the formation of the core when the earth was still forming). Sulfur is brought to the earth's crust and surface through volcanic activity. Once brought to the surface, it bonds with a wide range of proteins, minerals, and other compounds.

There are two components to the sulfur cycle—atmospheric and terrestrial—working in concert. According to the sulfur cycle framework, this element is released from minerals through the chemical weathering process. The sulfur is released into the atmosphere, where it bonds with oxygen to form sulfates. These sulfates return to the earth's surface in the form of precipitation, where they are consumed by plants, which in turn are consumed by animals along the food chain. When the animals and plants die and decompose, the sulfur is released back into the air and soil. Sulfur also returns to the oceans and other waterways and moves through marine ecosystems. The sulfur that is not absorbed by this ecosystem settles on the marine floor, where it combines with other sediment.

Although the sulfur cycle is invaluable to the myriad life forms on Earth, it also constitutes a danger when the delicate balance this cycle maintains is altered. For example, when a volcano erupts, it spews great volumes of sulfur high into the atmosphere. The sulfur that does not immediately return to the earth can block the sun's rays. A large number of volcanic eruptions can eventually lead to climate change, triggering global warming and global cooling.

As humans continue to develop industrial capabilities, the sulfur cycle is again taken out of balance because of the enormous volume of industrial pollution (which includes sulfur) released into the atmosphere. Scientists are concerned that sulfates (a greenhouse gas) will contribute to global warming.

MINING AND SULFIDE ORES

Sulfur deposits are generally found underground, most commonly in areas near inactive volcanoes and salt domes (protrusions in sedimentary basins in which large salt masses have been pushed outward). Because elemental sulfur melts at relatively low temperatures, the preferred method for mining this element involves the pumping of hot saltwater into a mine and using air pumps to extract the melted sulfur.

In this process the sulfur can be allowed to return to solid form or remain as a liquid, depending on the preferences of those who are purchasing it. This method has been dubbed the Frasch process. Its inventor, Herman Frasch, developed this mining practice in 1891 for the Standard Oil Company, enabling the launch of the Union Sulfur Company in 1892.

Sulfur also is found in ores (solid masses from which certain minerals may be extracted). For example, sulfide ore is a solid deposit of sulfide (a compound of sulfur with a negative charge, allowing it to bond with other minerals and elements) and other metals. Often, mining companies will extract sulfide ore to obtain other metals that are part of the deposit. This "involuntary" sulfur is then refined and sold. The trend of mining involuntary sulfur while pursuing other metals accounts for more than 70 percent of the sulfur produced in the world, which means that the Frasch process, in terms of sedimentary mining, is being replaced.

SULFUR AND CLIMATE CHANGE

In addition to its many implications for life on Earth, sulfur plays a significant role in climate change. Indeed, one of the most voluminous gases spewed from a volcanic eruption is sulfur dioxide.

Sulfur dioxide may travel as high as the stratosphere, where it blocks the sun's warming radiation, triggering a drop in the earth's temperature by as much as 0.5 degree Celsius (0.9 degree Fahrenheit). Such events occur frequently; they take place nearly once every eighty years. However, in the earth's past, such occurrences were more frequent and happened every few to a dozen years. Scientists attribute a number of prehistoric ice ages to sulfur dioxide emissions from high levels of volcanic activity.

Another form of sulfur emission is threatening to create global climate change in the modern era. By the 1960's, the emissions caused by fossil fuel use amounted to the equivalent of a volcanic eruption every 1.7 years. Since that period, however, better awareness of the issue has led to the reduction of fossil fuel use through the increased use of more fuel-efficient vehicles and technologies and by the introduction of alternative energy sources.

Sulfur and its compounds also are involved in climate change as indicators. In a 2010 study of lakes in the Canadian Arctic and in Norway, researchers found evidence of sulfur accumulation that could be caused by an increase in algae development. When normally cold seasons are shortened, the continuing warmth causes an increase in algae growth. This growth kills other vegetation, and the resulting degradation of the plants leads to the release of sulfur. This sulfur is returned to the water, soil, and air as part of the aforementioned sulfur cycle. It is feared that an increase in the pace of this cycle could hasten major climate change. Scientists are carefully watching such trends.

FUTURE IMPLICATIONS

Sulfur and sulfates remain important factors scientists consider when analyzing the ongoing issue of global warming and climate change. For example, later studies point to sulfur in the atmosphere binding in time with soot from industrial pollution. Because sulfur absorbs sunlight, this combination is considered by some scientists to be a possible remedy against global warming trends. This research is at an early stage, although it is clear that sulfur remains on the minds of scientists in pursuit of answers to the issues of global climate changes.

Similarly, sulfur is likely to remain an important natural resource for industry. Sulfuric acid is one of the most-used commodities in the industrial world. Some 200 million tons of this sulfuric compound is consumed annually. Despite the trials of the recent global economy, experts believe demand for sulfuric acid will continue, especially from makers of fertilizer. In particular, markets are in search of acid that has been derived from natural sulfur deposits rather than that which has been produced from the petroleum refining process (which in comparison with natural deposit sources is a more inelastic source). Sulfur exploration for commercial purposes, therefore, is likely to continue for the foreseeable future.

Michael P. Auerbach

FURTHER READING

Davenport, William G., and Matthew J. King. *Sulfuric Acid Manufacture.* Boston: Elsevier Science, 2006. Describes the many uses of one of the most popular derivatives of sulfur: sulfuric acid. Discusses the regulatory aspects of sulfuric acid use in industrial circles and its development for commercial use.

Drevnick, Paul E., et al. "Increased Accumulation of Sulfur in Lake Sediments of the High Arctic." *Environmental Science and Technology* 44, no. 22 (2010): 8415-8421. Reports an increase in the amount of inorganic sulfur (an industrial by-product) in Canadian and Norwegian lakes. Cautions that an acceleration of the sulfur cycle could have major implications for the planet's geochemical processes.

Jez, Joseph. *Sulfur: A Missing Link Between Soils, Crops, and Nutrition.* Madison, Wis.: American Society of Agronomy, 2008. An overview of the relationship between sulfur and animal, human, and crop health. Discusses the sulfur cycle and the connections between the dietary needs of humans and other animals and elemental sulfur.

"Mineral Resource of the Month: Sulfur." *Earth* 55, no. 6 (2010): 26-27. Discusses the nature of sulfur and its industrial uses. Describes the Frasch process and how it has evolved.

Ward, Peter L. "Sulfur Dioxide Initiates Global Climate Change in Four Ways." *Thin Solid Films* 517, no. 11 (2009): 3188-3203. Reviews the incidents in Earth's history in which volcanic activity triggered global climate change. These occurrences have spurred mass extinctions and major cooling periods. Also discusses how sulfur emissions from industrialized civilizations have contributed to an ongoing trend of global warming.

Wessel, G. R., ed. *Native Sulfur: Developments in Geology and Exploration.* Englewood, Colo.: Society for Mining Metallurgy, 1992. Among the topics covered are potential deposit locations in Europe and the temperature conditions that contribute to the formation of certain sulfur deposits.

See also: Earth Science and the Environment; Geologic Settings of Resources; Geothermal Power; Hydrothermal Mineralization; Quartz and Silica; Silicates.

T

TIDAL FORCES

Tidal forces are caused by the interaction of the gravitational fields of astronomical objects, and tides are caused by physical deformations of astronomical bodies subjected to gravitational fields. The cyclic relationships between the orbits of the moon, Earth, and sun cause periodicity in the cycles of tidal movements on Earth. Tidal power is generated by utilizing the kinetic energy contained within moving tidal waters. This power generates electrical energy for human use.

PRINCIPAL TERMS

- **bathymetry:** the measure of water depth at various points on Earth; also, the study of components affecting water depth
- **gravitational field:** the area surrounding an astronomical body affected by the deformations in space and time caused by the mass of that body
- **heliosphere:** area of outer space affected by the sun
- **hydrosphere:** all the waters on the earth's surface, including seas, oceans, lakes, rivers, and the water vapor in the lower atmosphere
- **lithosphere:** outer portion of the earth consisting of the earth's crust and part of the upper mantle
- **lunar cycle:** the time it takes for the moon and the earth to return to any single position relative to each other as dependent on the combination of the rotation of the earth and the orbit of the moon; approximated to 29.5 rotations of the earth on its axis
- **syzygy:** alignment of three astronomical bodies with regard to a certain direction of reference, such as when the sun, moon, and Earth align twice each lunar cycle
- **tidal force:** secondary effect of the gravitational forces between two objects that causes elongation of the objects along the axis of the line connecting their respective center points
- **tidal power:** utilized kinetic energy created by moving tidal currents; generates electrical energy
- **tidal turbine:** turbine housed within a tidal system and used to generate electrical energy from the kinetic energy of tidal movement

THE PHYSICS OF TIDES

Every astronomical object, including stars, planets, and satellites, generates a gravitational field, which is defined by the way in which the object's mass deforms the surrounding space. Larger objects have more mass and therefore generate a larger gravitational field.

When two astronomical objects interact, they exert force on each other through their gravitational fields, a force sometimes described as a gravitational pull or attraction. This force grows stronger the closer the objects move toward each other. The earth orbits the sun because of the gravitational pull exerted by the sun; the earth and the moon orbit each other because of their mutual gravitational attraction. Together this constitutes the earth-moon-sun system, which is responsible for many physical phenomena that affect Earth's climate, weather, and physical cycles.

As the moon orbits the earth, the structure of both the earth and the moon deform slightly in response to the gravitational forces they exert on each other. This causes both the earth and the moon to elongate slightly as they pull toward each other. The pull of the moon is strongest on the side of the earth facing the moon. The moon's pull causes the sides of the earth at right angles to the moon to depress slightly. This occurs because the gravitational pull of the moon is unequally distributed over the spherical surface of the earth.

The side of the earth facing away from the moon receives the least gravitational pull and, therefore, also undergoes the least amount of deformation. This differential deformation causes the phenomena known as tidal forces, which are most clearly demonstrated in the cyclic movement of ocean currents.

Earth's hydrosphere, which includes oceans, seas, lakes, and rivers, is more responsive to tidal forces because of its fluid state. The waters closest to the moon are pulled toward the moon as the earth deforms, while the waters on the side of the earth farthest from the moon experience less pull; therefore, more of those waters are "left behind." This causes a high tide on both the side of the earth closest to the moon and on the side farthest from the moon. The sides of the earth located approximately at right angles to the moon's position experience intermediate pull from the moon and therefore exhibit low tides as the waters are either pulled toward the moon or toward the side opposite the moon.

TIDAL VARIATIONS

As the earth spins on its axis, the position of the moon changes relative to the earth, creating two high tides and two low tides at each location per day. However, the tides are also affected by the topography of the earth, as landmasses affect the overall direction and force of tidal flow. Therefore the bathymetry, or overall measure of water depth, depends on both the tidal cycle and the particularities of the lithosphere.

Many locations experience the standard semidiurnal cycle of two high and two low tides per day, while in other areas, landmasses blocking ocean currents can lead to a cycle—the diurnal—of only one high and one low tide per day.

The timing of the tide coincides with the rotation of the earth and the position of the earth and the moon as they orbit around each other. The lunar cycle is the time it takes the moon and the earth to return to the same position with regard to one another. For instance, if the moon is directly above St. Louis, Missouri, it will take one lunar cycle for the rotation of the earth and the orbit of the moon to realign to bring the moon to the same point above St. Louis.

The lunar cycle reflects the relationship between the 24-hour rotation of the earth combined with the 27.3-day orbit of the moon around the earth (called the moon's sidereal period). Because the earth rotates on its axis in the same direction that the moon travels around the earth, the relationship between the two lags by about thirty minutes each day. The combination of these factors creates a cycle of 29.5 days for the earth and the moon to return to the same position, sometimes called a lunation. The cumulative effect of this relationship is that the timing of the tides also changes on a daily cycle as the moon-earth relationship progresses through each lunation. High and low tides arrive approximately fifty minutes later each day.

While the moon exerts the most important influence over the tides, the position of the sun also has a powerful effect on tidal strength. The gravitational pull of the moon has a far more immediate effect on the earth because the moon is much closer. On average, the moon is approximately 384,403 kilometers (239 miles) from the earth, whereas the sun is approximately 150 million km (93 million mi) from the earth. However, because of the sun's enormous mass, its gravitational field extends beyond the edges of the solar system, throughout the entire heliosphere. Therefore, a full understanding of tidal forces must take into account the relative positions of all three bodies.

Twice each lunar cycle, the moon, the earth, and the sun fall into a line with respect to one another. This state of alignment, called syzygy, causes the gravitational pull of the sun and the moon to become complementary or opposing, depending on whether they are on the same side or the opposite side of the earth. In both cases, the earth experiences higher high tides and lower low tides, called the spring tides, because the differential pull of gravity exerted on the earth is stronger. When the moon and sun are at right angles to each other with respect to the earth, their respective gravitational fields interfere with each other, leading to periods with less variation between high and low tides, called the neap tides.

Spring and neap tides are often associated with the phases of the moon, which is a measurement of which part of the moon is visible from Earth. As the moon completes its 29.5-day lunation, the portion of the moon that is visible from Earth changes as the angle between the earth and the sun alters the portion of the moon that is hit by sunlight and therefore visible to an observer on the earth.

Spring tides coincide with the full moon, which occurs when the sun, moon, and earth align so that the sun and moon are on opposite sides of the earth; spring tides also coincide with the new moon, which occurs when the sun and moon align on the same side of the planet. Neap tides are associated with the quarter moon and three-quarter moon phases, when the moon and sun are at right angles to one another.

HARNESSING TIDAL ENERGY

Tidal energy is a type of alternative, renewable energy gained by harnessing the kinetic energy of tidal movement to create electricity. Kinetic energy is the energy possessed by an object by virtue of its movement through space. This energy can be used to perform such physical work as powering generators that transform motion into electrical signals.

Although not widely used, tidal energy offers a virtually limitless source of energy and does not produce atmospheric pollutants such as the carbon dioxide that results from burning fossil fuels. Because the movement of the tides follows a predictable cycle, generation of tidal energy constitutes a more predictable system of energy generation than many other forms of renewable energy, including solar and wind power.

One method of harnessing tidal energy is to create a tidal barrage, which is a dam built across an estuary or bay that experiences sufficient tidal variation. A tidal barrage works in a similar way to a hydroelectric dam, except that tidal barrages depend partially on tidal changes in water levels to push water into and out of the main reservoir of the system. In addition, tidal barrages tend to be heavier than hydroelectric dams because they must support not only the weight of the overlying water but also the force of tidal currents. Tidal range must be in excess of 5 meters (16 feet) per day for the barrage to be effective.

As water flows towards the barrage, it is channeled through gates called sluices and flows into the basin or estuary on the other side of the barrage, creating a body of water called the hydrostatic head. When the tide recedes, gates at the bottom of the barrage open, allowing the water to rush through the lower passage (driven by gravitational energy). The water channeled through these gates flows over a series of turbine engines, which are attached to an electric generator.

While tidal barrages produce little pollution, they require significant destruction of ecological habitat and may have long-term detrimental effects on ocean life. Because the installation of the barrage changes water levels on both sides, the lives of many animals may be altered. In addition, the barrage greatly restricts the movement of fish, crustaceans, and other oceanic organisms to and from the bay or estuary.

Another method of harnessing power from tidal energy is the tidal fence, which is similar to a tidal barrage but generally smaller in scope. Tidal fences are usually placed in areas where features of the surrounding landmasses cause a relatively rapid flow of tidal waters. Tidal fences also use turbines to generate energy from tidal movement. Tidal fences are less expensive, are easier to install, and cause less environmental damage overall, but they still affect the movement of sea creatures and cause localized environmental destruction.

Another system used to capture energy from tidal streams are tidal turbines, which are individual turbines mounted on the sea floor. Tidal turbines function similarly to air turbines used to capture energy from wind power. However, tidal turbines must be constructed more sturdily because water is more than eight hundred times denser than air and exerts a far greater pressure on the turbine structure. Tidal turbines cost less to construct than either tidal fences or tidal barrages and have less potential to cause environmental damage. In addition, tidal turbines can be installed under the water in such a way that they do not impede ocean travel or shipping. Tidal turbines may be the most promising method of harnessing tidal energy, but the technology is still in its infancy and few tidal turbine systems have been installed.

Energy generated by tidal kinetic energy can be stored as electric energy, and this energy can be channeled later to provide electricity for human use. One of the main challenges facing tidal energy and other types of renewable energy is how to store excess energy until it is needed. Current research programs are trying to refine lithium-ion battery technology to create methods for storing vast amounts of energy for later use. If this technology is successful it may be possible to generate most of the world's electrical energy needs from renewable sources, thereby reducing reliance on fossil fuels and reducing overall environmental pollution.

Micah L. Issitt

FURTHER READING

Boeker, Egbert, and Rienk von Grondelle. *Environmental Physics: Sustainable Energy and Climate Change.* New York: John Wiley & Sons, 2011. Introductory physics text focusing on the development of sustainable energy and factors influencing climate change. Contains a detailed discussion of tidal energy and its role in the creation of sustainable energy.

Cartwright, David Edgar. *Tides: A Scientific History.* New York: Cambridge University Press, 2001. Historical overview of information and scientific studies of ocean tides. Includes descriptions of the physics and geologic forces involved in tidal movements and discussions of the various uses of tidal energy.

Da Rosa, Aldo Viera. *Fundamentals of Renewable Energy Processes*. New York: Academic Press, 2009. Describes research into renewable energy projects, including detailed descriptions of the physics, chemistry, and economics of renewable energy. Examines the development and usage of tidal energy.

Grotzinger, John, and Thomas H. Jordan. *Understanding Earth*. New York: W. H. Freeman, 2009. Detailed description of earth's development and the function of geochemical cycles. Contains a description of the tidal cycles and its effect on Earth's climate and ecosystems.

Sorensen, Bent. *Renewable Energy: Its Physics, Engineering, Environmental Impacts, Economics, and Planning*. 4th ed. Burlington, Mass.: Academic Press, 2011. Complex introduction to renewable energy technology and research. Contains a description of tidal energy and the methods used to harvest it for human use.

Stanley, Steven M. *Earth System History*. New York: W. H. Freeman, 2004. Historical overview of the development of Earth's systems and research into geochemical cycles. Contains information on tidal systems and their relationship to geochemical processes.

See also: Earth System Science; Geologic Settings of Resources; Hydroelectric Power; Ocean Power; Unconventional Energy Resources; Water and Ice; Wind Power.

TOXIC MINERALS

Toxic minerals are those minerals that present a certain risk to public health. Among these minerals are asbestos and silica. In addition, other minerals contain toxic metals such as mercury and lead. The most common diseases caused by toxic minerals are cancer and neurological disorders. The toxic effects of some minerals, such as lead or mercury, can be irreversible if left untreated.

PRINCIPAL TERMS

- **aluminum:** the most abundant metal in the earth's crust
- **arsenic:** a chemical element with a metallic appearance
- **cadmium:** a soft element
- **chromium:** a lustrous hard metal
- **LD$_{50}$:** in toxicology, the abbreviation for the lethal dose for 50 percent of test animals; lethal dose is expressed in milligrams of poison per unit body weight of the animal
- **lead:** a soft metal
- **mercury:** the only liquid metal at room temperature
- **mineral:** a solid, inorganic natural substance with a distinct chemical composition and a specific crystalline structure; minerals are formed by geologic processes
- **selenium:** an element of the sulfur family; several forms of selenium are known, including a metallic selenium
- **toxicology:** the study of the effects of toxic compounds on humans and environments

TOXIC MINERALS AND THEIR USE

Toxic minerals are minerals that are hazardous to human health. A mineral by itself, or its components, can be toxic. Typical examples of the former are asbestos, selenium, and silica minerals; typical examples of the latter are minerals containing such toxic chemicals as arsenic, lead, mercury, or cadmium.

Six fibrous silicate minerals are defined as asbestos: crocidolite, amosite, chrysotile, tremolite, actinolite, and anthophyllite. The most hazardous of the asbestos minerals are crocidolite and amosite. The best known in the United States is chrysotile, a magnesium silicate.

Asbestos, an excellent insulator, was widely used in construction beginning in the nineteenth century; it has since been banned in Europe and is regulated in the United States. Before the discovery of asbestos's toxicity, it was a common component of drywall, fire blankets, fireproof clothing for firefighters, thermal pipe insulation, and gas mask filters.

The toxicity of other minerals such as talc and vermiculite is related to asbestos contamination.

Talc is a magnesium silicate mineral with the chemical formula $Mg_3Si_4O_{10}(OH)_2$. It is widely used in a form of talcum powder for preventing diaper rash and in surgical gloves.

Vermiculite is a mineral from the group of mica (sheet) minerals with approximate chemical formula Mg^{+2}, Fe^{+2}, $Fe^{+3})_3$ $[(AlSi)_4O_{10}] \cdot (OH)_2 \cdot 4H_2O$. It was widely used in construction for fire protection and for the insulation of pipes and roofs. In the United States alone, about 35 million homes contain vermiculite.

Other potentially hazardous minerals are silica and selenium. Silica, or silicon dioxide (chemical formula SiO_2), can be found in nature as mineral quartz or sand. Silica is a common food additive. For example, table salt contains silica. Silica also is a primary component of diatomaceous earth, which is used as a filter aid, commonly in backyard pools. According to the Centers for Disease Control and Prevention, nearly 1.7 million workers in the United States are exposed to silica in a variety of industries and occupations, including in construction and mining.

Selenium is both a chemical element and a mineral. This mineral is mainly used to manufacture glass, making it green or red. Selenium also is used in the electronics industry.

Several minerals contain such toxic metals as lead, mercury, cadmium, chromium, and arsenic. These metals, also called heavy metals, are poisonous. The main mineral containing lead is galena (PbS), which is up to 86.6 percent lead. Other common lead minerals are cerrusite ($PbCO_3$) and anglesite ($PbSO_4$). Lead, which can be easily extracted from its minerals, was known to ancient Egyptians and Babylonians and is mentioned in the Bible. Since Roman times, people have used lead for plumbing and as a food and drink preservative. During the Middle Ages, lead was used for roofing.

The major use of lead today is for batteries. Lead also is used as a protective shield against radiation. One lead mineral that was historically used as a natural dye is chrome yellow, which is toxic. Chrome yellow is a natural yellow pigment made of lead chromate ($PbCrO_4$).

Cinnabar (HgS) is the most common mineral of mercury. It has been known for some time and has been an object of fascination because of its appearance. Mercury was used since Roman times as a natural dye. Dentists use it today to make amalgams to fill cavities, although not without controversy.

Several rare minerals contain the metal cadmium. Cadmium is used in industry for making some types of paints and batteries and in electroplating. Chromium is mined as chromite ($FeCr_2O_4$). It is used in metallurgy; stainless steel usually contains chromium. Chromium also is used as a paint pigment (school buses are painted in chrome yellow) and for tanning leather (such as for sofas).

Another heavy metal is arsenic, which is not exactly a metal but has several metallic properties. Arsenic occurs in many minerals. The most common arsenic minerals are realgar, orpiment, and arsenopyrite, which are frequently sources of arsenic in industry. Realgar is known as the ruby of arsenic and was used in ancient Rome as a red paint pigment. It is now used occasionally for killing weeds, insects, and rodents. Orpiment also was used as a pigment in painting in ancient times. However, it also was used in poison arrows, as its toxic nature was well known.

There also exist minerals of lower toxicity and minerals with inconclusive evidence of toxicity; these include aluminum, tin, and zinc. Most common aluminum minerals are feldspars. They are aluminosilicates and make up to 60 percent of the earth's crust, or the rock shell that surrounds the planet. There also exist other aluminum minerals, such as beryl ($Be_3Al_2Si_6O_{18}$), cryolite (Na_3AlF_6), and corundum (Al_2O_3). Aluminum is widely used in the airplane and electronics industries and for making foil (wrapping material) and cooking pans.

THE HEALTH HAZARD OF TOXIC CHEMICALS

All toxic minerals are hazardous to humans. The effect of toxic chemicals on human health can be immediate (acute) or can occur through years (chronic).

Most common diseases caused by toxic minerals are cancer and neurological disorders. Toxic minerals act as carcinogens, chemicals that alter genetic mechanisms. Experts have not determined how toxic minerals cause cancer.

Asbestos causes several serious illnesses, including asbestosis, lung cancer, and mesothelioma. Inhalation of asbestos fibers causes asbestosis (pulmonary fibrosis) within a short time. Asbestosis is the formation of scar tissue (fibrosis) inside the lungs. Symptoms include chest pain, tightness in the chest, cough, and shortness of breath that worsens with time. Asbestos also can cause cancer, especially lung cancer or mesothelioma. Mesothelioma is a rare cancer of the lining of body cavities.

The majority of asbestos-related cancers are caused by the asbestos mineral crocidolite. In the past, before crocidolite mines were closed, about 18 percent of crocidolite miners died from cancer. Chrysotile is more common but appears less dangerous. Both talc and vermiculite minerals, which contain asbestos, also cause cancers, especially of the lung.

The most common disease caused by exposure to silica is silicosis. Other diseases are lung cancer, autoimmune disorders, and chronic renal disease. Silicosis is a disease marked by inflammation and lesions in the lungs. Selenium can cause selenosis if taken in excessive amounts. Symptoms of this disease include hair loss, gastrointestinal disorders, fatigue, irritability, and nervous system damage.

Heavy metals primarily inhibit the functioning of important enzymes in the human body. Many enzymes have sulfhydryl groups. Heavy metals attach to these groups, making enzymes inactive.

The toxicity of lead and lead mineral (galena) has been known since ancient times. Humans used galena to obtain metallic lead. Minerals containing lead are toxic to many tissues and organs, including the heart, bones, intestines, kidneys, brain, and reproductive system. Also, lead ions inhibit enzymes necessary for the synthesis of hemoglobin (the oxygen carrier in human blood). Lead poisoning causes mental retardation and neurological disorders. This damage can be irreversible if untreated. Historically, lead poisoning was a major problem in children, particularly those who lived in old homes.

Lead poisoning is treated with calcium salt of ethylenediaminetetraacetic acid (EDTA), with the lead-catching, chelating agent (from the Greek *chela* meaning "claw") being administered intravenously. In the human body, calcium ions of EDTA are replaced by lead.

Mercury toxicity is notorious, but pure metallic mercury is not particularly hazardous. Ingestion of small amounts of mercury from dental fillings will not produce any health problems. Mercury compounds, however, are dangerous. Mineral cinnabar is extremely toxic by ingestion. The mining of cinnabar in ancient times was regarded as a death sentence. The exposure to cinnabar can result in damage to the brain, kidney, and lungs; birth defects; and several other diseases, including insanity. Damaging effects are usually irreversible. As in the case of lead, antidotes exist for mercury poisoning and include chelating agents such as British anti-lewisite (BAL). Mercury ions can attach to it and be deactivated (prevented from attacking body systems).

Various other minerals containing heavy metals are extremely toxic to humans. These other minerals include those containing cadmium and chromium. Poisoning with cadmium leads to loss of calcium ions (Ca^+) from bones, making them fragile and easy to break. Cadmium poisoning also causes pain, vomiting, and diarrhea. The toxicity and carcinogenic properties of chromium are known as well. Chromium minerals also can cause allergic reactions in some people.

Arsenic mineral poisoning causes headaches, confusion, severe diarrhea, and drowsiness and usually affects the lungs, skin, kidneys, and liver. The final result of arsenic poisoning is death. Arsenic poisoning was one of the favorite methods for murder in the Middle Ages. Because of its slow effect, arsenic poisoning often went undetected. Today, arsenic poisoning is a global problem because of its natural contamination of groundwaters. Chelating agents are used to treat arsenic poisoning.

Compared with other toxic minerals and their components, is not extremely toxic. Data on its low toxicity appeared only recently. Some toxicity of aluminum is related to its deposition in bones, where it replaces calcium, contributing to reduced hardness of bones and, when occurring in infants, growth retardation. There also is some inconclusive evidence that aluminum exposure increases the risk of breast cancer and some brain diseases, such as Alzheimer's desease.

MANAGEMENT OF TOXIC MINERALS

Humans have been mining toxic minerals for thousands of years and have been poisoned by these minerals in the process. Several U.S. government agencies, including the Environmental Protection Agency (EPA) and the National Toxicology Program, have data on toxic minerals that are available to the public, health practitioners, scientists, and industry workers.

The EPA requires that containers and vehicles in which toxic minerals are stored or carried display signs identifying the hazard. Public fears of developing cancer from toxic minerals have led to regulations that request the minerals' removal from public and private places.

Scientists normally conduct tests on animals to identify toxic effects and hazardous doses (the level of exposure and the amount of toxic mineral). For example, LDL_0 (the lowest dose at which animal death occurs) for aluminum is 6207 mg per kilogram (a low toxicity chemical) compared with arsenic of 70 mg/kg (a high toxicity chemical). Humans are able to deal with low doses of toxic minerals.

The level in which toxic effects are not observed is the threshold level. Higher levels of toxic minerals can also be tolerated if the time of exposure is short. Toxicologists normally know the threshold level of acute diseases caused by minerals. Scientists, however, do not know the threshold level of toxic minerals for cancer.

Sergei A. Markov

FURTHER READING

Chang, Raymond, and Jason Overby. *General Chemistry: The Essential Concepts.* 6th ed. Columbus, Ohio: McGraw-Hill, 2010. A comprehensive book on general chemistry that includes discussion of minerals and heavy metals.

Dodson, Ronald P., and Samuel P. Hammar, eds. *Asbestos: Risk Assessment, Epidemiology, and Health Effects.* Boca Raton, Fla.: Taylor & Francis, 2006. Comprehensive book on asbestos and its health risks. Includes chapters on the history of asbestos, analysis methods, molecular and cellular responses to asbestos, pathology of asbestos-induced disease, epidemiology, clinical diagnosis of asbestos-related disease, recommendations for practicing physicians, and asbestos regulations.

Hewitt, Paul G., John Suchocki, and Leslie A. Hewitt. *Conceptual Physical Science.* San Francisco: Pearson/Addison Wesley, 2008. A classic book in physical science. Several chapters cover Earth science and the properties of minerals.

Hill, John W., and Doris K. Kolb. *Chemistry for Changing Times.* Upper Saddle River, N.J.: Prentice Hall, 2001. A textbook on popular chemistry that emphasizes chemistry's connection to everyday life. Covers minerals and their components.

Nebel, Bernard J., and Richard T. Wright. *Environmental Science: Towards a Sustainable Future.* Englewood Cliffs, N.J.: Prentice Hall, 2008. This popular textbook on ecology and environment describes human control of hazardous chemicals.

Olmsted, John A., and Gregory M. Williams. *Chemistry.* 4th ed. Hoboken, N.J.: Wiley, 2006. A good book on general chemistry. Several chapters present information on lead, mercury, cadmium, aluminum, and other chemical elements and on their use in industry.

See also: Aluminum Deposits; High-Pressure Minerals; Industrial Metals; Lithium; Minerals: Physical Properties; Minerals: Structure; Quartz and Silica; Radioactive Minerals; Silicates; Uranium Deposits.

U

ULTRA-HIGH-PRESSURE METAMORPHISM

Metamorphic rock is one of three major rock types and is formed when pressure, temperature, or environmental chemicals cause either an igneous or sedimentary rock to change in density and mineralogical structure. Metamorphosis can be caused by extraterrestrial impact, exposure to heat from the earth's core, exposure to chemical environments, and tectonic forces. Ultra-high-pressure metamorphism is a type of metamorphism that occurs at extremely high pressures and usually accompanies the collision of continental plates.

PRINCIPAL TERMS

- **diagenesis:** chemical and texture changes that affect sediment during and after its formation into sedimentary rock
- **lithification:** process by which sediment is compacted or cemented into solid rock
- **lithostatic pressure:** pressure exerted by the weight of sediment overlying the sediment sample in question
- **magma:** superheated rock that exists in a liquid phase
- **megapascal:** geologic unit of extreme pressure, equivalent to 1 million pascals and 145.037 pounds per square inch
- **petrology:** branch of geology that studies rocks and the conditions that lead to the formation of rocks
- **polycrystalline:** crystal structure consisting of crystallite fragments
- **recrystallization:** process that occurs at high pressure, causing the atoms, molecules, or grains of a crystal to reorganize into another type of crystal
- **subduction zone:** area in which two tectonic plates collide, forcing one plate to push beneath the other and driving the submerged plate into the earth's mantle

METAMORPHIC PETROLOGY

Metamorphic petrology is a branch of geology that studies transformations in mineral composition, texture, color, and other qualities that occur within rocks. These transformations occur when the rocks are subjected to pressures, chemical environments, and temperatures different from those in which the rock originally formed.

Metamorphism can be distinguished from diagenesis, which is another transformative process that affects certain types of rock after solidification. Diagenesis occurs at low temperatures and pressures, whereas the process affecting metamorphic rock typically occurs at temperatures higher than 200 degrees Celsius (392 degrees Fahrenheit) and at pressures greater than 300 megapascals. Metamorphic rocks are sufficiently unique in structure and chemistry that they are considered one of the three basic rock types.

Sedimentary rocks are formed from sediment, a layer of debris that forms at the top layer of the earth's crust and is composed of loose minerals, organic waste, and bits of other rocks that have eroded into smaller fragments. Through burial and compression, this surface material can become lithified—that is, cemented and compressed to form rocks.

Sedimentary rocks come in three types: clastic, chemical, and organic. Clastic rocks are composed almost entirely of bits of other rocks that have become compacted into solid structures. Chemical rocks are sedimentary rocks that form through some chemical process. Many types of chemical rocks result from situations in which a riverbed or shallow sea gradually dries, after which underwater sediment may become solidified through dehydration and other chemical processes to become sedimentary rock. Organic rocks are formed from the accumulation of organic debris in the sediment that becomes lithified through compaction. Coal is an example of an organic sedimentary rock that forms from the remains of wood and other organic material that has become buried and compacted through millions of years.

Igneous rocks form when molten earth originating within the earth's mantle rises close to the surface and then cools and solidifies to form solid bodies.

The formation of igneous rock involves a change of state from liquid to solid, which is accompanied by a reduction in heat.

Igneous rocks come in two main types: plutonic and volcanic. Plutonic rock forms when magma slowly cools in pockets beneath the surface, sometimes taking millions of years to form. Plutonic rock is one of the most common types of rock and commonly underlies mountains and other major geologic features. Volcanic rock forms when magma breaks through the surface as lava and then cools and hardens into solid structures. Magma and volcanism have played major roles in the formation of the lithosphere. Geologists estimate that the majority of the earth's crust is composed of igneous rock.

Metamorphic rocks may start their existence as either igneous or sedimentary rocks but are subjected to an environment such that they undergo a change in chemical and physical properties. Most metamorphic rocks form deep in the earth's crust or the upper layers of the mantle, where temperatures and pressures are sufficient to induce significant chemical and physical changes. Though metamorphosis generally begins at 200 degrees Celsius (degrees 392 Fahrenheit) and at 300 megapascals, it can occur at much higher temperatures and pressures. The limit of the metamorphic process occurs when the rock becomes hot enough to melt. At this point, the rock becomes magma. Any further structural transformations fall into the category of igneous rock formation.

Metamorphism always involves the loss of water and an overall reduction in density. Some of the most familiar metamorphic rocks are gemstones such as diamonds and rubies. Rubies and sapphires form from the metamorphosis of certain types of clay, which give rise to corundum. Further pressure and temperature increases will cause the corundum to develop glass-like mineral inclusions including rubies and sapphires. Diamonds also develop from the metamorphic process acting on coal that has been compressed at depths for millions of years.

There are two basic stages to the metamorphic process. First, the volume of the rock is reduced as the rock is compacted, which reduces empty air space and water from the sample. After this, further pressure will cause recrystallization, which is a process by which the relationships and position of the individual grains, molecules, or atoms of a rock change to form a new crystal structure with a reduced volume. Many common stones, including marble, result from recrystallization; marble is a recrystallized form of limestone that has been subjected to intense pressures.

TYPES AND GRADES OF METAMORPHISM

Low-grade metamorphism occurs at temperatures between 200 and 320 degrees Celsius (608 degrees Fahrenheit) and at relatively low pressures, where high-grade metamorphosis occurs at temperatures above 320 degrees Celsius and at higher pressures. Rocks formed from low-grade metamorphism tend to be hydrous minerals, meaning that the rocks have water incorporated into their crystal matrix.

As pressure and temperature increase, water molecules are destabilized and removed from the rock. High-grade metamorphic rocks are characterized by having low levels of water molecules in their structure.

Many different physical conditions can lead to metamorphosis. One of the most basic types is contact metamorphism, which occurs when magma rises from the mantle into the crust. Only some of the surrounding rock becomes hot enough to melt. Just outside of this melting zone is a layer of rock that is subjected to intense heat, which may be sufficient in some cases to bring about metamorphic changes in the rock.

In rare cases, metamorphic rocks result from extraterrestrial impact events, such as when a meteor hits the surface of the earth. An impact of this type can generate sufficient pressure over a small area to bring about metamorphosis in the impacted rock. Evidence of an extraterrestrial impact also may include the presence of shock lamellae, which are thin strips of glass that form in the opposite direction from the rock's original crystalline structure. Shock lamellae form only under conditions of rapid, intense pressure, such as those that accompany a meteor impact.

Most types of metamorphic rocks form when rock is buried at sufficient depths to induce metamorphic changes. At increasing depths, rocks undergo increased heating because of their proximity to the mantle and increased lithostatic pressure, which is caused by the weight of the crust overlying them. A special type of burial metamorphism, called regional metamorphism, occurs over vast areas when large volumes of rock are simultaneously buried and subjected to metamorphic influences.

One of the most common scenarios for this type of metamorphism is the collision of two continents. The force of the collision causes the crust on both sides to deform, often leading to orogenesis, the formation of mountains. When the crust buckles at the site of impact, some of the crust is thrust beneath the surface, where it is subjected to intense pressure. As this process continues, layers of crustal rock are forced deeper into the earth, leading to metamorphism as pressures and temperatures increase.

ULTRA-HIGH-PRESSURE METAMORPHISM

Ultra-high-pressure metamorphism (UHPM) is a type of metamorphism that occurs when rocks are subjected to extremely high pressures and temperatures. The conditions that result in UHPM appear to be rare, and only twenty sites containing rocks generated by this type of metamorphism had been found between its discovery in the 1980's and 2010. Existing evidence of UHPM environments are associated with regional metamorphism related to the collision of tectonic plates and the subduction of crustal rock.

When oceanic tectonic plates are driven together, the older and therefore denser of the two plates will typically drive under the opposing plate in a process called subduction. The subducted plate is driven into the earth's mantle, where it desolidifies and blends with the rock of the mantle. This desolidified rock may eventually become magma, some of which will eventually rise toward the surface and solidify into igneous rock.

Another type of subduction, usually called A-type subduction, involves the collision of continental plates. Because continental crust has a lower density than oceanic crust, the denser of the two plates begins to subduct, but then rebounds because the crustal rock is not dense enough to remain in the mantle. A-type subduction zones often lead to orogeny, or the formation of mountains, as the two plates buckle against each other, forcing Earth materials to rise near the impact site.

UHPM occurs where a portion of the continental crust has descended to depths greater than 100 kilometers (62 miles) before returning to the surface. Geologists detect the existence of UHPM by looking for characteristic minerals that are created only in extreme temperature and pressure regimes. Crust that is transported to sufficient depths rapidly begins to recrystallize; however, water contained within the rocks tends to stabilize the matrix, slowing the process of converting low-density rock to high-density rock. When the force causing the subduction of the rock abates, the submerged strata is far more buoyant than the surrounding rock that typically exists at this depth, causing the recrystallized rock to rise rapidly. This relatively rapid return to the surface (though still requiring thousands of years) preserves mineralogical evidence of UHPM in the rock, whereas a longer period of rising to the surface would allow metamorphosis to occur a second time, returning most of the metamorphic rock to its original form.

EVIDENCE OF UHPM

One of the most common types of rock produced by UHPM is coesite, which forms when high pressures and relatively high temperatures are applied to quartz. Quartz is one of the most abundant materials on the surface of the earth and is made from a combination of silicon and oxygen atoms arranged in a crystal lattice. Coesite is unstable at sea-level temperatures and pressures and will eventually decay in this environment, returning to quartz.

Samples of intact coesite are rarely uncovered because of coesite's extreme instability at sea level. However, geologists can utilize indirect evidence to determine if coesite was present in a recovered sample of quartz. As coesite returns to quartz form, it forms into a type of quartz called polycrystalline quartz, which is characterized by the presence of microscopic crystal fragments called crystallites. Coesite has 10 percent less volume than quartz; therefore, when coesite metamorphoses into quartz, it expands in such a way that it causes the development of characteristic microscopic fractures in the surrounding quartz. Geologists use the presence of these pressure fractures in samples of polycrystalline quartz to determine if the quartz was subjected to UHPM environments that led to the formation of coesite.

In rare cases, geologists have discovered microdiamond inclusions in rocks subjected to UHPM environments. Microdiamonds are diamonds that are typically less than 1 micrometer in extent and are formed only in extreme high-pressure environments. Microdiamonds have been found as inclusions in zircon, a common mineral that occurs in sedimentary and igneous rocks as small inclusions. Microdiamonds also have been uncovered from samples of kyanite, a silicate mineral that forms in metamorphic environments in which the pressure rises into the thousands of pascals. Geologists have found microdiamonds formed in rocks that have undergone pressures of more than 5,000 megapascals and temperatures exceeding 1,000 degrees Celsius (1,832 degrees Fahrenheit).

Geologists are working to enhance the ability to find and study UHPM zones and to refine models that explain the formation of these extreme lithographic environments. Research into UHPM gained widespread attention in the 1990's, and since that time UHPM environments have been discovered on every continent.

The process of subduction, followed by return to the crust, takes millions of years to complete, and most evidence of UHPM has been found in sediment from ancient environments. The best-known sites to uncover UHPM minerals include the Daba Shan mountains of eastern China and the Dora-Maira massif of the Italian Alps.

Micah L. Issitt

FURTHER READING

Bucher, Kurt, and Rodney Grapes. *Petrogenesis of Metamorphic Rocks.* 8th ed. New York: Springer, 2011. Complex discussion of mineral composition and the construction of metamorphic rocks. Includes a detailed discussion of ultra-high-pressure metamorphic processes.

Dobrzhinetskaya, Larissa, et al., eds. *Ultrahigh-Pressure Metamorphism.* Burlington, Mass.: Elsevier, 2011. Complex description of metamorphic processes and the related phenomenon of subduction written for advanced students of geologic sciences.

Frisch, Wolfgang, Martin Meschede, and Ronald Blakey. *Plate Tectonics: Continental Drift and Mountain Building.* New York: Springer, 2011. Introduction to the science of plate tectonics with detailed discussion of geologic processes. Contains a brief discussion of ultra-high-pressure metamorphism.

Grotzinger, John, and Thomas H. Jordan. *Understanding Earth.* New York: W. H. Freeman, 2009. Introduction to Earth sciences and Earth system science written for general readers. Discusses subduction and ultra-high-pressure rock formation.

Monroe, James S., Reed Wicander, and Richard Hazlett. *Physical Geology.* 6th ed. Belmont Calif.: Thompson Higher Education, 2007. College-level text covering issues in geologic processes and research. Contains a brief discussion of metamorphic rock formation and the processes involved in ultra-high-pressure geologic environments.

Wicander, Reed, and James S. Monroe. *The Changing Earth: Exploring Geology and Evolution.* 5th ed. Belmont, Calif.: Brooks/Cole, 2009. Introduction to geologic principles and research methods. Discusses plate tectonics and subduction and their relation to the formation of ultra-high-pressure rock formations.

See also: High-Pressure Minerals; Igneous Rock Bodies; Metamorphic Rock Classification; Metamorphic Textures; Orbicular Rocks; Sub-Seafloor Metamorphism; Ultrapotassic Rocks.

ULTRAPOTASSIC ROCKS

Igneous rocks, from which ultrapotassic rocks derive, form when magma from within the earth's mantle rises toward the earth's surface and cools to form solid rock. Magma pockets lead to a variety of igneous deposits, including batholiths, sills, dikes, and volcanic pipes. Ultrapotassic rocks are a type of igneous rock that form deep within mantle rock contained within a craton, a stable portion of the mantle.

PRINCIPAL TERMS

- **craton:** stable portion of the earth's mantle and associated crust that forms the inner core of the lithospheric components
- **diamond:** mineral formed from carbon atoms organized in a characteristic structure known as a diamond lattice
- **diamond stability zone:** area within the earth's mantle where pressures and temperatures are sufficient for the formation of diamonds from deposits of carbon
- **dike:** igneous rock formation that develops in a vertical or diagonal direction from the surface of the earth and often connects to a magma chamber within the earth's crust or mantle
- **igneous rock:** type of rock formed when magma that develops within the earth's mantle rises to the surface, where it cools and solidifies into solid rock structures
- **magma:** molten rock, gas, and liquid minerals that develop within the earth's mantle from decompression, molecular excitation, and heat from the earth's core
- **volcanic pipe:** deposit of igneous rock formed when a deep magma pocket produces an explosion of gas and magma that rapidly rises to the surface and forms a cone-shaped deposit
- **xenocryst:** crystal that differs from the surrounding rock found in igneous rock deposits
- **xenolith:** inclusion of rock in the body of a different type of rock; often occurs in igneous rock formations

INTRUSIVE IGNEOUS ROCKS

Igneous rocks are formed from magma, a mixture of molten rock, gases, and other trace chemicals that form in pockets within the earth's mantle. The mantle is a layer of rock heated by radioactive energy from the earth's core and pressurized by the weight of rock above it.

The mantle shifts in response to thermal currents from the core. As this occurs, localized disturbances cause part of the mantle to destabilize. The molecules within the disturbed mantle move freely, generating energy that melts the surrounding rock. Magma is lower in density than surrounding rock and therefore rises to the surface, pushing through the solid rock of the mantle and the crust.

When magma cools in the crust or on the surface, it solidifies into igneous rocks. Magma that breaks through the surface of the crust before cooling is called lava and gives rise to volcanic rock or extrusive igneous rock. When magma cools beneath the earth's surface, insulated within pockets of solid rock, it develops into intrusive igneous rock or plutonic rock. Plutonic rock cools slowly, in millions of years, which causes the development of larger crystals. For this reason, intrusive igneous rock tends to be characterized as phaneritic in texture, meaning that the grains of the rocks can be discerned by the naked eye.

Igneous rocks also are characterized by their chemical composition. The primary components of igneous rocks are silica, a molecular combination of silicon and oxygen, and feldspar, which is another silica-based crystal that contains a variety of included materials. Some igneous rocks, called felsic rocks, are composed primarily of silica (55 to 70 percent) and frequently contain a variety of potassium-rich feldspars. By contrast, mafic rocks have lower levels of silica (less than 50 percent) and are rich in iron and magnesium. Ultramafic rocks have even lower concentrations of silica (less than 45 percent) and correspondingly higher concentrations of included metals.

IGNEOUS ROCK DEPOSITS

Magma that cools within the earth's crust forms into plutonic rock deposits. The largest plutonic deposits are called batholiths, which result when large pockets of magma join as they cool. Batholiths may exist deep within the mantle, within the continental cratons, which are the deep, solid portions of the lithosphere least affected by the collisions between tectonic plates.

Batholiths give rise to dikes, which are vertical or diagonal segments of plutonic rock that rise through the surrounding strata toward the surface. Dikes also may give rise to horizontal layers of igneous rock, called sills, that form when magma infiltrates layers of rock parallel to the surface.

At times, magma pockets deep within the earth may erupt because of a buildup of volcanic gases, passing rapidly through the mantle and crust to erupt onto the surface. These formations, called volcanic pipes, result in vertical, cone- or carrot-shaped deposits of igneous rock connected to deep subterranean deposits. Volcanic pipes may result from magma pockets that develop hundreds of kilometers beneath the earth's surface.

ULTRAPOTASSIC IGNEOUS ROCKS

Ultrapotassic igneous rocks are rocks that contain high levels of potassium in the form of potassium oxide (K_2O). A rock is considered ultrapotassic if it contains more than 3 percent potassium oxide by weight and contains a minimum of two times more potassium oxide than sodium oxide (Na_2O).

In addition, ultrapotassic rocks must contain more than 3 percent magnesium oxide (Mg_2O) by weight. Ultrapotassic rocks are usually mafic or ultramafic in composition and are a relatively rare form of igneous rock.

The formation of potassic and ultrapotassic rocks begins with the enrichment of magma deposits with potassium oxide-rich minerals from the surrounding rock. When magma of this type is subjected to intense pressures, such as those that arise from the collision of continental plates, ultrapotassic rocks may develop. Potassic rock sediments uncovered in Mediterranean and Italian strata appear to have resulted from the collision of continental plates.

In rare occasions ultrapotassic rocks may result from magma that forms above an active subduction zone, an area in which one tectonic plate collides with and is forced under another tectonic plate. Ultrapotassic sediments from the Sunda arc in Indonesia, Papua New Guinea, and Fiji originated from this tectonic environment.

The most uncommon environment for ultrapotassic rock formation is through intraplate magmatism, or magma that forms somewhere in the central portion of a tectonic plate. This type of magmatism is known as a hot spot, where a deep magma pocket pushes through the mantle and crust to create a volcano or a volcanic pipe. Ultrapotassic rock deposits from South Africa and western Uganda developed from intraplate magmatism surrounding the development of a hot spot.

Ultrapotassic rocks are generally divided into three types: 1, 2, and 3. Type 1 ultrapotassic rocks, called lamproites, have relatively low percentages of calcium oxide (CaO), aluminum oxide (Al_2O_2), and sodium oxide (Na_2O_2) coupled with high percentages of potassium and magnesium oxides. Type 2 rocks, called kamafugites, have low levels of silicate and high levels of calcium oxide. Type 3 rocks typically occur in areas in which mountain formation has taken place and usually include high levels of calcium and aluminum oxide.

LAMPROITES

Lamproites are ultrapotassic rocks in which the ratio of potassium oxide to sodium oxide exceeds 5 percent by weight. Most forms of lamproite have between 6 and 8 percent potassium oxide by weight.

Lamproites generally have high levels of magnesium oxide and are composed of between 45 and 50 percent silica, which categorizes them as mafic igneous rocks. Lamproites have been discovered connected to dikes and sills and also to batholith deposits. One of the most common locations for lamproites is within volcanic pipe deposits.

Lamproites generally develop only when a certain set of specific geoformational processes have taken place. The first stage occurs in the Benioff zone, which is an area of deep seismic activity that occurs beneath a subduction zone. Millions of years later the minerals created by subduction over a Benioff zone are subjected to a process in which two continental plates collide, leading to orogenesis, or mountain formation. In these conditions, lamproites may develop in the deep sediment, near the edges of the continental craton. Generally, lamproites will form only when a portion of the mantle melts at depths greater than 150 kilometers (93 miles) and at temperatures of between 1,100 to 1,500 degrees Celsius (2,012 to 2,732 degrees Fahrenheit).

Lamproites are of particular interest in geology because of the number of rare minerals the rocks contain. As lamproite magma rises to the surface, it may carry with it a variety of xenoliths, which are incompatible minerals or pieces of other rock types that become included in the magma as it hardens. Inclusions in lamproites often include rocks that are found only in deep mantle strata. These mineral types include a variety of iron-rich and titanium-rich minerals. Deposits of aluminum, nickel, quartz, and chromium also are commonly found in lamproites.

KIMBERLITES

Kimberlites are ultrapotassic rocks derived from intraplate magma activity, generally confined to portions of the crust in which the plate overlies old portions of a continental craton. Kimberlites appear to be related to hot spots that develop within the core of a continental craton. Most kimberlite deposits occur within five degrees of a known hot spot, while others that are not close to a recent hot spot may have originated along the moving path of an ancient hot spot.

Kimberlites are volatile rocks because they contain high levels of carbon dioxide (CO_2) and water (H_2O) within their structure, leading them to fracture easily at sea-level pressure and temperature. Kimberlites have low sodium oxide levels and frequently contain a large variety of incompatible elements and minerals incorporated into their matrices.

Kimberlites occur in dikes and sills but are best known from large deposits uncovered from volcanic pipes. Most kimberlite deposits contain inclusions of ultramafic rocks from the upper mantle and may also contain xenoliths and xenocrysts, a type of xenolith consisting of individual crystals embedded within the matrix of another rock type.

Kimberlites are generally divided into two groups, based on the level of mica, a mineral silicate containing aluminum, potassium, magnesium, iron, and silicon. Both types of kimberlite may occur together in the same deposit.

Kimberlites are derived from magma pockets that originated at depths of between 100 and 200 km (62 to 124 mi) beneath the surface and are created by partial melting of mantle rock containing high levels of water and carbon dioxide. Kimberlites are usually formed in situations where intraplate magma rises to the surface rapidly, in less than ten years. Most known kimberlite deposits originated in strata that developed before the Cenozoic period (65 million years ago) indicating that the tectonic and geologic processes involved in the formation of kimberlite may have been more common in the ancient Earth's lithosphere.

ULTRAPOTASSIC ROCKS AND DIAMONDS

Ultrapotassic rocks are economically important because of the discovery of diamonds as xenocrysts in deposits of kimberlite and lamproite. Kimberlite is named after the town of Kimberly, in the Northern Cape region of South Africa, where diamonds were first discovered in kimberlite deposits in the 1870's. The first known lamproite diamond deposit was discovered in 1979 in Western Australia.

Lamproite diamonds are often of poor quality in comparison with kimberlite inclusions. However, lamproite deposits sometimes contain rare-colored diamond inclusions and a variety of other precious stones.

Diamonds form within cratons at depths of between 140 and 300 km (87 and 186 mi) beneath the surface, a depth that is known as the diamond stability zone. At deeper or shallower depths, diamonds will not form. Diamonds may therefore be included as xenocrysts in deposits of kimberlite and lamproite that developed from magma that originated within the mantle below or within the diamond stability zone.

Diamonds most frequently occur in kimberlite and lamproite that hardens within volcanic pipes. These explosions of magma and gas quickly rise to the surface, thereby dragging inclusions of various mantle rocks, xenoliths, and xenocrysts, which may include diamonds. Many of the world's most productive diamond mines are part of ancient kimberlite pipes.

Cratons occur in two general types, based on the age of material contained within the craton. Those composed of mantle rock and crust that are more than 2.5 billion years old are called Archean age cratons, while those arising from mantle rock that is between 1.6 and 2.4 billion years old are called Proterozoic age cratons. Diamonds are usually found in kimberlite and lamproite derived from Archean age cratons containing strata that is 2.4 billion years old or more.

Kimberlite and lamproite pipes, sills, and dikes are derived from magma chambers that develop deep within the center of a craton. These pipes, sills, and dikes, therefore, provide geologists with samples of rocks and minerals from the ancient earth in addition to precious and semiprecious stones. For this reason, kimberlite and lamproite are important from both an economic and a scientific perspective. Minerals and mantle rocks included within kimberlite and lamproite deposits give geologists data regarding the composition and environmental conditions of the ancient Earth at the time when these rocks first formed or became incorporated into the mantle.

Micah L. Issitt

FURTHER READING

Duff, Peter McLaren Donald. *Holmes' Principles of Physical Geology.* 4th ed. London: Chapman & Hall, 1994. Basic introduction to geologic science and research. Chapter 5 introduces igneous rocks and the processes that lead to their formation. Contains a brief discussion of ultrapotassic rocks.

Gill, Robin. *Igneous Rocks and Processes: A Practical Guide.* Hoboken, N.J.: Wiley-Blackwell, 2011. Detailed examination of the chemical and physical processes involved in igneous rock formation and development. Contains a discussion of ultrapotassic rocks and their formation.

Monroe, James S., Reed Wicander, and Richard Hazlett *Physical Geology.* 6th ed. Belmont Calif.: Thompson Higher Education, 2007. General text covering physical geology and geologic research. Contains a discussion of igneous rock formation, chemistry, and identification.

Montgomery, Carla W. *Environmental Geology.* 4th ed. Columbus, Ohio: McGraw-Hill, 2008. Introductory text covers aspects of geologic research, development, and theory with an emphasis on the environmental role of the lithosphere and the impact of lithospheric development.

Stanley, Steven M. *Earth System History.* New York: W. H. Freeman, 2004. Detailed introduction to Earth system science, placing the features and processes of the lithosphere within the context of their relationship to geochemical cycles and the earth's systems as a whole.

Wilson, Marjorie. *Igneous Petrogenesis: A Global Tectonic Approach.* New York: Springer, 2007. Highly detailed text aimed at helping geology students understand the detailed processes involved in the formation of igneous rocks. Contains a thorough discussion of ultrapotassic rocks.

See also: Batholiths; Crystals; Geologic Settings of Resources; High-Pressure Minerals; Igneous Rock Bodies; Orbicular Rocks; Rocks: Physical Properties; Ultra-High-Pressure Metamorphism.

UNCONVENTIONAL ENERGY RESOURCES

Unconventional energy resources are forms of energy harvested from sources other than those utilized for the majority of human energy needs. The vast majority of energy produced for human consumption comes from the combustion of fossil fuels, which are materials that form from the decomposition and fossilization of organic material. Unconventional energy resources include both new sources for potential energy generation and new ways to utilize existing sources of energy. Unconventional sources of fossil fuels include oil shale and tar sand, hydroenergy, the kinetic energy of ocean waves, biofuel, and nuclear energy.

PRINCIPAL TERMS

- **bitumen:** viscous, semiliquid material produced as a by-product of the decay of organic material; typically called tar
- **fossil fuel:** fuel source derived from the fossilized remains of organic matter, including coal, petroleum oil, and natural gas
- **gasification:** process that converts carboniferous material to methane and carbon gases by subjecting the material to extreme heat in an environment that prevents combustion
- **hydroelectricity:** electricity produced by converting the kinetic energy within moving water
- **kerogen:** complex fossilized form of organic matter that forms oil and natural gas
- **kinetic energy:** the energy possessed by an object because of its motion
- **nuclear fission:** the process of inducing the nucleus of an atom to split into lighter nuclei, which produces free neutrons and energy
- **oil shale:** sedimentary rock that contains solid deposits of kerogen as inclusions within the rock matrix
- **renewable energy:** energy source that does not depend on a finite, exhaustible resource
- **turbine:** engine that uses the viscosity and movement of a fluid medium, such as a liquid or gas, to perform mechanical work, turning a rotary mechanism that can be used to generate other types of energy

UNCONVENTIONAL ENERGY RESOURCES

Fossil fuels are materials generated by the decomposition of organic matter through millions of years. The remains of dead animals and plants become buried under successive levels of sediment and are subjected to increasing pressure from the weight of material overlying them and to increasing heat generated by chemical decomposition and thermal heat from the earth's core.

Through millions of years this mixture of organic material will give rise to kerogen, a tar-like material that will further develop into petroleum and natural gas with even more temperature and pressure. Further compression at great depths leads to the formation of coal, a type of igneous rock.

Coal, petroleum oil, and natural gas are mined from pockets within the earth and then shipped to facilities in which they are subjected to combustion. Heat produced by the combustion of fossil fuels is then filtered through turbine engines to generate electricity. Between 80 and 90 percent of global consumer energy is derived from the combustion of fossil fuels.

The major drawbacks to fossil fuels include the production of environmental pollutants, such as carbon dioxide and methane, which leads to global warming and other types of ecological damage. Fossil fuels are a limited resource because it takes millions of years for new deposits to form from organic matter.

As conventional supplies of fossil fuels have begun to dwindle, scientists are working to develop new resources for energy development, often called unconventional or alternative energy resources. Some unconventional energy resources are based on finding new ways to obtain fossil fuels, while others focus on renewable energy, which is energy derived from potentially limitless sources.

UNCONVENTIONAL FOSSIL FUELS

Several methods have been developed to harvest fossil fuels from unconventional sources or to convert less desirable forms of fossil deposits into usable fuels. Although these unconventional deposits are available in large quantities, they are more difficult and expensive to harvest, transport, and process than are conventional deposits.

Oil sands and tar oils. Vast quantities of fossil fuel deposits are mixed with sand and other minerals, often called oil sands or tar oils. Fossilized sediments in tar oils are in the form of bitumen, a thick, semiliquid material often called tar; only 10 percent of tar sand is composed of bitumen. The harvest of this fossil fuel requires vast amounts of raw material. For each barrel of oil, 2 to 4 tons of tar sand and four barrels of water must be used, creating a significant amount of waste. It is estimated that tar-sand processing produces 300 percent more waste than traditional oil production. In addition, harvesting oil sand requires mining, which leads to environmental damage to vast ecological areas.

Oil shales. Fossil fuels also can be harvested from oil shales, which are sedimentary rock deposits that contain solid kerogen. When oil shale is mined, processed, and subjected to intense heat, petroleum oil can be extracted. Oil shale is more complex than conventional oil to process because the rocks must be crushed and heated before oil can be extracted—a process called retorting. Oil shales have been harvested in small quantities since the 1970's but are not preferred because of the increased cost and difficulty of processing.

Harvesting oil shale has a number of drawbacks in addition to high cost, including significant environmental impact. Harvesting shale requires more raw material than harvesting raw oil and therefore causes higher levels of environmental damage during the mining process. Oil shale harvesting has been linked to increased levels of water pollution and increased levels of carbon dioxide emissions. In addition, though oil shale accounts for trillions of barrels of potential oil, it is a finite resource that will eventually be exhausted if oil consumption continues to increase at its current rates.

HYDROENERGY

The movement of water within oceans, rivers, and streams produces kinetic energy that can be harnessed and used to generate electric energy. A number of methods are used to capture hydroenergy, most of which rely on the same basic mechanism—utilizing moving water to rotate turbine engines—which fuels the generation of electric power.

Hydroelectric power. Hydroelectric power is the most common type of hydroenergy, accounting for more than 50 percent of renewable energy production and more than 15 percent of electricity generated in many countries. To harvest hydroelectricity, engineers construct a dam such that there is a significant difference in elevation between the water on one side of the dam and that on the other. Water is prevented from falling into the lower level by a series of gates. When these gates are opened, water flowing through channels passes over turbines, which spin to generate electric current. At night, when human power consumption` is typically at its lowest, water is pumped from the lower basin into the upper basin, thereby preparing the dam to produce more energy for the next day.

Hydroelectric dams provide a renewable energy source, but the installation and operation of dams causes environmental and ecological damage. Because dams alter water depth, they destroy habitat for many types of birds, aquatic mammals, fish, reptiles, and amphibians. In addition, dams block passageways that are used by certain animals for daily or migratory movements.

Tidal energy. Another way to generate hydroenergy is to use the kinetic energy of tidal movement. In bays and estuaries, tidal energy is harvested using a special kind of dam, a tidal barrage, that functions similarly to a hydroelectric dam but utilizes the movement of tidal currents to push water through the turbine channels. Because tidal barrages utilize the motion of tidal currents, they require less energy to operate than do hydroelectric dams. However, barrages pose a similar risk to the environment by blocking travel for many types of species and by altering water depth, which affects the surrounding habitat.

Another method of harnessing tidal energy is to use a series of turbine engines placed just under the surface of the water. Called tidal turbines, these machines are useful in areas where steady, reliable tidal-current levels dominate. Tidal turbines are similar to wind turbines, but because water is more viscous than air and because tidal currents are more constant than air currents, tidal turbines provide a more reliable source of continuous energy. Tidal turbines are more affordable and less complex than are tidal barrages, but the technology is still in its infancy, and most tidal turbine programs are in the experimental stage.

Tidal turbines produce little pollution and do not require significant environmental damage during installation, but research indicates that they may disrupt the movements of fish and other aquatic animals. Research is ongoing to investigate the best ways to implement tidal turbine technology with the lowest level of ecological impact.

Wave energy. Another relatively new and still experimental method for capturing hydroenergy involves utilizing the kinetic energy in ocean waves to drive electric generators. As wind passes over the surface of the ocean, it transfers a certain amount of kinetic energy to the water, thus generating waves. The strength of surface waves therefore varies according to meteorologic and atmospheric conditions.

Several different models of wave power are being investigated for commercial potential. One of the most promising techniques involves the use of a long, floating, tube-like structure with hinged joints that allow the structure to flex and bend in response to the undulating motion of ocean waves. As the sections of the tube bend, hydraulic rams between the sections are forced to compress, pushing liquid through a hydraulic turbine and thereby generating electricity.

Experimental evidence suggests that wave-power generators pose a low level of environmental threat because most of the technology is mounted at the surface of the ocean. The primary area of concern for environmental scientists is that the wave generators may cause noise pollution in the aquatic environment. Wave technology is a relatively recent field, and research is ongoing to investigate the potential environmental effects of long-term operation.

BIOFUELS

Biofuels are alternatives to fossil fuels and are made from living plant matter rather than from fossilized remains. The concept of biofuels was developed in the early twentieth century, but the relative ease of fossil fuel mining and production largely stalled research and development until the end of the century. The two most common biofuels in development are ethanol and biodiesel.

Ethanol. Ethanol is an alcohol that is produced in a similar way to beer or spirits. Carbohydrates from plant matter are fermented through the addition of organic acids; fermentation transforms the carbohydrates into alcohols. Most modern biofuels are made using the sugars and starches from corn, but it also is possible to use the cellulose from plant waste to manufacture ethanol. Alternatively, ethanol can be produced by a process called gasification, in which plant matter is subjected to high heat in an atmosphere devoid of oxygen; this converts the plant matter into synthesis gas, or syngas, a hydrocarbon-rich blend of gases. Syngas can then be utilized in separate processes to create ethanol or other fuels.

Biodiesel. Biodiesel is another type of biofuel made by combining alcohol with vegetable or animal oils. Biodiesel is typically manufactured from ethanol and vegetable oil derived from corn. However, methods exist to create biodiesel using recycled oil, such as used cooking oil. Biodiesel can be used directly in cars manufactured to run on diesel but cannot be used as a sole fuel source for petroleum-burning engines.

Both ethanol and biodiesel are used as additives, which are blended by traditional gasoline to more efficient fuels that lower levels of pollutants in their emissions. Support for biofuel increased rapidly in the 1990's, largely because of the potential for biofuel to create a lucrative new market for agricultural crops. Government subsidies were established to support farmers who dedicated a certain portion of their growing efforts to producing corn for biofuel production. In addition, biofuel can potentially be generated using plant waste and other recycled materials, which could mean a significant reduction in solid waste on a global scale.

One of the primary criticisms of biofuels is that production methods utilize petroleum gas in the manufacturing process, mitigating the ultimate goal of reducing fossil fuel reliance. In addition, biofuels have been used to bolster the corn market, and increased corn production can lead to increased use of nitrogen fertilizers, which also constitute a destructive pollutant. Some estimates indicate that increased nitrogen pollution may be more harmful than the carbon dioxide emissions produced by fossil fuel combustion. In addition, the increased agricultural area needed for corn production can lead to further deforestation, for example, thereby hastening environmental collapse in some ecosystems.

NUCLEAR POWER

Nuclear power uses energy produced by nuclear fission to generate heat, which is then used to generate electricity. Fission is the process by which the nucleus of an atom is induced to split into smaller nuclei, producing light and heat as a by-product of the reaction.

Nuclear power can come from the fission of uranium, plutonium, and thorium, though most modern nuclear reactors utilize uranium. Uranium is mined from sediments, where it occurs in two varieties, usually blended together within the matrix of sedimentary rock. Uranium-235 can be used for fission, while the more common uranium-238 is not generally appropriate for use as reactor fuel.

To induce fission, samples of uranium are bombarded with a stream of free neutrons. When one of the neutrons strikes an atom of uranium-235, it causes the atom to split, thereby releasing heat. Heat from this reaction is used to convert water into water vapor, which rises with force and spins the blades of large turbine engines to generate electricity. In essence, then, though nuclear fission involves the release of vast amounts of atomic energy, this atomic energy must be converted to kinetic energy before it is used to generate electricity. The basic method of utilizing kinetic energy from rising currents to operate a turbine engine is similar to the method used to generate electricity from fossil fuel combustion.

Approximately 82 percent of the energy produced through nuclear fission is converted to kinetic energy, making nuclear energy a more efficient method of producing energy than combustion. Research indicates that fission of one atom of uranium produces 10 million times more energy than the combustion of an atom of coal. In addition, nuclear fission does not produce carbon dioxide as a by-product; instead, it utilizes atmospheric carbon dioxide to cool the reactor environment, thereby contributing little to atmospheric pollution.

While nuclear power generates little atmospheric pollution, the process produces significant amounts of hazardous waste that pose an environmental and ecological threat. Nuclear reactors around the world together generate between ten and twenty tons of radioactive waste each year. Nuclear waste produces heat and radiation, which can be harmful or lethal to nearby organisms. Nuclear waste also can pollute soil and groundwater, spreading the harmful effects of the radiation over a large area. Safely disposing of the waste is an expensive and dangerous process that involves burying waste in large, cooled concrete storage containers filled with glass particles. These waste disposal containers must be monitored for leaks and to keep the radioactive material safe from tampering.

There are additional concerns regarding nuclear energy—namely, the potential for nuclear waste or nuclear fuel to be stolen and used to manufacture weapons. Given security and monitoring processes, the risk of nuclear theft is extremely low, yet this fear has contributed to negative public opinion toward nuclear power.

In addition, though technology allows for the safe operation of nuclear power plants, the potentially catastrophic effects of a nuclear plant meltdown have further reduced popular support. Because of these drawbacks both in waste and in public opinion, the use of nuclear fission is declining on a global scale, despite nuclear energy being a cleaner, more sustainable, and more affordable way to manufacture energy.

Micah L. Issitt

FURTHER READING

Boeker, Egbert, and Rienk von Grondelle. *Environmental Physics: Sustainable Energy and Climate Change.* New York: John Wiley & Sons, 2011. Detailed text on environmental development as it relates to sustainable energy development, with a focus on climate change. Written as a reference for students of physics and environmental sciences.

Da Rosa, Aldo Viera. *Fundamentals of Renewable Energy Processes.* New York: Academic Press, 2009. Introduction to renewable and alternative energy research and technology. Contains data on nuclear energy, tidal energy, wind energy, and solar power.

Fanchi, John R., and Christopher J. Fanchi. *Energy in the Twenty-First Century.* 2d ed. Hackensack, N.J.: World Scientific, 2011. Introduction to energy resources and use addressing the relationship between energy resources and politics. Contains sections discussing alternative energy resources and the future of energy management.

Ferrey, Steven, and R. Anil Cabral. *Renewable Power in Developing Countries: Winning the War on Global Warming.* Tulsa, Okla.: Pennwell Books, 2006. Addresses issues in global warming and its relationship to energy management in developing Asian countries. Contains information on nuclear, hydroelectric, and wind energy production.

Galarraga, Ibon, Mikel Gonzalez-Eguino, and Anil Markandya. *Handbook of Sustainable Energy.* Northhampton, Mass.: Edward Elgar, 2011. Detailed text on sustainable energy resources and the technology used to develop alternative energy programs. Includes a detailed discussion of climate change and the economics of alternative energy.

Sorensen, Bent. *Renewable Energy: Its Physics, Engineering, Environmental Impacts, Economics, and Planning.* 4th ed. Boston: Academic Press, 2011. Reference work containing information on alternative energy technologies, including solar, tidal, and nuclear power. Also contains information on energy cycling and the environmental impact of current energy trends.

See also: Geologic Settings of Resources; Hydroelectric Power; Nuclear Power; Ocean Power; Oil and Gas Distribution; Radioactive Minerals; Solar Power; Tidal Forces; Uranium Deposits; Water and Ice; Wind Power.

URANIUM DEPOSITS

Uranium is a radioactive element that naturally occurs on Earth. The element is used in the manufacture of weapons and as a fuel source for power plants. Uranium has proved to be a valuable economic and political resource, and its extraction and use have become a pressing international concern.

PRINCIPAL TERMS

- **alpha particle:** essentially a helium nucleus, a particle consisting of two protons and two neutrons, emitted by radioactive atoms during alpha decay
- **beta particle:** an electron emitted from the nucleus of an atom during the process of beta decay
- **gamma ray:** a high-energy photon, with no mass or charge, emitted when atomic nuclei are broken down
- **half-life:** the amount of time it takes for half of a sample of a radioactive element to decay
- **highly enriched uranium:** uranium that has been processed to alter its composition of uranium isotopes, making it more useful as fuel or for nuclear weapons
- **ionizing radiation:** radiation that creates ions by breaking down atomic nuclei
- **isotopes:** atoms of the same element with differing numbers of neutrons in their nuclei
- **nuclear fission:** a reaction in which atomic nuclei are broken down; generates a large amount of energy
- **plutonium:** a radioactive element with ninety-four protons in its nucleus; can be used as an energy source in the same way as uranium
- **radiation:** unstable isotopes of an element decay in various ways; the emission of particles and energy from nuclei is known as radiation
- **uranium:** an element with ninety-two protons in the nucleus and with only radioactive isotopes
- **uranium ore:** uranium as it is typically found in nature, as shiny black rocks that may be found combined with a variety of other elements
- **yellowcake:** uranium bonded with oxygen that has been milled from the uranium ore found in nature; has a yellow, powdery appearance

CHEMICAL AND PHYSICAL PROPERTIES

Uranium is an element containing ninety-two protons in the nucleus. It is the heaviest element found in any significant quantity on Earth. (Larger elements, such as plutonium, are found only in small quantities. Most of the heavier elements exist only when created in laboratories.) Uranium's uses and properties can vary based on the number of neutrons present.

Three main isotopes of uranium are found in nature. More than 99 percent of naturally occurring uranium is uranium-238. All isotopes of uranium are radioactive.

Uranium-238, which contains 146 neutrons, takes the longest of the isotopes to decay, with a half-life of roughly 4.5 billion years. In a lengthy chain of decay, uranium-234 is one of the eventual by-products, with a half-life of 240,000 years. Uranium-238 cannot sustain the chain reactions used by nuclear fission to generate energy. With the addition of a neutron to the nucleus, however, which is done in a nuclear reactor, uranium-238 can be converted to plutonium-239, which can be used for nuclear fission. Uranium-235 is the only isotope of uranium that can sustain a chain reaction necessary for nuclear fission. As a general rule, for isotopes to be usable for nuclear fission, they must have an odd number of neutrons in the nucleus.

Radioactive nuclei can decay in different ways. In alpha decay, which is what occurs with all three isotopes of uranium, an alpha particle is emitted from the nucleus. When the nucleus has an odd number of neutrons, a significant amount of energy is released in the form of a gamma ray. This energy allows the breakdown of the nucleus to continue further, particularly when a large number of odd-neutron–numbered nuclei are present. Enrichment of uranium, which involves increasing the ratio of uranium-235 to uranium-238 in a sample, is the focus of the primary applications for the element.

A HISTORY OF URANIUM

Uranium was officially discovered as an element in the late eighteenth century, but some of its features have been known for much longer. For example, the coloring effects of uranium have been found in the stained glass of the ancient Romans.

Uranium was found largely by accident. Originally, the shiny black rocks of uranium ore were considered the waste product of a silver mining operation in the town Joachimsthal, in modern-day Czech Republic. The material was called pitchblende; it is now known as uranium ore. The silver mining operation also led to illness among the workers, illness that was not immediately diagnosed by doctors of the time.

The discovery of uranium itself is credited to German chemist and pharmacist Martin Klaproth, who, in 1789, isolated the metal from uranium ore mined in Joachimsthal that had been sent to him. Klaproth named uranium for the planet Uranus, which was the most recently discovered planet at the time. Uranium originally was intended to be a temporary name.

Even after the discovery of uranium, its most important properties remained unknown. In the nineteenth century it was used in artwork. This began to change in the twentieth century, soon after French physicist Antoine Henri Becquerel discovered radiation with a sample of the uranium. At the time, theories of emission tied radiation to light from the sun. In 1896, Becquerel left a sample of the uranium compound on a sheet of photograph paper on a cloudy day. The image generated from the compound indicated that the radiation came from the uranium rather than the sun.

The discovery of radiation unleashed a number of scientific discoveries, including the work of Pierre and Marie Curie, who discovered the radioactive elements polonium and radium from those same pitchblende rocks from Joachimsthal. In 1903, Becquerel and the Curies were awarded the Nobel Prize in Physics for the discovery of radiation. (Marie Curie received the Nobel Prize in Chemistry in 1911 for the discovery of the two elements.)

Discoveries followed that would cast further light on the radioactive nature of uranium. Working together at McGill University in Montreal, Canada, Ernest Rutherford and Frederick Soddy discovered the alpha particle, the key component in the radioactive decay of uranium. Rutherford received the Nobel Prize in Chemistry in 1908, while Soddy received the award in 1921.

The research that ultimately brought uranium to the forefront of the public consciousness was the U.S. government's Manhattan Project, directed by physicist J. Robert Oppenheimer. The project involved the development, construction, and detonation of the world's first atomic bombs, near the end of World War II.

On July 16, 1945, the United States tested the first atomic bomb, in New Mexico. On August 6, an atomic bomb was dropped on the Japanese city of Hiroshima. The immediate death toll was estimated at seventy thousand people, with that number doubling in the next year as a result of radiation exposure. A second bomb was dropped on Nagasaki, Japan, on August 9. Fighting stopped the next day, and Japan surrendered on August 14, ending World War II. Within the next few years, both the United Kingdom and the Soviet Union successfully tested nuclear bombs of their own, changing the landscape of geopolitics for the nuclear age.

Civilian uses for nuclear power soon followed. In 1951, a working nuclear reactor was built in southeastern Idaho. By the end of the decade, full-scale nuclear power plants had been built. In the years to come, people would get a glimpse of the strong potential for nuclear power, and its destructive capabilities. Many of the issues that would arise continue into the twenty-first century and will shape the future of nuclear power use.

In 1963, discussions began regarding guidelines for nuclear weapon testing and usage. The framework was thus begun for the Treaty on the Non-Proliferation of Nuclear Weapons (1970). The treaty called for inspections to prevent development of further nuclear weapons beyond the five countries that had built the bomb (the United States, the Soviet Union [now Russia], the United Kingdom, France, and China). Many other nations signed the treaty; the only major notable holdouts were Israel, India, North Korea, and Pakistan.

In part as a result of this framework, nations have been reducing their stockpiles of weapons-grade uranium, both for reasons of utility and for reasons of security. Outside the scope of the treaty, nations have gained access to uranium through existing stockpiles.

MINING URANIUM

Sources of uranium have been found throughout the world. Uranium has been mined in the American Southwest. The three nations mining the most uranium are Kazakhstan, Canada, and Australia. In World War II, much of the uranium used for the Manhattan Project came from Shinkolobwe, in the Belgian Congo (now the Democratic Republic of the Congo).

Mined uranium is milled to form yellowcake uranium (U_3O_8), which is the form in which it is shipped. Yellowcake uranium is then enriched for use in power plants or weapons. In uranium enrichment, the percentage of uranium-235 in the sample is increased. Low-enriched uranium (LEU) involves uranium composed of 3 to 5 percent uranium-235, which is useful as a fuel for power plants. Highly enriched uranium (HEU), which is enriched to be 90 percent uranium-235, is used for nuclear weapons.

Enriching uranium involves the use of fluorine gas, which binds to the uranium to form uranium hexafluoride (UF_6). The uranium-235 is then separated using diffusion of a gas or, more recently, using centrifuges. Separating by weight produces uranium oxide (UO_2), which contains a greater proportion of uranium-235. The uranium from which it was taken, now called depleted uranium, is disposed of, while the uranium oxide becomes fuel.

URANIUM'S USES

Uranium has a number of uses. Historically, it was used in artwork for its shine and coloring. After discovery of its atomic properties, it became a fuel source for atomic weapons and nuclear power plants.

HEU also can be used to produce technetium-99. Fission of uranium-235 is used with molybdenum-98 to produce molybdenum-99, which decays to technetium-99. This isotope is used for diagnostic medical imaging. Because of the desire to reduce available stocks of HEU, ongoing efforts have focused on ways to produce technetium-99 from LEU. The first such stocks were produced at the end of 2010, but efforts are ongoing to ensure the transition without a supply shortage.

While other applications involve the processed form of the element, uranium ore (because of its radioactivity) is useful for radiometric dating, a method geologists use to determine age. In time, uranium-238 decays into lead-206. Knowing the half-life of uranium allows scientists to calculate the age of the sample based on the ratio of presence of these two elements. (Because lead has multiple isotopes—uranium-235 becomes lead-207, for example—the presence of other lead in the sample will not necessarily interfere with this calculation.)

HEALTH RISKS OF URANIUM

Uranium ore is relatively safe. The alpha particles released by the decay of uranium-238 can be blocked by a sheet of paper and do not penetrate the skin. While uranium in its natural form may have fewer health risks from radiation, it still can cause significant problems if dust that contains uranium particles is inhaled into the lungs. High rates of lung cancers have been found among people who work in uranium mines; as a result some uranium mining operations have been shut down.

A much greater health risk arises when uranium is enriched. The gamma radiation from uranium-235 can penetrate much deeper, including through human tissue. This can lead to radiation poisoning (which led to the death of Marie Curie in 1934). The energy from an atomic blast will cause death instantly.

POLITICS OF URANIUM

Because of its chemical properties and its use as an engine for both power and warfare, uranium has been a source of controversy since the early twentieth century. Mining operations present a health risk, and the most publicized early use of uranium was as history's most deadly weapon. Furthermore, power plants designed to enhance civilian life hold the potential for large-scale destruction.

While scientists created the atomic bomb, the term "atomic bomb" was not created by a researcher on the Manhattan Project but much earlier by science fiction writer H. G. Wells, who, having read a scientific paper about the potential of uranium, created the term in the book *The World Set Free* (1914), a novella about an element with the potential to destroy the world. In the story, Wells foreshadowed uranium's destructive potential more than thirty years before the first atomic bomb was detonated.

In addition to the fears raised by the possibility of the world's destruction from atomic bombs, particularly during the Cold War, concerns have been raised after accidents at nuclear power plants around the world. A 1979 accident at Three Mile Island in Pennsylvania led to a loss of coolant. The heat of the energy from the plant began to cause the metal containing the uranium to melt. While it was brought under control and no deaths were caused by the accident, the meltdown led to public fear. A 1986 accident at the Chernobyl nuclear plant in Ukraine led to two deaths from the immediate explosion and another twenty-eight deaths from radiation poisoning in the weeks that followed. Others have claimed the disaster has had long-term health effects that continue into the twenty-first century.

Following damage to the Fukushima Daiichi nuclear plant from the major earthquake and tsunami in Japan in 2011, a review of possible health risks was published in one of the world's leading medical journals. In response, a group of doctors noted that the initial article did not include the psychological and social impact from the fallout. Not only would anxiety about health risks and an inability to return home have a large health impact, but people in affected areas might also be considered contaminated by others.

Technically, nuclear plants present less of an environmental hazard than, for example, coal, which emits tons of pollutants into the atmosphere. However, while the risks from nuclear power plants remain relatively small from a quantitative perspective, the severity of an adverse outcome, in the minds of many, leads those risks to outweigh the potential benefits. Despite its potential, civilian uses for uranium face an uncertain future.

Joseph I. Brownstein

FURTHER READING

Brumfiel, Geoff. "Iran's Nuclear Plan Revealed: Report Paints Detailed Picture of Nation's Intention to Build a Warhead." *Nature* 479 (2011): 282. Discusses Iran's nuclear ambitions, explains how Iranians might have figured out a way to construct a more powerful uranium warhead, and discusses why they would choose uranium over plutonium as the warhead's power source.

Christodouleas, John P., et al. "Short-Term and Long-Term Health Risks of Nuclear-Power-Plant Accidents." *New England Journal of Medicine* 364 (2011): 2334-2341. Examines the potential health consequences that can result from nuclear fallout. Published following damage to the Fukushima Daiichi nuclear power plant, resulting from a major earthquake and tsunami in Japan.

National Research Council Committee on Medical Isotope Production Without Highly Enriched Uranium. *Medical Isotope Production Without Highly Enriched Uranium.* Washington, D.C.: National Academies Press, 2009. Available at http://books.nap.edu/catalog.php?record_id=12569. Given the desire to reduce stocks of highly enriched uranium, this study was organized to determine ways of getting isotopes for medical imaging usage without such uranium.

Reiche, Ina, et al. "Development of a Nondestructive Method for Underglaze Painted Tiles—Demonstrated by the Analysis of Persian Objects from the Nineteenth Century." *Analytical and Bioanalytical Chemistry* 393, no. 3 (2009): 1025-1041. Following its initial discovery but preceding the finding of its full potential, uranium was used in a variety of artworks. In this paper researchers detail a method for determining the composition of some of these art pieces.

Stockton, Peter, and Ingrid Drake. "From Danger to Dollars: What the US Should Do with Its Highly Enriched Uranium." *Bulletin of the Atomic Scientists* 66, no. 6 (2010): 43-55. Despite nuclear nonproliferation, the United States still has a significant amount of highly enriched uranium. This article discusses "downblending" the stock unneeded by the military to make it suitable for power plants.

Zoellner, Tom. *Uranium: War, Energy, and the Rock That Shaped the World.* New York: Viking Penguin, 2009. This book for general readers provides a history of uranium and details how it has affected politics, largely in the past one hundred years.

See also: Lithium; Nuclear Power; Radioactive Minerals; Toxic Minerals; Unconventional Energy Resources.

W

WATER AND ICE

Water is a molecule formed from two hydrogen atoms covalently bonded to a single oxygen atom. The chemical properties of water are unique among molecular substances on Earth and have had a dominant effect on the evolution of life.

PRINCIPAL TERMS

- **covalent bond:** powerful bond between two atoms that involves the sharing of a pair of electrons
- **evaporation:** type of vaporization, the transition of a liquid to gaseous state; occurs on the top layer of a liquid
- **groundwater:** water that penetrates the surface of the lithosphere and gathers in layers beneath the crust of the earth
- **hydroelectricity:** methods used to convert the kinetic energy in moving water to electric energy
- **hydrogen bond:** electrochemical bond between a hydrogen atom and an electropositive atom such as those that occur between water molecules in liquid water and ice
- **hydrosphere:** the collective waters of the earth contained in the oceans, rivers, seas, lakes, and the lower level of the atmosphere
- **precipitation:** rain, sleet, hail, snow, or other forms of liquid or solid water that fall from the atmosphere to the hydrosphere or lithosphere
- **solvent:** substance that dissolves the chemical structure of another substance upon contact
- **specific heat:** the heat at which a liquid will transition to a gaseous state
- **surface tension:** measure of the potential for the surface of a liquid to resist external force

CHEMICAL PROPERTIES OF WATER

Water (H_2O), also called dihydrogen monoxide, is a naturally occurring molecule that can exist on Earth in solid, liquid, and gaseous phases. Water is the most abundant molecule on Earth, covering approximately 70 percent of the earth's surface.

The hydrosphere, which is composed of the collective waters of the seas, oceans, rivers, lakes, and lower atmosphere, consists of more than 1.3 billion cubic kilometers (312 million cubic miles) of water.

The chemical properties of the water molecule are in large part responsible for the nature of life on Earth. The bodies of most living organisms are more than 60 percent water, and some living cells contain more than 90 percent water by volume.

The molecular structure of water consists of a single oxygen atom joined with two hydrogen atoms through covalent bonding, which involves the sharing of electrons between two atoms. Water molecules are often described as having a bent shape, because the pull of oxygen on its shared electrons is stronger than the pull of hydrogen. For this reason, the hydrogen atoms tend to gather toward one side of the water molecule, while lone electrons are pulled to the other side. This chemical symmetry means that one side of the molecule has a slight negative charge and the other (the side with the hydrogen atoms) has a slight positive charge. Water is therefore a polar molecule, which is attracted to other water molecules and to other molecules and ions that have a positive or negative charge.

Because of their slight positive and negative charges, water molecules bond to one another through hydrogen bonding, which is the momentary attraction of the negative end of one water molecule with the positive end of a nearby molecule. In liquid water, molecules are constantly moving, and the hydrogen bonds between them are frequently broken and re-formed. The average hydrogen bond lasts for less than one trillionth of a second before being broken. However, the millions of hydrogen bonds forming across a body of water provide a collective strength that gives water unique structural and adhesive properties.

The polarity of water molecules makes water an excellent solvent, meaning it can cause the dissolution of other molecules and substances, as water molecules form bonds with other polar molecules they encounter. Any compound with a negative or positive net charge will dissolve in liquid water. Water's function as a solvent is important in a variety of biological processes and also allows water to dissolve minerals and other elements from the lithosphere of the earth, returning them to the hydrosphere and biosphere and thereby acting as the most important catalyst in many geochemical cycles.

Hydrogen bonding gives water high surface tension, which is the capacity of the layer on the surface of a liquid to resist external pressure. Therefore, certain substances and materials will float on top of a body of water rather than penetrate the surface. In addition, hydrogen bonds cause water to tend to gather into drops, rather than spreading evenly, which is important to plants that use water's natural adhesion to help distribute water throughout the plant's tissues.

The collective energy of hydrogen bonding also means that water has a high specific heat, which is the heat at which the liquid form will give rise to a gas. Water therefore must absorb a large amount of heat energy before it will change temperature, and it also tends to release heat energy slowly back into the environment. These properties mean that water is a natural regulator of climate, tending to absorb excess heat when available and to release heat later when the ambient heat has reduced.

PROPERTIES OF ICE

Most substances become more dense as they cool because heat energy is lost, causing collisions between molecules to decrease and thereby allowing molecules to settle closer to one another. As water cools, its density increases until the temperature reaches 4 degrees Celsius (39 degrees Fahrenheit), reaching the maximum density for liquid water. If water continues to cool beyond this point, the vibrational energy of individual water molecules decreases to such a point that the hydrogen bonds between molecules become fixed, rather than alternate between bonds. At this point, water begins to decrease in density as the molecules align into a matrix, simultaneously increasing in volume.

At approximately 0 degrees Celsius (32 degrees Fahrenheit) liquid water becomes solid ice. The organized arrangement of atoms in ice allow for more space between molecules, thereby increasing the volume by more than 9 percent. This property is unusual among liquids, but it also occurs in the common mineral silica (SiO_2) as it transitions from liquid magma to solid silica rock. Because ice is less dense than water, it does not sink and will rise to the surface of the liquid, further buoyed by the surface tension of the liquid. In areas where freezing occurs, this creates a layer of frozen ice that insulates the liquid water beneath. This insulation prevents further water from reaching the freezing point and thereby allows organisms to survive under the surface of frozen bodies of water. Similarly, the preservation of this liquid water means that chemical reactions can continue under the surface of the ice.

THE HYDROLOGIC CYCLE

The hydrologic or water cycle is a geochemical cycle that moves water molecules between the environmental spheres of the earth. The lithosphere is the combination of rocky minerals, sediment, and molten rock that make up the earth's crust and part of the earth's mantle.

The atmosphere is an envelope of gaseous elements that surround the earth in layers and differentiate the planetary environment from that of outer space. The biosphere is the sum of all living organisms on the planet and their interactions with the other environmental spheres. Geochemical cycles are the processes that move various elements through the environmental spheres through interconnected sets of chemical and physical reactions.

The hydrologic cycle begins with evaporation, which is the process by which liquid water transitions into a gaseous form and enters the earth's atmosphere. Approximately 80 percent of evaporation occurs over the ocean, with the remainder occurring over lakes, rivers, and water temporarily deposited on the lithosphere. Water rises into the atmosphere as water vapor and, as it gains in altitude, the temperature of the atmosphere decreases, causing water vapor to condense. Hydrogen bonds cause water vapor to form into clouds, which are pockets of water droplets or ice crystals that move according to the development of thermal currents in the atmosphere.

Water vapor in clouds and in more diffuse gaseous form throughout the atmosphere is transported around the earth on thermal currents and eventually returns to the earth as precipitation, which is the return of condensed water to the hydrosphere and lithosphere in the form of rain and snow. Precipitation occurs when the density of water vapor in the atmosphere reaches a certain critical level, often triggered by temperature changes in the atmosphere that are related to windstorms that develop over the ocean.

A portion of the water delivered to the lithosphere as precipitation returns to the atmosphere through evaporation, because the temperature of the lithosphere is generally warmer than the temperature of the hydrosphere; this causes water to more easily vaporize into gas. Some of the precipitation over the lithosphere penetrates the minerals of the earth and becomes groundwater, which is water that saturates and remains within the mineral layers of the earth. Flooding occurs when groundwater levels rise to a point that the soil is oversaturated and can no longer absorb additional water.

Groundwater eventually returns to the hydrosphere as it filters into streams and lakes that are connected to the ocean. Additional groundwater enters the ocean as seepage in coastal areas, where the lithosphere and the ocean meet. Water that filters back into the hydrosphere through the lithosphere carries a variety of minerals and other trace elements because of the solvent properties of water. These minerals enrich water sources, thereby contributing to the nutrient cycle in aquatic environments.

The organisms of the biosphere also absorb water into their tissues, using it to fuel and to carry the products of chemical reactions. Water from within organisms is returned to the atmosphere during respiration and as a component of biological waste. Water returned to the hydrosphere through the biosphere also carries a variety of minerals and other elements that are thereafter transferred through the hydrosphere into the other environmental spheres.

HYDROELECTRICITY

Hydroelectricity is the process of using water to generate electricity for human consumption. The basic method is to harness the kinetic energy of moving water and then to transition this energy into electric currents that can be stored and used to power electric devices.

Generally, the generation of hydroelectricity involves using running water to power turbine engines, devices with spinning axles attached to blades that move in response to the flow of a fluid through the blades. As the blades turn, they power an electric generator, which forces a flow of electrons through a circuit and generates electric currents.

Hydroelectric generators can utilize the gravitational flow of liquid water, which is the tendency for water to flow downward with the force of gravity from an area of higher elevation or depth. To harness this energy, engineers control the flow of water and filter the water through channels containing turbines.

A number of different methods generate hydroelectricity, depending on environmental conditions. Hydroelectric dams function by blocking the flow of water along the path of a river or other waterway. This creates an artificially generated imbalance in water depth on either side of the dam. The higher water on one side of the dam can then be used to power the flow of water through the turbines at the bottom of the dam. Another way of artificially generating water flow is to utilize a series of pumps to control or enhance the flow of water in a certain area. In this case, hydroelectricity can be generated continuously, rather than relying on naturally occurring gravitational imbalances to generate flow.

Turbines can be installed in rivers to capture energy from the flow of the river current. Similarly, turbines can be installed in oceanic areas, where they derive energy from the tidal movement of ocean currents. Tidal power and wave power are two of the most recent forms of hydroelectricity and have developed into major areas of research. Energy present in the kinetic movement of tides provides a consistent and predictable source of energy generation but is utilized in relatively few areas.

All forms of hydroelectric energy offer the advantage of being renewable, as they are not based on the supply of a limited ingredient. However, this type of energy usage often carries environmental consequences because the environment must be altered to install hydroelectric generators and other equipment.

Micah L. Issitt

FURTHER READING

Barry, Roger, and Thian Yew Gan. *The Global Cryosphere: Past, Present, and Future.* New York: Cambridge University Press, 2011. Detailed academic text presenting the formation and function of ice in relation to the global geochemical and physical cycles. Contains a detailed discussion of the chemical and physical properties of both water and ice.

Brutsaert, Wilfried. *Hydrology: An Introduction.* New York: Cambridge University Press, 2005. Concise introduction to the field of hydrology, which studies the distribution, quality, and environmental processes of water over the earth's surface. Discusses chemical properties and the generation of hydroelectric energy.

Gosnell, Mariana. *Ice: The Nature, the History, and the Uses of an Astonishing Substance.* New York: Alfred A. Knopf, 2005. General-interest investigation of ice and its relationship to culture. Discusses the physical and chemical properties of ice and its shifting patterns in the environment.

Jacobson, Michael C., et al. *Earth System Science: From Biogeochemical Cycles to Global Change.* San Diego, Calif.: Academic Press, 2006. Introductory text covering Earth system science. Discusses the hydrologic cycle and its relationship to the other environmental and biogeochemical cycles active on Earth.

Macdougall, Douglas. *Frozen Earth.* Berkeley: University of California Press, 2006. General-interest text covering glaciation and its role in shaping Earth processes and physical properties.

Tagare, Digamber M. *Electricity Power Generation: The Changing Dimensions.* Hoboken, N.J.: Wiley-Blackwell, 2011. Advanced scientific text presenting aspects of various systems used to generate electricity. Chapter 2 discusses past and present methods used to generate hydroelectricity, including their economic and environmental impact.

Wallace, John M., and Peter V. Hobbs. *Atmospheric Science: An Introductory Survey.* 2d ed. Burlington, Mass.: Academic Press, 2006. Detailed introduction to the study of atmospheric processes, including discussions of the cycling of oxygen, nitrogen, and water in the atmosphere.

See also: Chemical Precipitates; Clathrates and Methane Hydrates; Desertification; Evaporites; Geologic Settings of Resources; Hydrothermal Mineralization; Manganese Nodules; Sulfur; Tidal Forces; Unconventional Energy Resources.

WELL LOGGING

Reservoir rock data obtained by well logging are of vital importance to the petroleum industry. With these data, the production potential of a well can be determined and many problems involving the structure, environment of deposition, and correlation of rock strata can be solved.

PRINCIPAL TERMS

- **conductivity:** the opposite of resistivity, or the ease with which an electric current passes through a rock formation
- **correlation:** the tracing and matching of rock units from one locality to another, usually on the basis of lithologic characteristics
- **gamma radiation:** electromagnetic wave energy originating in the nucleus of an atom and given off during the spontaneous radioactive decay of the nucleus
- **hydrocarbons:** organic compounds consisting predominantly of the elements hydrogen and carbon; mixtures of such compounds form petroleum
- **lithology:** the mineralogical composition of a rock unit
- **neutron:** an uncharged, or electrically neutral, particle found in the nucleus of an atom
- **permeability:** measured in millidarcies, the capacity of a rock unit to allow the passage of a fluid; rocks are described as permeable or impermeable
- **petroleum:** a natural mixture of hydrocarbon compounds existing in three states: solid (asphalt), liquid (crude oil), and gas (natural gas)
- **porosity:** the volume of pore, or open, space present in a rock
- **reserves:** the measured amount of petroleum present in a reservoir rock that can be profitably produced
- **reservoir:** any subsurface rock unit that is capable of holding and transmitting oil or natural gas
- **resistivity:** a measure of the resistance offered by a cubic meter of a rock formation to the passage of an electric current
- **sonde:** the basic tool used in well logging; a long, slender instrument that is lowered into the borehole on an electrified cable and slowly withdrawn as it measures certain designated rock characteristics

LOGGING PROCEDURE

A well log is a continuous record of any rock characteristic that is measured in a well borehole. The log itself is a long, folded paper strip that contains one or more curves, each of which is the record of some rock property. Since the first "electric log" was run in a well in France for the Pechelbronn Oil Company in 1927, well logs have been the standard method by which well data have been displayed and stored. Well logs and the information they record can be classified in two ways: by type (radioactive, sonic, electric, or temperature) and by purpose (lithology, porosity, or fluid saturation determination).

After a well has been drilled, it is standard procedure to log it. Logging has been compared to taking a picture of the rock formations penetrated by the borehole. The technique consists of lowering the logging tool, or "sonde," to the bottom of the borehole on the end of an electric cable that is attached to a truck-mounted winch at the surface. The truck also contains the instruments for recording the logged data. The sonde, which is 4.5 to 6 meters long and has a diameter of 7.5 to 13 centimeters, is then pulled up the borehole at a constant rate, measuring and recording the data of interest. Measurements are recorded coming "uphole" rather than going "downhole" because it is easier to maintain a constant sonde velocity by pulling it up. On the downward course, the sonde has a tendency to "hang up" on numerous irregular surfaces in the borehole. It is essential to run logs to evaluate petroleum potential before the borehole is lined with steel pipe, because the well completion process is expensive and will be done only if economically justified. Data obtained from the well by logging are used to determine such rock parameters as lithology, porosity, and fluid saturation.

RADIOACTIVE AND SONIC LOGS

Of the logging curves that can be run for lithology identification and correlation, the most useful are the gamma-ray, spontaneous potential, and caliper. The gamma-ray log records the intensity of natural gamma radiation emitted by minerals in the rock formations during radioactive decay. One advance in gamma-ray technology has been the development of the gamma-ray spectrometry tool, a device that measures the energies of the gamma rays and makes possible the identification of individual minerals. The spontaneous potential (SP) log measures small natural potentials (voltages) caused by the movement of fluids within the formations. These currents largely arise as a result of salinity differences between the pore waters of the formations and the mud in the borehole. Although the borehole is drilled with a bit of a particular size, its diameter is never constant from top to bottom because the rotating drill pipe wears away the rocks along the hole. The caliper log provides a continuous measurement of borehole diameter by means of spring-activated arms on the sonde that are pressed against the wall of the borehole. Softer rocks have larger-diameter boreholes, and harder rocks have smaller diameters because of their difference in resistance to wear.

Porosity is determined by using, singly or in combination, the sonic, neutron, and density logs. The standard sonic tool has an arrangement of two transmitters, each with its own signal receiver. The transmitters send out sound waves, which are detected back at the receivers after passing through the rock. The neutron log tool bombards the formation with fast neutrons. These neutrons are slowed by collisions with ions in minerals and fluids. Because a hydrogen ion has approximately the same mass as does a neutron, collisions with hydrogen ions are most effective in slowing the neutron for the same reason that a billiard ball is slowed more by a collision with another billiard ball than it is by a collision with the rail of the table. The slowed neutrons are deflected back to the tool to be counted and recorded. Like the neutron log, the density log is a nuclear log. The density sonde bombards the formations with medium-energy gamma rays. The gamma rays collide with electrons in the formation, causing the gamma-ray beam to be scattered and its intensity reduced before it returns to the detector on the sonde.

ELECTRIC AND TEMPERATURE LOGS

Most logs used to determine water and oil saturations in the formations employ some method of measuring the passage of an electric current through the rock. The electric logs can be subdivided into induction logs and electrode logs. The induction log measures the conductivity of the formation and is the most commonly used device. The induction-logging sonde generates a magnetic field that induces a current deep in the formation. The passage of this current is measured by the logging tool. In the electrode-log system, electrodes on the sonde put current directly into the borehole fluid or the formation. The resistance to the flow of the electric current through the formation is measured as the formation resistivity. The short normal log, microlog, and microlaterolog measure resistivity immediately adjacent to the borehole, while the laterolog and guard log measure resistivity deep in the formation. Deep readings are made by narrowly focusing the electric current beam and directing it straight into the formation rather than letting it diffuse through the mud and into the adjacent formations. The laterolog and microlaterolog are most commonly used when the borehole mud has a base of saltwater rather than freshwater.

The temperature log, a nonelectric log, continuously records borehole temperature and can also be used for fluid identification. The dipmeter log is a resistivity device run with three or four electrodes arranged around the perimeter of the sonde. If the rock layers are inclined at any angle to the horizontal, this inclination, or dip, can be detected, because the electrodes will encounter bed boundaries at slightly different times on different sides of the borehole. An online computer converts these differences to angle of dip. The cement-bond log is a sonic device that measures the degree to which cement has filled the space between the steel pipe, or casing, that lines the inside of the borehole and the formations behind it (complete filling is desired).

DETERMINATION OF LITHOLOGY

Lithology refers to the mineralogical composition of the rock unit, or formation. Oil and natural gas occur almost exclusively in the sedimentary rocks sandstone, limestone, and dolomite; the latter two are known as carbonate rocks. Shale, the most abundant of all sedimentary rocks, is never a reservoir rock for hydrocarbons because it is impermeable. An important purpose of well logs is to determine the lithology of the rock formations and thus, to identify those that possess suitable permeability to serve as reservoir rocks. The principal radioisotopes (thorium, uranium, and potassium), from which most natural gamma radiation emanates, are usually found in minerals in clays and shales. Therefore, the gamma-ray log is used to differentiate shales from sandstones and limestones and to calculate the amount of clay that might be present in some sandstones. Since most fluid movement is in or out of porous and permeable formations, the SP curve may be used to identify such rocks. These rocks are usually sandstones—hence, the identification of lithology. While permeable zones can be located, it is not possible to calculate actual permeability values. Next to permeable formations, the diameter of the borehole is reduced by the buildup of mud cake on the borehole wall. Borehole diameter also changes dramatically through shales, because shale is weak and crumbles, or "caves," enlarging the borehole. The caliper log can therefore be used as a lithology log to identify permeable sandstones and "caving" shale.

In addition to assessing rock units for their petroleum content and environmental information, well logs are used extensively for correlation—that is, the matching and tracing of rock units from one locality to another. Since it is lithology rather than porosity and fluid saturation that is geologically the most significant factor in rock identification, the lithology logging curves are most commonly used for this purpose. Correlation may be accomplished either by matching log curves "by eye" or by statistical and computer analysis.

DETERMINATION OF POROSITY

Porosity is a measure of the total open space in a rock unit that is available for the storage of hydrocarbons. Such space is normally expressed as a percentage of the total rock volume. Knowledge of formation porosity is necessary to determine the total petroleum reserves in the formation or oil field. The sonic log has historically been the most widely used porosity tool. The time, in microseconds, required for a sound wave to travel through one meter of the rock is continuously plotted on the log. This travel time is the reciprocal of velocity, so a wave that has a high velocity has a short travel time. The use of a dual transmitter-receiver system for modern sonic logs eliminates the effects of changing borehole diameter and deviations of the borehole from the vertical. Formation travel times are functions of lithology and porosity. If the formation lithology is known, the porosity can be calculated. Because a sonic wave travels faster through a solid than through a liquid or gas, increasing porosity causes greater travel times. The presence of shale in the rock formations, however, can also cause unusually high travel times and

erroneously high porosity calculations. The neutron tool principally senses the hydrogen ions present in the formation fluids, which, in turn, are found in the pore spaces. This log is affected by lithology, because clays in shales have water within their crystal structure, and the tool senses this water as if it were pore water. The density log measures electron density, which is directly related to the overall, or bulk, formation density. The greater the bulk density, the lower the porosity, because mineral matter is denser than fluid-filled pore space. A related log is the variable-density log, which is used to locate rock zones that are highly fractured and thus, potential reservoir rocks. The fractures in the rock have the effect of lowering the bulk density.

Because accurate interpretation of the data is dependent on a knowledge of lithology, it is common practice in the petroleum industry to run porosity tools in combination, particularly the neutron and density logs. Cross-plotting the readings from the two logs provides both lithology and porosity information. In addition, neutron and density logs respond oppositely to the presence of natural gas in a formation. Density porosity readings increase, whereas neutron porosity readings decrease. Therefore, this log combination will detect the presence of gas-bearing zones by the separation of the two curves.

DETERMINATION OF FLUID SATURATION

Fluid saturations—water and oil—are the most important quantities to be determined from well-log analysis. Because the water and oil saturations together must equal 100 percent, knowing one necessarily determines the other. The significance of these values is clear: If the rock unit of interest is not oil-bearing, or if it contains hydrocarbons in quantities that are not economically feasible to produce, the well will be abandoned rather than completed.

To understand the quantitative assessment of formation-fluid saturation, one must first understand what occurs within the borehole and in the formations that are penetrated by the well. Because high temperatures are generated by friction as the drill bit grinds its way through solid rock, specially formulated "drilling mud" is continuously circulated down the borehole to cool the bit. In addition, the mud clears the borehole by bringing to the surface the pulverized rock material, or cuttings. The drilling mud is usually a water-based fluid with various mineral additives. Within the borehole, there is a tendency for the fluid portion of the mud to separate from the mineral fraction. The fluid, or mud filtrate, seeps into the permeable rock formations, completely flushing out and replacing the natural formation fluids adjacent to the borehole. This area is the "flushed zone." Some of the filtrate moves deeper into the formation, where it continues to displace the natural fluids, creating a partially flushed area, or "invaded zone." The solid portion of the mud that has separated from the filtrate forms a "mud cake," lining the inside of the borehole on the surfaces of the permeable formations.

SPECIAL-PURPOSE LOGS

Within a reservoir, the rock matrix, freshwater, and hydrocarbons act as electrical insulators. Any electric current that passes through the rock is carried by dissolved ions in saltwater in the pore spaces of the formation. Therefore, where a current flows readily, the pore fluid is saltwater. Where electrical resistance is high, it is likely that hydrocarbons occupy the pore spaces. An induction log is used to assess electrical conductivity. The principal advantage of the induction log is that electrical currents largely bypass the high-resistance invaded zone where rocks have been penetrated by well fluids, and give a better picture of the true formation resistivity (the inverse of the conductivity) deep in the formation. Even so, the flushed zone and the invaded zone are still sampled to some extent, and corrections must be made to obtain the true resistivity of the uncontaminated formation. This can be accomplished by running logs that sample the formations only immediately adjacent to the borehole and, therefore, read either the flushed zone or the invaded zone resistivity.

In addition to corrections for flushed zone and invaded zone resistivities, other corrections must be applied to the log readings to allow for the effects of changing borehole diameter and bed thickness. When the true resistivity value and formation porosity are known, one can calculate the percentages of water and oil in the formation and make a quantitative determination of the total volume of hydrocarbons. Formations that contain natural gas rather than oil can be readily identified with the temperature log. As gas moves out of the formation and into the borehole, it is under less pressure and expands. As it expands, it cools, and the borehole temperature

opposite the gas-bearing formation is significantly lowered.

Other special-purpose logs are available to the petroleum industry. One of these, the dipmeter log, provides information on the angle of dip of the formations encountered. This information can be used to determine environments of deposition, because rock layers in channel deposits, reefs, offshore bars, and other sedimentary features will have some unique pattern of dip. Increasingly, geologists have been using other well-log curves to refine their environmental interpretations. This is done by examining the curve patterns within the sedimentary rock units and noting whether the log parameters increase or decrease downward, for example, or change gradually or abruptly. The log, in effect, is measuring rock properties, such as particle size, that are controlled by the physical conditions within the site of deposition.

IMPORTANCE TO THE PETROLEUM INDUSTRY

Well logging is a little-known but necessary part of the petroleum industry. It is the method whereby geologists and petroleum engineers obtain the information on petroleum reservoir rock characteristics that allow them to make decisions about the economic potential of an oil well—that is, whether it can be completed as a "producer" or must be plugged and abandoned as a "dry hole." Such decisions involve millions of dollars and cannot be made without an examination of all the available relevant data. They must usually be correct, or the oil company will not survive financially.

The 1990's brought increased use of directional drilling, the downhole tool, and a nonrotating drill stem. A different type of well logging called "measure while drilling" (MWD) is well suited to this method of drilling. MWD uses a sonde as part of the drill stem near the tool. The sonde carries the usual logging instruments, along with magnetometers and accelerometers to help in establishing the drill stem's position and the orientation of the hole. High-capacity batteries power the equipment. A transmitter assembly creates a series of pulses in the pressure of the drilling mud by opening and closing the valve. This allows small amounts of mud to pass directly from the drill stem into the borehole, thereby bypassing the drill tool. These pressure pulses are monitored by sensors at the top of the borehole. Computers in the sonde and in the logging truck communicate by means of these pulses and thereby provide logging information continuously while drilling. The logging information is therefore immediately available, and it is easier to interpret than with the older system because the sonde is in a relatively fresh borehole so the drilling mud has had little time to penetrate into the surrounding rock and modify its properties. Based on the logging data, decisions can be made regarding whether the well should be completed, the direction of the borehole, and which rock zones should be tested for their fluid content and producibility. As in many business endeavors, time saved in the decision-making process can be turned into money earned.

Donald J. Thompson

FURTHER READING

Berg, Robert R. *Reservoir Sandstones.* Englewood Cliffs, N.J.: Prentice-Hall, 1986. Includes a good summary of well logs and logging procedures. Discusses various types of log, emphasizing how they are used in specific formation evaluation and sedimentological problems. Appropriate for general readers in the context of the basic discussion of well logging.

Brock, Jim. *Analyzing Your Logs.* Vol. 1, *Fundamentals of Open Hole Log Interpretation.* 2d ed. Tyler, Tex.: Petro-Media, 1984. A fundamental discussion of the basic characteristics of petroleum reservoir rocks, followed by a description of the theory and use of each of the major types of well log. Includes diagrams and graphs and correction charts for use with each type of log; questions and problems accompany each section. Designed for a short course in well log evaluation.

Coates, George R., Lizhi Xiao, and Manfred G. Prammer. *NMR Logging Principles and Applications.* Houston: Halliburton Energy Services Publication, 1999. Discusses the applications, benefits, physics, and fundamentals of NMR logging. Touches on MRI logging as well.

Gorbachev, Yury I. *Well Logging: Fundamentals of Methods.* New York: John Wiley & Sons, 1995. An introduction to all aspects of geophysical well logging. Includes numerous useful illustrations and a short bibliography.

Hyne, Norman J. *Dictionary of Petroleum Exploration, Drilling, and Production.* Tulsa, Okla.: PennWell, 1991. Covers all terms associated with petroleum and the petroleum industry. A great resource for the beginner in this field. Illustrations and maps.

_____. *Nontechnical Guide to Petroleum Geology, Exploration, Drilling, and Production.* 2d ed. Tulsa, Okla.: PennWell, Corporation, 2001. Provides a well-rounded overview of the processes and principles of gas and oil drilling. Covers foundational material.

Krygowski, Daniel, George B. Asquith, and Charles R. Gibson. *Basic Well Log Analysis.* 2d ed. Tulsa, Okla.: American Association of Petroleum Geologists, 2004. Discusses the theory of well logging and of the application of major log types. Includes descriptions, examples, problems, and comprehensive case studies. Well illustrated, with a comprehensive bibliography. Intended for the working geologist but appropriate for readers with minimal geologic training.

Luthi, S. *Geological Well Logs.* New York: Springer-Verlag, 2001. Discusses characterization of geological properties, as well as such tools as borehole imaging and nuclear magnetic resonance. Written for researchers and graduate students.

Pirson, Sylvain J. *Geologic Well Log Analysis.* 3d ed. Houston: Gulf Publishing, 1983. Emphasizes the use of well logs for sedimentary environment interpretation, structural analysis, facies analysis, and hydrogeology rather than the evaluation of hydrocarbonate-bearing rock units.

Rider, Malcolm H. *The Geological Interpretation of Well Logs.* 2d ed. Houston: Gulf, 1999. Overview of oil well logging and analysis. Discusses openhole and borehole logging. Illustrations and index.

Selley, Richard C. *Elements of Petroleum Geology.* 2d ed. San Diego: Academic Press, 1998. Contains a section on the theory and application of well logging, with a discussion of each of the major types of log. Appropriate for general readers. Includes excellent diagrams and an extensive bibliography.

Serra, Lorenzo. *Well Logging: Data Acquisition and Applications.* Paris: Technip Editions, 2004. Discusses physical principles and practical measurements of logging, as well as such factors as radioactivity, porosity, texture, and nuclear magnetic resonance. Appropriate for engineers and geologists.

See also: Earth Resources; Offshore Wells; Oil and Gas Distribution; Oil and Gas Exploration; Oil and Gas Origins; Oil Chemistry; Oil Shale and Tar Sands; Onshore Wells; Petroleum Reservoirs; Strategic Resources.

WIND POWER

Power-generating devices that use wind as an energy source are the oldest known devices engineered specifically to produce power. From ancient grain mills to today's complex wind turbines, wind has provided humanity with a power source. As the world's demand for energy increases at an accelerating rate, the use of wind power as a legitimate alternative energy source will also increase.

PRINCIPAL TERMS

- **aerodynamic lift:** a measure of the degree of wind force acting on a rotor, which is translated into power to a generator in a wind-power machine
- **axis system:** the vertical or horizontal orientation of the power shaft of a wind-power machine
- **power flux:** the amount of energy that can be obtained from a cross-sectional area of the wind at a given velocity
- **power grid:** the mechanical power distribution system of a social structure, and its independent power distribution systems
- **rotor:** the bladed system on a wind-power machine that is set into motion by the wind and translated into power to a generator
- **step-up gearing:** a gear system that increases the revolutions of the downstream shaft (generator) over the revolutions of the upstream shaft (rotor)

EARLY USES

Wind farms—large arrays of windmills on towers—have become a common feature of the landscape in the twenty-first century. A sailboat was the first device designed to use wind power to replace human energy; sailing craft have been in use since prehistoric times. It was obvious long before the development of science and engineering that the wind offered available energy. If the wind could push huge sailing craft over the known oceans of the world, then it could perform other tasks as well. A full thousand years after the first sailing craft, the first wind systems were planned for use in Babylonia to pump water through an irrigation system, although there is no evidence that they were ever built. The construction techniques to bring about such systems did not fully materialize until about 1000 C.E., when wind power was used widely in the Middle East.

The first widespread use of the wind for power was to grind grain, a task usually reserved for slaves, draft animals, or domestic servants. Along with this discovery came the realization that the wind was equally good at pumping water, and wind-powered pumps were put to work lifting water from wells and irrigation canals and to drain fields (most famously in Holland). Just before the beginning of the Industrial Revolution, wind-powered devices were in use across the world, being employed to pump water, grind grain, saw lumber, and even turn carousels. Yet the limits of using wind power were well known, and its use typically was neglected in favor of more reliable energy sources, as available. Water power, for example, which operated with few interruptions, was usually more consistent than wind power. Water-powered devices were generally smaller and required less investment in equipment. Because the wind shifts direction and velocity, windmills needed to be engineered to operate under a wide range of conditions. Even with its limitations, wind was in many cases the only choice for power, and tens of thousands of wind-powered systems were constructed around the world.

REEVALUATION AS ENERGY SOURCE

Just before the turn of the nineteenth century came the discovery of new and widespread uses for electricity. One of those devices was the electrical generator. Not long after, Moses G. Farmer was issued a patent for a wind-powered generator. After four thousand years of providing power for humankind, the wind finally had become linked with electricity. Yet, with the onset of the Industrial Revolution, steam power became the energy means of choice. Soon thereafter, petroleum, cheaply priced oil, and internal combustion engines all but drove to extinction the use of wind power, with the exception of the iconic farm windmill, which continued to be used for pumping water.

Beginning in the early 1970's, wind power was recognized as a potentially important world power source. The price of the world's energy had increased when the Organization of Petroleum Exporting Countries (OPEC) forced world oil prices up from $3 to $32 per barrel in seven years. Most nations of the West realized how dependent they were on foreign energy sources and how closely energy prices were linked to the economic health of nations. Indeed, the free and abundant flow of cheap energy was recognized as directly related to the economic vitality and well-being of any country. Therefore, measures were taken to reduce dependency on foreign energy sources. The most significant step taken was the encouragement of energy conservation, and second to it was the improvement and development of alternative and renewable energy sources, which include wind power, solar energy, hydroelectric power, synthetic fuels, and ocean and geothermal energy.

In the United States, President Jimmy Carter described the energy crisis of the late 1970's as the "moral equivalent to war." Congress responded and passed the Wind Energy Systems Act of 1980, which would provide an eight-year, $900 million program to develop cost-effective wind power systems across the United States. California responded immediately and launched a full-fledged crusade to harness the power of the wind. By 1990, wind power contributed 1 percent of California's energy needs. This number represented more than 1,000 megawatts (million watts), about the output of a single average-sized nuclear reactor. That power, which amounted to three-quarters of the world's total wind energy output, could supply 400,000 households and save 4 million barrels of oil each year. California's example is being followed in Hawaii, where constant trade winds are being harnessed. The California experience, however, represents only a fraction of the possible wind-generating power capacity in the United States. Some of the best sites are in North and South Dakota, Montana, and Wyoming. Because they lie far from population centers, power distribution costs are high, and for that reason alone it may never be practical to fully develop them. Yet, if their wind-generating capacity were fully developed, these states alone might produce 20 percent of the electricity needed in the United States.

Because of supportive government policies in several European countries, and because of technical improvements, wind energy became the fastest-growing energy source during the latter part of the 1990's. By 1998, the cost of wind-generating capacity had dropped to one-third of what it was in 1981 and became more competitive with other energy sources. In 1995, worldwide wind energy-generating capacity stood at 4,990 megawatts, but by the end of 2008 it had increased to 120,800 megawatts. A 1998 Danish study concluded that wind energy may provide 10 percent of the world's electricity by the 2020's, and the Worldwatch Institute optimistically predicted that wind-generated power would eventually exceed hydroelectric power, which supplied more than 20 percent of the world's electricity in 1999.

WIND-POWERED GENERATORS

The laws of physics determine the capabilities and efficiencies of wind-powered generators. The physical realities of wind power are as follows: Wind force varies with the square of its velocity; the power, however, varies with the cube of the velocity; the wind's ability to do work is limited by its flux, or the amount of energy that can be created by a cross-sectional area of wind traveling at a given velocity; and a perfect system can extract only 60 percent of the total wind power available. Realistically, after electrical conversion systems and (in some systems) storage, the conversion efficiency drops to between 20 and 50 percent. Therefore, one needs to garner a large area of wind to produce significant power. That equates directly into a very large wind turbine required to produce a relative modicum of electrical power.

A typical wind generator sits atop a tower about 35 meters off the ground and sports a blade approximately 60 or 70 meters in length. This device will generate about 300 to 500 kilowatts of power. Wind farms may occupy hundreds of acres, populated with many (perhaps several hundreds) of these wind turbines. By 2007, California's optimal wind generation capacity—consisting of more than a thousand generators over thousands of acres—produced over 30 percent of the world's total wind power, and supplied 2.3 percent of California's total power.

The typical wind generator consists of a tower, a generator, gears, the rotor, the axis system, and speed control. Some wind generators have batteries for power storage and inserters to convert power states. Wind generators must be placed on towers to enable them to capture optimal wind states. Generators placed too close to the ground will be adversely influenced by obstructions to a steady wind flow located on the ground, such as buildings, hills, and trees. Obviously, the tower must be high enough to allow clearance for large, efficient rotors. The rotors of the modern wind generator (also known as propellers, blades, or turbines) are quite unlike the windmills of the past. They come in many forms, depending on

A wind farm with giant turbines at Altamira, California. (PhotoDisc)

the task they are required to perform. The generator task itself is principally determined by the kind of generator mounted on the tower. Typically, the rotors are based on one of two designs. The most common wind-power-generator design is the horizontal axis rotor system. In this system, the shaft of the rotor is aligned horizontally with the ground, with the rotor blades mounted perpendicular to the rotor shaft. This design requires that the generator be mounted at the top of the tower. The disadvantages are obvious: Installation and maintenance must be done high above the ground. These rotors look much like aircraft propellers—in fact, they are designed along some of the same aerodynamic principles. The other type of design is called the vertical axis system. In this design, the shaft is mounted vertically with the rotors placed alongside the shaft. The rotor designs for these wind generators are not propeller types; one design looks like an eggbeater and is called exactly that. In these designs, the wind rotates the rotor and shaft, which are mounted vertically.

That allows the generator to be placed at the bottom of the assembly on the ground. Often there are no requirements for a tower to be constructed. Other vertically mounted designs are called paddle vanes and s-rotors.

REFINEMENTS TO GENERATORS

With few exceptions, in wind-power systems, the wind cannot turn the generator shaft fast enough to generate power. Therefore, gear systems are required to "step up" the system from the turning rotor to the generator. A typical blade rotation speed of 200 to 400 rotations per minute (RPM) must be geared up to 1,800 to 3,600 RPM for the generator to deliver adequate power. These gearing systems require that more force be delivered to the turning rotor than if there were no gears between the rotor and the generator shaft. Consequently, gear systems limit the minimum amount of wind necessary to turn the rotor and deliver power from the generator.

The wind-power generator must operate in a very wide range of conditions, considering a constantly variable wind speed and direction. To deliver a consistent energy output, the generator shaft should ideally turn at a more or less constant RPM. In most commercial generators, that is accomplished in a variety of ways. The pitch of the rotor is selectively changed along the entire length, or the rotor is selectively pitched at the tip. Some rotors have flaps, much like an aircraft wing, to change the degree of aerodynamic lift provided by the rotor. The degree of aerodynamic lift ultimately determines the power output of the rotor. Some generators change the angle at which the system is facing the wind in order to increase or decrease the amount of impinging wind energy. As important as it is to maintain a maximum effective energy output by regulation, it is also necessary to allow energy to go to waste if the wind speed exceeds the levels the system can tolerate. There are also systemic control mechanisms relating to aspects other than speed control. Among them are braking systems to slow or stop the generator in the event of dangerous winds or system failure. Wind systems need to be shut down in the event that the wind speed is too low, a condition that can deliver unacceptably low or inconsistent power to the regulatory mechanisms.

Wind systems have been designed for home use, and there are many homes around the world whose sole or partial power source is wind power. These homes rely on small wind-power-generation systems nearly identical to their commercial counterparts. Typically, they rely on direct current (DC) generators that feed banks of lead acid batteries storing the power from the wind generator, which allows for a constant supply of power to the home even when the wind is not blowing. These homes use direct current for many purposes, and many have special DC appliances. Some appliances do not operate on DC, so these homes have devices called inverters, which change direct currents to alternating currents (AC). During the storage and conversion process, however, there is energy loss so that the wind-power system becomes less efficient at each stage. The typical cost of equipping a home with a wind-power and storage system is about the cost of a new automobile.

DISADVANTAGES AS AN ENERGY SOURCE

The same problems associated with wind-power generation encountered by prehistoric humans still exist. Generating power from wind flow is fraught with myriad difficulties. Consistency and direction have already been mentioned. In addition, the storage of wind power is necessary for use in windless conditions or at times when the wind is so strong that the generators have to be shut down for their protection. In a large utility system, the power demand varies considerably from hour to hour and day to day; therefore, the utility must maintain backup generating capacity. Pacific Gas and Electric gets up to 7 percent of its power from California wind turbines and has had no problem adjusting to their varying power output. Environmental concerns also plague the use of modern wind-power devices. Whereas many environmentalists applaud the benign and clean nature of wind-power generation, many complain that the devices themselves are a blight on the landscape. Other complaints lodged against wind farms include the dangers posed to migrating birds and noise from the rotors.

ADVANTAGES AS AN ENERGY SOURCE

Although wind power is unlikely to replace other, higher-level energy sources, such as petroleum, coal, or nuclear fission or fusion energy, it remains an important energy source. The relatively low level of technology required is somewhat offset by the initial cost of the equipment. Yet, if wind energy is utilized efficiently by public utilities, it clearly offers a great number of social advantages. It is renewable and nonpolluting, and it is available to a variable degree nearly everywhere. The equipment demands few further technical advances in order to be mass-produced, at which time the cost per unit would decline dramatically. The additional development of standard home interfaces would effectively decrease domestic use of other, nonrenewable energy resources, pending the likely requirement of a technological advance in power storage technology. A combination of locally available energy such as wind and solar power with maximal use of conservation techniques could result in a decline or stabilization in the energy requirement by the private sector of the economy. As the free flow and abundance of energy is the life's blood of any economy, such developments serve to benefit society as a whole.

Dennis Chamberland

FURTHER READING

Andrews, John, and Nick Jelley. *Energy Science: Principles, Technologies, and Impact.* New York: Oxford University Press, 2007. Discusses various forms of energy; environmental and socioeconomic impacts; and principles of energy consumption, including information on the generation, storage, and transmission of energy. A strong mathematics or engineering background is required for the mathematical examples. Contains useful information for undergraduates and professionals interested in energy resources.

Boyle, Godfrey, ed. *Renewable Energy.* 2d ed. New York: Oxford University Press, 2004. Provides a complete overview of renewable energy resources. Discusses wind energy and other renewable energy sources, as well as the basic physics principles, technology, and environmental impacts. Includes references and a further reading list with each chapter.

Clark, Wilson. *Energy for Survival.* Garden City, N.Y.: Doubleday, 1974. Discusses all energy alternatives at the disposal of the United States. Includes a very detailed discussion of the use of wind energy in its various forms. Illustrated and indexed.

Gipe, Paul. *Wind Energy Comes of Age.* New York: John Wiley, 1995. A look at the evolution of wind power and the procedures and protocol involved in the conversion process. An introduction for the general reader. Illustrations, index, and bibliography.

Letcher, Trevor M., ed. *Future Energy: Improved, Sustainable, and Clean.* Amsterdam: Elsevier, 2008. Considers the future of fossil fuels and nuclear power. Discusses renewable energy supplies, including solar, wind, hydroelectric, and geothermal power. Focuses on potential energy sources and currently underutilized energy sources. Covers new concepts in energy consumption and technology.

Marier, Donald. *Wind Power for the Homeowner.* Emmaus, Pa.: Rodale Press, 1981. The definitive text for homeowners wanting to install their own wind-energy systems. Details the practical and technical side of installation of wind-energy systems. Discusses cost and legal issues. Well illustrated. Includes relevant tables, references, and addresses of equipment manufacturers.

Naisbitt, John. *Megatrends: Ten New Directions Transforming Our Lives.* New York: Warner Books, 1984. Details ten directions that are transforming the lives of Americans, from the evolving trends in technology to politics and the economy. Discusses the significance of energy and power sources. Covers the dynamics of a constantly shifting social structure and the role that world energy sources play in it.

Patel, Mukund R. *Wind and Solar Power Systems.* Boca Raton, Fla.: CRC Press, 1999. Provides a thorough examination and clear explanation of wind and solar energy plants. Discusses their operations, protocol, and applications. Suitable for the reader without prior knowledge of alternative energy sources. Illustrations, maps, index, and bibliography.

Torrey, Volta. *Wind-Catchers: American Windmills of Yesterday and Tomorrow.* Brattleboro, Vt.: Stephen Greene Press, 1981. A detailed, meticulous recounting of the windmill (a broad category in which the author includes wind generators) throughout world history and its ultimate infiltration into the United States. Describes the scientific as well as the historical details. Covers futuristic wind generators and contains some rather exciting photos of futuristic Danish wind-generator designs.

Vogel, Shawna. "Wind Power." *Discover* 10 (May, 1989): 46-49. Up-to-date account of the current state of wind power in the world, providing an excellent sketch of existing wind-power systems. Discusses the current attitude toward wind energy, foreign investments, and the state of world energy in relation to wind-power economics.

Walker, John F., and Nicholas Jenkins. *Wind Energy Technology.* New York: John Wiley & Sons, 1997. Deals with the history of wind energy, the applications of the resource, and the potential for converting wind into electricity. Illustrations and index.

Weiner, Jonathan. *Planet Earth.* New York: Bantam Books, 1986. Based on the television series of the same name. Details the energy equation on the planet, its balance, and the methods for extracting it. Provides an excellent look at the balance sheet for world energy, including wind power and its place in the future of world power needs.

See also: Coal; Geothermal Power; Hydroelectric Power; Nuclear Power; Nuclear Waste Disposal; Ocean Power; Oil and Gas Distribution; Oil and Gas Origins; Solar Power; Strategic Resources; Unconventional Energy Resources.

XENOLITHS

Xenoliths are blocks of preexisting rocks within magma. Consequently, xenoliths provide a sampling of materials through which the magma has traversed on its rise toward the surface. Some xenoliths originate within the mantle and provide the only means of obtaining samples of this elusive material.

PRINCIPAL TERMS

- **assimilation:** the absorption of the chemical components of wall rock or xenoliths into a magma
- **country rock:** rocks through which a magma is intruding; also known as "wall rock"
- **diatreme:** a pipelike conduit in the crust of the earth filled with fragmented rock produced by gas-rich volcanic eruptions
- **eclogite:** rock composed principally of garnet and pyroxene that formed at high pressures associated with great depths
- **felsic rocks:** igneous rocks rich in potassium, sodium, aluminum, and silica, including granites and related rocks
- **mafic rocks:** igneous rocks rich in magnesium and iron, including gabbro, basalt, and related rocks
- **magma:** a naturally occurring silicate-rich melt beneath the surface of the earth
- **mantle:** the intermediate zone between the crust and the core of the earth
- **peridotite:** a class of ultramafic rocks made up principally of pyroxene and olivine, with subordinate amounts of other minerals
- **segregation:** the concentration of early-formed minerals in a magma by crystal settling or crystal floating
- **ultramafic:** a term for any rock consisting of more than 90 percent ferromagnesium minerals, including olivine and pyroxene

OCCURRENCE OF XENOLITHS

Xenoliths are fragments of preexisting rocks that have been incorporated in a magma as it makes its way into higher levels of the crust. The term "xenoliths" is derived from the Greek roots *xeno* and *lith*, meaning "strange or foreign" and "rock," respectively. These rock fragments are pieces of previously formed rocks that become incorporated into the magma and perhaps removed from their source as the magma moves. The xenoliths may retain their original identity with minor alteration, or they may be greatly altered by attendant heat and fluids present in the magma. Xenolithic inclusions may be preserved near to their original sources along the borders of an intrusive magma, or they may be carried for great vertical distances from where they originated. In this manner, fragments of deep crust and mantle material from as much as 200 to 300 kilometers below the earth's surface have been brought to the surface by volcanic eruptions.

Xenoliths represent fragments of the rocks through which a magma has moved to its site of final emplacement and crystallization. They may be found in products of explosive volcanism such as volcanic tuff and breccia, within crystalline igneous rocks as in lava flows, and within shallow and deep-seated igneous rocks. Explosive volcanic materials are ejected by highly gas-charged eruptions that produce diatremes and maar-type volcanoes. These are volcanic craters in the form of inverted conelike or dishlike depressions in the surface surrounded by a rim of ejected deposits. Xenoliths are found as angular or rounded blocks embedded in ash tuffs or volcanic breccia in the rim and the pipelike conduit underlying the crater. In certain types of basalt and related magmas that originate deep in the mantle, such as kimberlite, rare fist-sized fragments of mantle material are transported upward from near the source of the magma origin. Fragments may also be collected from rocks traversed by the magma along its path of vertical ascent through the crust. Xenoliths in crystalline igneous rocks are embedded within the rock and are not exposed until erosion exposes the xenolith by removing overlying material. Granite rocks typically contain large xenoliths of metamorphic or sedimentary rocks. Such xenoliths reflect the typical intrusive process that produces granites. In this process, sub-surface magma chambers expand and move upward by physically plucking country rocks from the wall and roof.

WALL-ROCK XENOLITHS

Three basic varieties of xenoliths are recognized: wall-rock xenoliths, cognate xenoliths, and mantle xenoliths. Wall-rock xenoliths are represented by blocks and pieces of the adjacent country rock that have been incorporated into the magma. Cognate xenoliths are inclusions of the chilled margins of the magma or comagmatic segregations—that is, masses of previously solidified magma that break loose and are later incorporated in more energetic magmatic motions. Mantle xenoliths are presumed to be pieces of the mantle that become incorporated in magmas that are formed by partial melting deep within the earth.

Magmatic intrusions make room for themselves by three processes: forceful injection, stoping, and assimilation. Stoping occurs when the magmatic front advances by injection into fractures and surrounding blocks of country rock. These blocks may sink in the magma, and they may be slightly altered or totally assimilated within the magma depending on the characteristics of the magma and the wall rock. Most wall-rock xenoliths in felsic magmas do not move far from their source. Xenoliths are abundant near the margins of most intrusions. Because the margins are more likely to be losing heat and cooling at rates faster than the interior of the intrusion, the magma is more viscous, and xenoliths are less likely to move very far from the source. In contrast, fast-moving magmas in volcanic conduits often carry a wide variety of xenoliths from country rock traversed by the magma. In this way, a wide variety of crustal rocks cut by the volcanic vent may be brought to the earth's surface.

Xenoliths usually show effects of the high temperatures to which they are exposed. Preexisting minerals within a xenolith react to form new minerals that are in equilibrium with the magma. Thus, most xenoliths are brought to a high-grade metamorphic state unless they are composed of high-temperature refractory minerals to begin with, or unless they are exposed to high temperatures for short periods of time. The German term *Schlieren* is used to describe hazy, ill-defined streaks of xenoliths that are almost completely assimilated. Materials caught up in low-temperature felsic magmas are more likely to be altered by reactive assimilation, in which there is an exchange of ions between the xenolith and the magma. Inclusion of large blocks of country rock may alter the composition of magma by enriching it with elements that were not originally abundant.

One process that is not igneous in nature but creates xenoliths as well is the subterranean flow of salt. Buried salt beds are capable of plastic flowage in response to the weight of overlying rocks. The salt is not molten but nevertheless flows under pressure. Often the salt, which has a lower density than the enclosing sedimentary rocks, will rise thousands of feet through overlying sediments to form salt domes or salt plugs. The rise of large masses of salt is very much like the rise of magma. Pieces of wall rocks and sub-salt rocks may be incorporated into the salt as xenoliths. In this manner, salt domes in the Persian Gulf have brought up blocks of sedimentary, igneous, and metamorphic rocks from great depths in the crust. There are even ultramafic igneous xenoliths in the Weeks Island salt dome in southern Louisiana that are thought to be fragments of mantle-derived ultramafic intrusions emplaced along fault zones prior to deposition of the salt.

COGNATE XENOLITHS

Cognate xenoliths, also called "autoliths," are xenoliths from parts of the magma that have previously crystallized. Magmas solidify over a wide range of temperatures. Large bodies of magma may require tens of thousands of years to crystallize fully. Material on the outer edge of the magma will cool and crystallize more rapidly than that of the interior, resulting in chilled margins. Elsewhere within the magma, early formed crystals will either float or sink depending on the specific gravity differential. Feldspar crystals tend to float and collect near the top of the magma chamber, and mafic minerals such as olivine or pyroxene tend to sink to the bottom of the chamber. Some magmas undergo energetic degassing because of the reduction in confining pressure as they approach shallow levels in the crust. The rapid evolution of dissolved volatiles may disrupt previously crystallized portions of magma (chilled margins or crystal segregations) and mix solid cognate xenoliths with the mobile fluid phase.

Amphibolite xenoliths in medium-grained pink granite, west of Whirlwind Lake, Northwest Territories. (Geological Survey of Canada)

MANTLE XENOLITHS

Perhaps the most exotic xenoliths are those that originate within the mantle. The mantle lies at depths of five to forty kilometers below the surface and extends down to the top of the outer core nearly three thousand kilometers beneath the surface. No drill has penetrated to the mantle; therefore, these materials are completely inaccessible for direct sampling. The probable composition and mineralogical makeup of the mantle is postulated from calculations of the density, pressure, and temperatures that exist at mantle depths and by comparisons with meteorites, which are pieces of asteroids that have been fragmented by collisions with other asteroids. The interior of asteroids are thought to reproduce conditions similar to those of the mantle. Fortunately, pieces of the upper mantle are delivered to the surface of the earth as xenoliths in some magmas that originate by partial melting deep within the earth. For these rocks to make it to the surface without significant alteration by the host magma, they must be delivered to the surface in a fairly short period of time. Thus, it is not surprising that mantle xenoliths are found in volcanic rocks associated with rift zones and magmatic zones between plates, which allow for the rapid rise of gas-rich magmas to the surface.

Typical mantle xenoliths are composed of peridotite (olivine and pyroxene-rich rocks) incorporated in mafic volcanic rocks such as basalt. Basalts are formed from magma that originates by the partial melting of mantle materials. They often incorporate xenoliths from their place of origin as well as fragments of crustal rocks torn from walls of the conduit along which they are rising. Most notable of these magmas are varieties of peridotites known as "kimberlites" and "lamproites." Kimberlite (mica peridotite) is a potassic ultramafic rock that occurs in intrusive pipes and plugs, and explosively formed volcanic craters that overlay them. Lamproite is a porphyritic, ultrapotassic ultramafic rock that occurs in dikes and small intrusions. Kimberlite and lamproite magmas typically contain up to 75 percent xenoliths and xenocrysts. Kimberlites contain abundant xenoliths of lherzolite—a mantle peridotite with magnesian olivine, pyroxene, and minor calcium-plagioclase, spinel, or garnet.

The range of xenoliths in basalt is more restricted than that in kimberlites because basalts form at shallower levels of the mantle. Thus, basalts incorporate less of the upper mantle on their way to the surface. Spinel lherzolites are common in alkali basalts such as those found in Hawaii, Arizona, the Rio Grande Rift, and Central Europe. Basalts also contain xenoliths of harzburgite, dunite, and eclogite. Harzburgite (a peridotite with magnetite and spinel) and dunite (an ultramafic rock that consists of mostly olivine with minor chrome-bearing spinel as an accessory mineral) probably represent residual melts following fractionation of lherzolite.

The occurrence of mantle xenoliths in volcanics is limited to intraplate magmatic environments such as oceanic islands (Hawaii and Tahiti) and continental volcanic provinces or rifts (the southern Colorado Plateau and the Eifel District, Germany). Kimberlite intrusions favor old, stable, thick continental crust (South Africa and central North America). These environments are characterized by simple plumbing systems in which magmas rise rapidly to the surface; otherwise, xenoliths would sink in the host magma. Xenoliths are more rare in complex systems found in interplate environments, such as collisional magmatic provinces or island arcs.

SCIENTIFIC VALUE

The chief scientific value of xenoliths rests in their being samples of materials collected as magma ascended through the earth's mantle and crust. Wall-rock xenoliths provide samples of country rock that remain close to their original source. More importantly, xenoliths in rapidly ascending, volatile-rich volcanic magmas may provide samples of crustal rocks from the walls of the conduit throughout its entire path. These materials are brought to the surface in relatively unaltered states. In addition to mantle xenoliths, some localities—such as Kilbourne Hole, New Mexico, and Williams, Arizona—contain significant xenoliths of

granite gneiss representing crustal basement rocks. Cognate xenoliths provide information on the earliest parts of the magma to crystallize. Basalt, kimberlite, and lamproites contain xenoliths from the mantle. Distribution of nodule occurrences is not random. Siliceous basalts rarely contain nodules, whereas alkali basalts commonly contain eclogite and spinel peridotites. Kimberlites contain abundant nodules, including garnet peridotite. These differences suggest different depths of origin for the different magmas. Mantle xenoliths are the only means of obtaining samples of the mantle. Study of these materials can lead not only to determining the composition of the mantle but also to developing an understanding of its physical state and some of its processes.

From a study of these rare rocks, various processes and conditions of the lower crust and upper mantle can be inferred. The mineralogical combinations serve as geobarometers and geothermometers. Controlled crystallization studies at a variety of temperatures and pressures are employed to characterize the stable mineral assemblage within differing crustal and mantle environments. A mixture of minerals or rocks exposed to elevated temperatures, as found in the lower crust and upper mantle, recrystallize into a mineral assemblage that is in equilibrium with the higher temperatures. Alternatively, by examining altered rocks such as some altered xenoliths, it is possible to determine their prior state before metamorphism by the magma.

ECONOMIC VALUE

Wenoliths and xenocrysts find their principal economic value as a source of diamonds formed deep within the mantle and, to a much lesser extent, the olivine gem peridot. Diamonds occur as xenocrysts in kimberlite, lamproite, and alluvial gravels derived from kimberlites. Diamonds form at very high pressures. Minimum conditions required are pressure greater than 40 kilobars, which is equivalent to depths greater than 120 kilometers, and temperatures of approximately 1,000 degrees Celsius. Diamonds are associated with magnesia garnet-bearing lherzolites and coesite.

Diamonds make up only one part in twenty million of a typical diamond-bearing kimberlite, and many kimberlites are devoid of diamonds.

By far, most diamond production is from kimberlites in West Africa, South Africa, and Siberia, but diamonds are also mined in Australia, Brazil, and India. Small concentrations of diamonds have been found in kimberlite and lamproite bodies in Arkansas, Wyoming, Montana, Michigan, and Canada. After unsuccessful attempts to develop mining operations at the lamproite body at Murfreesboro, Arkansas, the area has been turned into the Crater of Diamonds State Park, where several small diamonds are found by tourists each year.

René De Hon

FURTHER READING

Best, Myron G. *Igneous and Metamorphic Petrology.* 2d ed. Malden, Mass.: Blackwell Science Ltd., 2003. A college-level textbook that should be accessible to most readers. Part 1 contains comprehensive treatments of all major plutonic rock bodies, complete with drawings, diagrams, and photographs detailing the essential features of each pluton type. Chapter 1 contains a section on how petrologists study rocks. The appendix contains chemical analyses of plutonic rocks and descriptions of important rock-forming minerals. One of the best books available for the serious student of igneous and metamorphic rocks.

Dawson, J. B. *Kimberlites and Their Xenoliths.* New York: Springer-Verlag, 1980. A technical treatment of xenoliths and their scientific significance as samples of the mantle.

Faure, Gunter. *Origin of Igneous Rocks: The Isotopic Evidence.* New York: Springer-Verlag, 2010. Discusses chemical properties of igneous rocks, and isotopes within these rocks formations. Specific locations of igneous rock formations and the origins of these rocks are provided. Diagrams, drawings, and an overview of isotope geochemistry make this accessible to undergraduate students as well as professionals.

Grapes, Rodney. *Pyrometamorphism.* 2d ed. New York: Springer, 2010. Discusses the formation of fused rocks and basaltic intrusions such as xenoliths. Describes heating and cooling sequences. The second edition includes additional references and illustrations.

Jerram, Dougal, and Nick Petford. *The Field Description of Igneous Rocks.* 2d ed. Hoboken, N.J.: Wiley-Blackwell, 2011. Begins with a description of field skills and methodology. Contains chapters on lava flow and pyroclastic rocks. Designed for student and scientist use in the field.

Legrand, Jacques. *Diamonds: Myth, Magic, and Reality.* Rev. ed. New York: Crown, 1985. Provides comprehensive coverage of diamonds, including their worldwide occurrence, geology, crystallography, mining, and cutting.

Mitchell, Roger H. *Kimberlites, Orangeites, and Related Rocks.* New York: Plenum Press, 1995. Provides a good introduction to the study of kimberlites and related rocks, including an extensive bibliography that will lead the reader to additional information.

_____. *Kimberlites, Orangeites, Lamproites, Melilitites, and Minettes: A Petrographic Atlas.* Thunder Bay, Ontario: Almaz, 1997. Covers the worldwide distribution of kimberlites and related rock types. Color illustrations and bibliography.

Morris, E. M., and J. D. Pasteris. *Mantle Metasomatism and Alkaline Magmatism.* Boulder, Colo.: Geological Society of America, 1987. This collection of papers presented at the Symposium on Alkali Rocks and Kimberlites provides a technical discussion of the chemistry, mineralogy, and petrology of the mantle as determined from mantle xenoliths.

Nixon, Peter H., ed. *Mantle Xenoliths.* New York: John Wiley & Sons, 1987. Covers a wide range of topics related to kimberlites. Considerable attention is paid to regional kimberlite occurrences and to the foreign rocks that are brought up by the kimberlite diatremes. Several of the articles are general enough to suit a beginning reader, but the work is best suited for the undergraduate and graduate student.

Raymond, L. A. *Petrology: The Study of Igneous, Sedimentary, and Metamorphic Rocks.* 2d ed. Long Grove, Ill.: Waveland Press, 2007. Provides comprehensive coverage of the field of petrology. Xenoliths are discussed in the sections on igneous rocks.

Sinkanka, John. *Gemstones of North America.* Tucson, Ariz.: Geoscience Press, 1997. Written for the amateur collector, the text describes gem-hunting localities in North America and several diamond localities in the United States.

See also: Andesitic Rocks; Anorthosites; Basaltic Rocks; Batholiths; Carbonatites; Diamonds; Granitic Rocks; Igneous Rock Bodies; Kimberlites; Komatiites; Orbicular Rocks; Platinum Group Metals; Plutonic Rocks; Pyroclastic Rocks; Rocks: Physical Properties.

APPENDIXES

GLOSSARY

abyssal plain: vast, flat, underwater plains.

accretionary prism: the complexly deformed rocks in a subduction zone that are scraped off the descending plate or eroded off the overriding plate.

aerodynamic lift: a measure of the degree of wind force acting on a rotor, which is translated into power to a generator in a wind-power machine.

aggregate: a mineral filler such as sand or gravel that, when mixed with cement paste, forms concrete.

A horizon: the surface soil layer; also known as topsoil.

albedo: the fraction of visible light of electromagnetic radiation that is reflected by the properties of a given type of surface.

algae: diverse group of autotrophic organisms that live in aqueous and subaqueous environments.

alkali metals: soft, shiny, highly reactive metals that readily lose their one valence electron; group 1 of the periodic table of elements.

allotrope: form of an element with different physical and chemical properties.

alloy: a metal composed of two or more elements.

alpha particle: a helium nucleus emitted during the radioactive decay of uranium, thorium, or other unstable nuclei.

alpha rays: combination of two protons and two neutrons released from a radionuclide during the radioactive decay process.

alumina: sometimes called aluminum sesquioxide, alumina is found in clay minerals along with silica; tricalcium aluminate acts as a flux in cement manufacturing.

aluminum: the most abundant metal in the earth's crust.

amalgam: an alloy of mercury and another metal; gold and silver amalgams occur naturally and have been synthesized for a variety of uses.

amethyst: a variety of quartz noted for its violet to purple color produced by irradiated molecules of iron contained within its crystalline structure.

amphiboles: a group of generally dark-colored, double-chain silicates crystallizing largely in the orthorhombic or monoclinic systems and possessing good cleavage in two directions intersecting at angles of about 56 and 124 degrees.

angle of repose: the maximum angle of steepness that a pile of loose materials such as sand or rock can assume and remain stable; the angle varies with the size, shape, moisture, and angularity of the material.

anhydrous: having little or no water in its physical structure.

anion: an atom that has gained electrons to become a negatively charged ion.

anorthosite: a light-colored, coarse-grained plutonic rock composed mostly of plagioclase feldspar.

aphanitic: igneous rocks with fine grains that can be distinguished only with visual aid.

aplite: a light-colored, sugary-textured granitic rock generally found as small, late-stage veins in granites of normal texture; in pegmatites, aplites usually form thin marginal selvages against the country rock but may also occur as major lenses in the pegmatite interior.

aqueous: solution that contains water acting as a solvent or medium for reactions.

aqueous solution: synonymous terms for fluid mixture that is hot and has a high solvent capacity, permitting it to dissolve and transport chemical constituents; it becomes saturated upon cooling and may precipitate metasomatic minerals; also known as a hydrothermal fluid or intergranular fluid.

aquifer: a porous, water-bearing zone beneath the surface of the earth that can be pumped for drinking water.

aragonite: a carbonate with the orthorhombic crystal structure of the calcium carbonate compound; it forms in marine water or under high-pressure, metamorphic conditions.

arkose: a sandstone in which more than 10 percent of the grains are feldspar or feldspathic rock fragments; also called feldspathic arenite.

aromatic hydrocarbons: ring-shaped molecules composed of six carbon atoms per ring; the carbon atoms are bonded to one another with alternating single and double bonds.

arsenic: a chemical element with a metallic appearance.

asbestosis: deterioration of the lungs caused by the inhalation of very fine particles of asbestos dust.

ash: fine-grained pyroclastic material less than 2 millimeters in diameter.

assimilation: the absorption of the chemical components of wall rock or xenoliths into a magma.

augite: an essential mineral in most basalts, a member of the pyroxene group of silicates.

aureole: a ring-shaped zone of metamorphic rock surrounding a magmatic intrusion.

authigenic minerals: minerals that formed in place, usually by diagenetic processes.

avalanche: any large mass of snow, ice, rock, soil, or mixture of these materials that falls, slides, or flows rapidly downslope.

axis system: the vertical or horizontal orientation of the power shaft of a wind-power machine.

backscattering: a reflection of energy waves by an object.

banded iron formations: the type of iron deposit most commonly mined; composed of iron ores in alternating layers with chert.

bar: a unit of pressure equal to 100 kilopascals and very nearly equal to 1 standard atmosphere.

barrel: the standard unit of measure for oil and petroleum products, equal to 42 U.S. gallons or approximately 159 liters.

basalt: a dark rock containing olivine, pyroxene, and feldspar, in which the minerals often are very small.

basic: a term to describe dark-colored, iron- and magnesium-rich igneous rocks that crystallize at high temperatures, such as basalt.

basin: area containing thick sedimentary rock; usually rich in hydrocarbons.

batholith: the largest type of granite/diorite pluton, with an exposure area in excess of 100 square kilometers.

bathymetry: the study of underwater depth.

belt: a geographic region that in some way is distinctive.

benzene: a carcinogenic aromatic hydrocarbon found in crude oil; an important ingredient in gasoline.

beta particle: an electron emitted from the nucleus of an atom during the process of beta decay.

beta rays: electrons or positrons released from a nucleus during the radioactive decay process

B horizon: the soil layer just beneath the topsoil.

binary cycle: the process whereby hot water in the primary cycle gives up heat in a heat exchanger; a fluid such as isobutane in the secondary cycle absorbs heat, is pressurized, and drives a turbine generator.

biodegradation: biological processes that result in the breakdown of a complex chemical into simpler building blocks.

biomarkers: chemicals found in oil with a chemical structure that definitely links their origin with specific organisms; also called geochemical fossils.

biotite: type of layered, silicate mineral composed of silica molecules bonded with magnesium and iron and appearing in dark brown and black varieties.

bitumen: a generic term for a very thick, natural semisolid; asphalt and tar are classified as bitumens.

black smokers: active hydrothermal vents along seafloor ridges, which discharge acidic solutions of high temperature, volume, and velocity, charged with tiny black particles of metallic sulfide minerals.

blowout preventer: a massive device used at a rig wellhead to function as a shutoff valve if the pressure in a reservoir gets out of control.

blue ground: the slaty blue or blue-green kimberlite breccia of the South African diamond pipes.

boom: a metal or plastic barrier used to prevent the spread of oil during a spill.

bort: a general term for diamonds that are suitable only for industrial purposes; these diamonds are black, dark gray, brown, or green in color and usually contain many inclusions of other minerals.

boule: a large, synthetically made single crystal with many industrial and technological applications.

Bowen's reaction principle: a principle by which a series of minerals forming early in a melt react with the remaining melt to yield a new mineral in an established sequence.

British thermal unit (Btu): the amount of heat required to raise the temperature of 1 pound of water by 1 degree Fahrenheit at the temperature of maximum density for water (39 degrees Fahrenheit or 4 degrees Celsius).

by-product: a mineral or metal that is mined or produced in addition to the major metal of interest.

cable-tool drilling: a repetitive, percussion process of secondary use in the boring of relatively shallow oil and gas wells.

cadmium: a soft element.

calcite: the main constituent of limestone, a carbonate mineral consisting of calcium carbonate.

caldera: an underground magma chamber.

cap rock: an impervious rock unit, generally composed of anhydrite, gypsum, and occasionally sulfur, overlying or capping any salt dome or salt diaper.

carbonate: ionic salt formed from the interaction of carbonic acid with basic elements

carbonate rock: a rock composed mainly of calcium carbonate.

carbonate rocks: the general terms for rocks containing calcite, aragonite, or dolomite.

carbonates: a large group of minerals consisting of a carbonate anion (three oxygen atoms bonded to one carbon atom, with a residual charge of two) and a variety of cations, including calcium, magnesium, and iron.

carbonatite: silicate mineral consisting of reduced levels of silica and rich in carbonate molecules formed from carbonic acid.

carbon cycle: a natural cycle by which carbon is absorbed by plants through the soil and air; plants in turn are eaten by animals, which exhale the carbon as carbon dioxide into the atmosphere and which decompose after death, returning carbon to the soil.

carbon sequestration: a chemical process whereby carbon dioxide is removed from an energy source and isolated in a secure chamber to prevent its release into the atmosphere.

casing: the metal pipe that provides the structure for a well.

catalyst: a chemical substance that speeds up a chemical reaction without being permanently affected by that reaction.

cation: an atom that has lost electrons to become a positively charged ion.

cation exchange capacity: the ability of a clay to adsorb and exchange cations (positively charged ions) within its environment.

cellulose: the substance forming the bulk of plant cell walls.

cementation: the joining of sediment grains, which results from mineral crystals forming in void spaces between the sediment.

ceramics: nonmetal compounds, such as silicates and clays, produced by firing the materials at high temperatures.

chain silicates: a group of silicates characterized by joining of silica tetrahedra into linear single or double chains alternating with chains of other structures; also known as inosilicate.

chalcedony: a form of cryptocrystalline quartz formed from microcrystals of quartz and other included minerals.

chalcogenides: nonmetallic elements that are known to form ores with certain metal elements; also

called ore formers.

chemical weathering: a change in the chemical and mineralogical composition of rocks by means of reaction with water at the earth's surface.

chert: multicolored, fine-grained sedimentary rock composed of silica and formed in the ocean.

chromium: a lustrous hard metal.

clastic rock: a sedimentary rock composed of broken fragments of minerals and rocks; typically a sandstone.

clathrate gun hypothesis: a theory that states that the Permian extinction was caused by a sudden release of methane from gas clathrates.

clay: a term with three meanings—a particle size (less than 2 microns), a mineral type (including kaolin and illite), and a fine-grained soil that is like putty when damp.

clay mineral: a group of minerals, commonly the result of weathering reactions, composed of sheets of silicon and aluminum atoms.

clay stone: a clastic sedimentary rock composed of clay-sized mineral fragments.

cleavage: the tendency for minerals to break in smooth, flat planes along zones of weaker bonds in their crystal structure.

clinker: irregular lumps of fused raw materials to which gypsum is added before grinding into finely powdered cement.

coal: dark brown to black sedimentary rock formed from the accumulation of plant material in swampy environments.

coastal wetlands: shallow, wet, or flooded lowlands that extend seaward from the freshwater-saltwater interface; they may consist of marshes, bays, lagoons, tidal flats, or mangrove swamps.

coastal zone: coastal waters and lands that exert a measurable influence on the uses of the sea and its ecology.

coefficient of thermal expansion: the linear expansion ratio (per unit length) for any particular material as the temperature is increased.

coesite: a mineral with the same composition as quartz (silicon dioxide) but with a dense crystal structure that forms only under very high pressures.

coherent texture: an arrangement allowing the minerals or particles in a rock to stick together.

cohesion: the strength of a rock or soil imparted by the degree to which the particles or crystals of the material are bound to one another.

completion procedures: all methods and activities necessary in the preparation of a well for oil and gas production.

compound: a chemical combination of elements with distinct properties that may differ from the elements from which it formed.

compressive strength: the ability to withstand a pushing stress or pressure, usually given in pascals or pounds per square inch.

computer-aided design: software that generates high-resolution images that may be manipulated by the user.

conchoidal fracture: fracturing of the type exhibited by materials that do not break along planes of separation.

concrete: a composite construction material that consists of aggregate particles bound by cement.

conductivity: the opposite of resistivity, or the ease with which an electric current passes through a rock formation.

contact metamorphic facies: zones of contact metamorphic effects, each of which is characterized by a small number of indicator minerals.

contact metamorphism: metamorphism characterized by high temperature but relatively low pressure, usually affecting rock in the vicinity of igneous intrusions.

continental margin: the offshore area immediately adjacent to the continent, extending from the shoreline to depths of approximately 4,000 meters.

convection: fluid circulation produced by gravity acting on density differences arising from unequal temperatures within a fluid; the principal means of heat transfer involving fluids of low thermal conductivity.

correlation: the tracing and matching of rock units from one locality to another, usually on the basis of lithologic characteristics.

country rock: rocks through which a magma is intruding; also known as "wall rock."

covalent bond: a chemical bond characterized by electron sharing.

craton: a stable portion of the earth's mantle and associated crust that forms the inner core of the lithospheric components.

creep: the slow and more or less continuous downslope movement of Earth material.

crust: the outermost layer of the earth; crust may be continental or oceanic.

crust: the veneer of rocks on the surface of the earth.

crystal: a material with a regular, repeating atomic structure.

crystal growth: the second stage of crystallization; a transfer of heat, matter, or both drives the crystal to expand from the nucleation site by adding molecules, atoms, or ions to the lattice.

crystal habit: the external appearance of a crystal, including its shape and visible physical features.

crystalline: solid structure composed of atoms arranged in a regular, repeating pattern to form a repeating three-dimensional structure.

crystal-liquid fractionation: physical separation of crystals, precipitated from cooling magma, from the coexisting melt, enriching the melt in elements excluded from the crystals; this separation, or fractionation, leads to extreme concentration of incompatible elements in the case of pegmatite magma.

crystallization: the formation and growth of a crystalline solid from a liquid or gas.

crystallography: the scientific study of crystals and how they form.

crystal structure: the regular arrangement of atoms in a crystalline solid.

crystal system: any one of six crystal groups defined on the basis of length and angular relationship of the associated axes.

deformation: geologic process whereby the shape and size of rocks and minerals are altered because of the application of pressure, temperature, or chemical forces.

density: mass per unit volume.

deposition: the settling and accumulation of sediment grains after transport.

depositional environment: the environmental setting in which a rock forms; for example, a beach, coral reef, or lake.

depth sounding: determining the depth of a specific point underwater; the resulting data are used in bathymetry.

derivative maps: maps that are prepared or derived by combining information from several other maps.

derrick: apparatus from which a drill string and casing are suspended.

desalinization: the process of removing salt and minerals from seawater or from saline water occurring in aquifers beneath the land surface to render it fit for agriculture or other human use.

detrital minerals: minerals that have been eroded, transported, and deposited as sediments.

detrital rock: a sedimentary rock composed mainly of grains of silicate minerals as opposed to grains of calcite or clays.

diagenesis: the physical and chemical changes that occur to sedimentary grains after their accumulation

diamond: a high-pressure, high-temperature mineral consisting of the element carbon; it is the hardest naturally occurring substance and is valued for its brilliant luster

diamond stability zone: area within the earth's mantle where pressures and temperatures are sufficient for the formation of diamonds from deposits of carbon

diatreme: a volcanic vent or pipe formed as the explosive energy of gas-charged magmas breaks through crustal rocks.

differential movement: the unequal movement of various parts of a building or pavement in response to swelling or shrinkage of the underlying soil.

diffusion: process whereby atoms move individually through a material.

dike: a tabular igneous rock formed by the injection of molten rock material through another solid rock

directional drilling: the controlled drilling of a borehole at an angle to the vertical and at an established azimuth

direct or single flash cycle: the process whereby hot water under great pressure is brought to the surface and is allowed to turn, or "flash," to steam, driving an electrical turbine generator.

dispersant: a chemical that breaks spilled oil into droplets that will sink instead of remain on the surface of the water.

disposable battery: a type of battery, also called a primary cell battery, in which the electrochemical reaction is irreversible and the battery cannot be recharged.

divalent ion: an ion with a charge of 2 because of the loss or gain of two electrons.

divergent boundary: an area where two tectonic plates are spreading apart instead of crashing together or sliding against one another.

dodecahedron: a three-dimensional geometric configuration with twelve faces.

dolomite: a mineral composed of calcium magnesium carbonate.

double flash cycle: the process whereby two flash vessels are employed in cascade, each operating a turbine to extract more power.

downhole tool: a drill bit and motor mounted on the end of the drill string; fluid pumped into the drill string drives the downhole tool, and the drill string is not rotated.

dragline: a large excavating machine that casts a rope-hung bucket, collects the excavated material by dragging the bucket toward itself, elevates the bucket, and dumps the material on a spoil bank, or pile.

drilling fluids: a carefully formulated system of fluids used to lubricate, clean, and protect the borehole during the rotary drilling process.

drilling mud: substance used to lubricate a drill and to prevent oil from flowing into the well during drilling.

drilling rig: the collective assembly of equipment, including a derrick, power supply, and draw-works, necessary in cable-tool and rotary drilling.

drill string: the length of steel drill pipe and accessory equipment connecting the drill rig with the bottom of the borehole.

dry hole: a well drilled for oil or gas that had no production.

dye: a colored solution.

dyke: igneous rock formation that develops in a vertical or diagonal direction from the surface of the earth and often connects to a magma chamber within the earth's crust or mantle.

earthflow: a term applied to both the process and the landform characterized by fluid downslope movement of soil and rock over a discrete plane of failure; the landform has a hummocky surface and usually terminates in discrete lobes.

eclogite: rock composed principally of garnet and pyroxene that formed at high pressures associated with great depths.

ecological succession: the process of plant and animal changes, from simple pioneers such as grasses to stable, mature species such as shrubs or trees.

ecosystem: a self-regulating, natural community of plants and animals interacting with one another and their nonliving environment.

elasticity: the maximum stress that can be sustained without suffering permanent deformation.

electricity: a flow of subatomic charged particles called electrons used as an energy source.

electrolysis: process by which liquid or dissolved metals are separated by electromagnetic attraction.

electromagnetic radiation: energy emitted and absorbed by charge particles; travels in a wave-like manner.

electronegativity: a measure of how tightly an atom holds its electrons.

electrowinning: a process by which metals are pulled from solution or a liquid.

electrum: a term commonly used to designate any alloy of gold and silver containing 50 to 80 weight percent gold.

element: a pure substance that cannot be broken down into anything simpler by ordinary chemical means.

El Niño/Southern Oscillation: a reversal in precipitation patterns, ocean upwelling, and thermocline geometry that is accompanied by a weakening of the trade winds; the phenomenon typically recurs every three to seven years.

energy: the capacity for doing work; power (usually measured in kilowatts) multiplied by the duration (usually expressed in hours, sometimes in days).

enhanced oil recovery: the use of heat, chemicals, or gases to loosen oil from reservoir rocks.

equilibrium: geologic state in which temperature and pressure, along with other environmental conditions, are in balance.

erosion: the movement of soil and rock by natural agents such as water, wind, and ice, including chemicals carried away in solution.

estuarine zone: an area near the coastline that consists of estuaries and coastal saltwater wetlands.

estuary: a zone along a coastline, generally a submerged valley, where freshwater system(s) and river(s) meet and mix with an ocean.

evaporation: type of vaporization, the transition of a liquid to gaseous state; occurs on the top layer of a liquid.

evaporite: type of mineral resulting from concentration within an evaporating aqueous medium.

evaporite minerals: that group of minerals produced by evaporation from a saline solution; for example, rock salt.

evapotranspiration: the movement of water from the soil to the atmosphere in response to heat, combining transpiration in plants and evaporation.

exsolve: the process whereby an originally homogeneous solid solution separates into two or more minerals (or substances) of distinct composition upon cooling.

extrusion: igneous rocks that are formed after liquid lava cools on the surface of the earth.

extrusive rock: igneous rock that has been erupted onto the surface of the earth.

facies: a part of a rock, or a group of rocks, that differs from the whole formation in one or more properties, such as composition, age, or fossil content.

feldspar: a silicate rock that forms in igneous environments and is the most common mineral in the earth's crust.

felsic: mineral group characterized by high proportions of silicate minerals, including quartz, feldspar, plagioclase, and micas.

felsic rocks: igneous rocks rich in potassium, sodium, aluminum, and silica, including granites and related rocks.

ferric: describes iron compounds with an oxidation number of +3.

ferrous: describes iron compounds with an oxidation number of +2.

fiber optics: flexible transparent fiber made of fused quartz or silica glass; used in computer and telecommunications applications.

fiber-reinforced composites: materials produced by drawing fibers of various types through a material being cast to produce a high weight-to-strength ratio.

field: one or more pools (petroleum reservoirs); where multiple, the pools are united by some common factor.

fineness: a measure of the purity of gold or silver expressed as the weight proportion of these metals in an alloy; gold fineness considers only the relative proportions of gold and silver present, whereas silver fineness considers the proportion of silver to all other metals present.

fixed carbon: the solid, burnable material remaining after water, ash, and volatiles have been removed from coal.

flame test: a chemistry test that detects the presence of some metal ions based on the color of the flame when the material in question is held in it; lithium, for example, turns a flame crimson.

flint: a type of microcrystalline quartz that exists in large quantities within the earth's crust and is notable for its glass-like characteristics.

flocculation: the process by which particles are released from a colloid.

flow rate: the amount of water that passes a reference point in a specific amount of time (liters per second).

fluid inclusions: microscopic drops of parental fluid trapped in a crystal during growth; inclusions persist indefinitely unless the host crystal is disturbed by deformation or recrystallization.

fluidization: process whereby the rocks and minerals contained in a volcanic eruption are vaporized within a pyroclastic flow.

foliation: a texture or structure in which mineral grains are arranged in parallel planes.

fossil fuel: fuel source derived from the fossilized remains of organic matter, including coal, petroleum oil, and natural gas.

fracking: hydraulic fracturing; uses water and chemicals to break up shale and release natural gas.

fumarole: a gas- and steam-producing vent at volcanic craters.

gabbro: a coarse-grained, dark-colored plutonic igneous rock composed of plagioclase feldspar and pyroxene.

gamma radiation: electromagnetic wave energy originating in the nucleus of an atom and given off during the spontaneous radioactive decay of the nucleus.

gamma ray: a high-energy photon, with no mass or charge, emitted when atomic nuclei are broken down: such an electromagnetic ray may be released from a variety of sources, including the decay of unstable radionuclides; it is a form of ionizing radiation and is therefore potentially harmful to biological tissues.

gasification: process that converts carboniferous material to methane and carbon gases by subjecting the material to extreme heat in an environment that prevents combustion.

gem: a cut and polished stone that possesses the durability, rarity, and beauty necessary for use in jewelry and therefore of value.

gemstone: any rock, mineral, or natural material that has the potential for use as personal adornment or ornament.

generator: a machine that converts the mechanical energy of the turbine into electrical energy.

geographic information system: a series of data collected and stored in an organized manner in a computer system.

geomembrane: a synthetic sheet (plastic) with very low permeability used as a liner in landfills to prevent leakage from the excavation.

geophysics: the quantitative evaluation of rocks by electrical, gravitational, magnetic, radioactive, seismological, and other techniques.

geopolymer: a large molecule created by linking together many smaller molecules by geologic processes.

geotherm: a curve on a temperature-depth graph that describes how temperature changes in the subsurface.

glass: a solid with no regular periodic arrangement of atoms; that is, an amorphous solid.

goethite: a yellowish-brown or reddish-brown iron oxide-hydroxide.

grade: the classification of an ore according to material content, or according to value.

grain boundary: the interface between the small crystals (crystallites) that make up a larger polycrystal material.

grains: the individual particles that make up a rock or sediment deposit.

granite: an igneous rock that is known for its hardness and durability; in modern times, it has been used on the exterior of buildings, as it is able to resist the corrosive atmospheres of urban areas.

granitic/granitoid: descriptive terms for plutonic rock types having quartz and feldspar as major mineral phases.

granitization: the process of converting rock into granite; it is thought to occur when hot, ion-rich fluids migrate through a rock and chemically alter its composition.

granules: small grains or pellets.

graphite: a crystalline variety of the element carbon, characterized by its softness and ability to cleave into flakes; the carbon atoms are arranged in sheets that are weakly bonded together.

gravitational field: the area surrounding an astronomical body affected by the deformations in space and time caused by the mass of that body.

greenhouse effect: a process whereby carbon dioxide, methane, and other types of gas are released into the atmosphere, retaining the sun's heat and causing global warming.

greenhouse gas: an atmospheric gas capable of absorbing electromagnetic radiation in the infrared part of the spectrum.

greywacke: a sandstone in which more than 10 percent of the grains are mica or micaceous rock fragments; also called lithic arenite.

grid: a pattern of horizontal and vertical lines forming squares of uniform size.

groundmass: the fine-grained material between phenocrysts of a porphyritic igneous rock.

groundwater: water found below the land surface.

guano: fossilized bird excrement, found in great abundance on some coasts or islands.

guest molecule: a molecule contained inside a clathrate.

gypsum: a natural mineral, hydrated calcium sulfate; it helps control the setting time of cement.

gyre: an ocean-scale surface current that moves in a circular pattern.

half-life: the amount of time it takes for half of a sample of a radioactive element to decay.

halide: salt formed from the reaction between a halide ion, including chloride, bromide, and fluoride, with a basic material in the environment.

halite: mineral form of rock salt or common salt formed from a reaction between a chloride ion and sulfur atoms in an aqueous environment.

halogen: one of a group of chemical elements including chlorine, fluorine, and bromine.

hardness: the resistance to abrasion or surface deformation.

head: the vertical height that water falls or the distance between the water level of the reservoir above and the turbine below.

heavy metal: one of a group of chemical elements including mercury, zinc, lead, and cadmium.

heliosphere: area of outer space affected by the sun.

hematite: an iron oxide that is usually gray or black, but can also be red; contains 70 percent iron.

heterogeneous mixture: a nonuniform mixture.

high-level wastes: wastes containing large amounts of dangerous radioactivity.

highly enriched uranium: uranium that has been processed to alter its composition of uranium isotopes, making it more useful as fuel for nuclear weapons.

horizon: a layer of soil material approximately parallel to the surface of the land that differs from adjacent related layers in physical, chemical, and biological properties.

hornblende: type of silicate rock characterized by its dark color, which is the result of inclusions of iron and magnesium within the rock matrix.

hornfels: the hard, splintery rocks formed by contact metamorphism of sediments and other rocks.

hot spot: a column of magma that rises from the mantle, remaining in one place as the lithospheric plate moves over it; also known as a mantle plume.

humic acid: organic matter extracted by alkalis from peat, coal, or decayed plant debris; it is black and acidic but unaffected by other acids or organic solvents.

hummocky: a topography characterized by a slope composed of many irregular mounds (hummocks) that are produced during sliding or flowage movements of earth and rock.

hydration: a process whereby, when soils become wet, water is sucked into the spaces between the particles, causing them to grow several times their original size.

hydraulic cement: any cement that sets and hardens under water; the most common type is known as Portland cement.

hydraulic fracturing: the underground splitting of rocks by hydraulic or water pressure as a means of increasing the permeability of a formation.

hydrocarbon: an organic compound consisting of hydrogen and carbon atoms linked together.

hydroelectricity: methods used to convert the kinetic energy in moving water to electric energy.

hydrogen bond: electrochemical bond between a hydrogen atom and an electropositive atom such as those that occur between water molecules in liquid water and ice.

hydrologic cycle: the cycle of water movement on the earth from ocean to land and back.

hydrolysis: the breakdown of water by energy into its constituent elements of water and hydrogen.

hydrosphere: all the waters on the earth's surface, including seas, oceans, lakes, rivers, and the water vapor in the lower atmosphere.

hydrostatic: state of fluid pressure equilibrium.

hydrostatic pressure: pressure within a fluid at rest, exerted at a specific point.

hydrothermal: characterizing any process involving hot groundwater or minerals formed by such processes.

hydrothermal solution: a watery fluid, rich in dissolved ions, that is the last stage in the crystallization of a magma.

hydrothermal vent: a fissure through which magmatically heated fluids escape to the earth's surface.

hydroxl: a combination of an oxygen and hydrogen atom, which behaves as a single unit with a negative charge.

hyperspectral imaging: a remote sensing technique that detects electromagnetic radiation across a wide spectrum of wavelengths, including visible light.

hypersthene: a low-calcium pyroxene mineral.

hyperthermal field: a region having a thermal gradient many times greater than that found in nonthermal, or normal, areas.

igneous rock: a major group of rocks formed from the cooling of molten material on or beneath the earth's surface.

ignimbrite: a pumice-filled pyroclastic flow caused by the collapse of a volcanic column and accelerated by gravity.

ijolite: a dark-colored silicate rock containing the minerals nepheline (sodium aluminum silicate) and pyroxene (calcium, magnesium, and iron silicate).

immiscible liquids: liquids not capable of being mixed or mingled.

improved oil recovery: the process of pumping oil out or injecting gas and water to increase pressure in a reservoir, thereby forcing oil out.

inclusion: a foreign substance enclosed within a mineral; often very small mineral grains and cavities filled with liquid or gas; a cavity with liquid, a gas bubble, and a crystal is called a three-phase inclusion.

incompatible elements: chemical elements characterized by odd ionic properties (size, charge, electronegativity) that tend to exclude them from the structures of common minerals during magmatic crystallization.

index mineral: an individual mineral that forms under a limited or very distinct range of temperature and pressure conditions.

insolation: the radiation from the sun received by a surface; generally expressed in terms of power per unit area, such as watts per square meter.

intermediate rock: an igneous rock that is transitional between a basic and a silicic rock, having a silica content between 54 and 64 percent.

intrusion: liquid rock that forms beneath the earth's surface.

inverse problem: determining information about an object or phenomenon by analyzing observed measurements about the object or phenomenon.

ion: an atom that has a positive or negative charge.

ionic bond: the strong electrical forces holding together positively and negatively charged atoms.

ionizing radiation: radiation that creates ions by breaking down atomic nuclei.

isograd: a line on a geologic map that marks the first appearance of a single mineral or mineral assemblage in metamorphic rocks.

isostructural: having the same structure but a different chemistry.

isotopes: atoms of an element that contain the same number of protons but different numbers of neutrons.

isotropic: having properties that are the same in all directions—the opposite of anisotropic, having properties that vary with direction.

I-type granitoid: granitic rock formed from magma generated by partial melting of igneous rocks in the upper mantle or lowermost crust.

karat: a unit of measure of the purity of gold (abbreviated "k"); pure gold is 24 karat.

karst: a region formed by the weathering of underlying rock, typically limestone or dolomite.

kerogen: a waxy, insoluble organic hydrocarbon that has a very large molecular structure.

kick: a "belch" during uncontrolled pressure in a reservoir; can lead to a blowout or a gusher.

kilojoule: a unit of electrical energy equivalent to the work done to raise a current of electricity flowing at 1,000 amperes for 1 second (1,000 coulombs) by 1 volt; equivalent to 4.184 calories; approximately the energy needed to raise 100 kilograms one meter.

kilowatt: 1,000 watts; a unit of measuring electric power.

kimberlite: an unusual, fine-grained variety of peridotite that contains trace amounts of diamond.

kinetic energy: the energy possessed by an object because of its motion.

lahar: combination of water, ash, and mud that creates tremendous mudslides in a pyroclastic flow.

land: in the legal sense, any part of the earth's surface that may be owned as goods and everything annexed to that part, such as water, forests, and buildings.

landforms: surface features formed by natural forces or human activity, normally classified as constructional, erosional, or depositional.

landscape: the combination of natural and human features that characterize an area of the earth's surface.

landslide: a general term that applies to any downslope movement of materials; includes avalanches, earthflows, mudflows, rockfalls, and slumps.

land use: the direct application of a tract of land.

land-use planning: a process for determining the best use of each parcel of land in an area.

lapilli: pyroclastic fragments between 2 and 64 millimeters in diameter.

laterite: a deep red soil, rich in iron and aluminum oxides, and formed by intense chemical weathering in a humid tropical climate.

lattice: an infinite array of discrete points upon which a crystal is built; the crystal's arrangement of atoms, molecules, or ions is repeated at each lattice point.

layered igneous complex: a large and diverse body of igneous rock formed by intrusion of magma into the crust; it consists of layers of different mineral compositions.

LD_{50}: in toxicology, the abbreviation for the lethal dose for 50 percent of test animals; lethal dose is expressed in milligrams of poison per unit body weight of the animal.

leach: to dissolve from the soil.

leachate: water that has seeped down through the landfill refuse and has become polluted.

leaching: the dissolving or removal of soluble materials from a soil horizon by percolating water.

lead: a soft metal.

lease: a permit to explore for oil and gas on specified land.

level: all connected horizontal mine openings at a given elevation; generally, levels are 30 to 60 meters apart and designated by their vertical distance below the top of the shaft.

light detection and ranging: a remote sensing technique involving the emission of beams of light.

lignin: a family of compounds in plant cell walls, composed of an aromatic nucleus, a side chain with three carbon atoms, and hydroxyl and methoxyl groups, plus the molecule that binds cellulose fibers together.

lime: a common name for calcium oxide; it appears in cement both in an uncombined form and combined with silica and alumina.

limestone: a sedimentary rock composed mostly of calcium carbonate formed by organisms or by calcite precipitation in warm, shallow seas.

limonite: an iron oxide-hydroxide of varying composition; includes hematite and goethite.

lithic fragment: a grain composed of a particle of another rock; in other words, a rock fragment.

lithification: process by which sediment is compacted or cemented into solid rock.

lithology: the general physical type of rocks or rock formations.

lithosphere: outer portion of the earth consisting of the earth's crust and part of the upper mantle.

lithospheric plates: giant slabs composed of crust and upper mantle; they move about laterally to produce volcanism, mountain building, and earthquakes.

lithostatic pressure: pressure exerted by the weight of sediment overlying the sediment sample in question.

loading: an engineering term used to describe the weight placed on the underlying soil or rock by a structure or traffic.

lode deposit: a primary deposit, generally a vein, formed by the filling of a fissure with minerals precipitated from a hydrothermal solution.

low-level wastes: wastes that are much less radioactive than high-level wastes and thus less likely to cause harm.

luminescence: the emission of light by a mineral.

lunar cycle: the time it takes for the moon and the earth to return to any single position relative to each other, as dependent on the combination of the rotation of the earth and the orbit of the moon; approximated to 29.5 rotations of the earth on its axis.

luster: the reflectivity of the mineral surface; there are two major categories of luster: metallic and nonmetallic.

lysimeter: a simple pan or porous cup that is inserted into the soil to collect soil water for analysis.

macronutrient: a substance that is needed by plants or animals in large quantities; nitrogen, phosphorus, and potassium are macronutrients.

mafic rocks: igneous rocks rich in magnesium and iron, including gabbro, basalt, and related rocks.

mafic/ultramafic: compositional terms referring to igneous rocks rich (mafic) and very rich (ultramafic) in magnesium- and iron-bearing minerals.

magma: a body of molten rock typically found at great depths, including any dissolved gases and crystals.

magnetite: an iron oxide that is usually black; contains 72 percent iron.

manganese nodules: rounded, concentrically laminated masses of iron and manganese oxide found on the deep-sea floor.

mantle: layer of hot, solid rock between the earth's core and outer crust.

marble: a metamorphic rock that has been used since Grecian times as a preferred building stone; it is known for its ability to be carved, sculptured, and polished.

marine current energy conversion: power from the transfer of kinetic energy in major ocean currents into usable forms, such as electricity.

massif: a French term used in geology to describe very large, usually igneous intrusive bodies.

mass wasting: the downslope movement of earth materials under the direct influence of gravity.

megapascal: geologic unit of extreme pressure, equivalent to 1 million pascals and 145.037 pounds per square inch.

megawatt: 1 million watts; a unit of measuring electric power.

mercury: the only liquid metal at room temperature.

mesodesmic: molecular bond in which one of the molecule's oxygen ions is capable of bonding with other ions.

metal: an element with a metallic luster, high electrical and thermal conductivity, ductility, and malleability.

metallic ore: sedimentary rock containing minerals rich in metals and often used in the harvest and derivation of metal for industrial processes.

metamict rock: mineral that has lost its defined crystal structure during the process of radioactive decay, generally forming into an amorphous mineral.

metamorphic: rock that was formed by heat and pressure; tectonic, or mountain-building, forces of the earth's crust create and alter the mineral composition and texture of the original rock material.

metamorphic facies: an assemblage of minerals characteristic of a given range of pressure and temperature; the members of the assemblage depend on the composition of the protolith.

metamorphic grade: the degree of metamorphic intensity as indicated by characteristic minerals in a rock or zone.

metamorphic rocks: rocks formed by the effects of heat, pressure, or chemical reactions on other rocks.

metamorphic zone: areas of rock affected by the same limited range of temperature and pressure conditions, commonly identified by the presence of a key individual mineral or group of minerals.

metamorphism: changes in the structure, texture, and mineral content of solid rock as it adjusts to altered conditions of pressure, temperature, and chemical environment.

metasomatism: chemical changes in rock composition that accompany metamorphism.

metastable: the state of crystalline solids once they are outside of the temperature and pressure conditions under which they formed; thus, diamond forms at very high pressures within the earth but is metastable at the earth's surface.

meteoric water: water that originally came from the atmosphere, perhaps in the form of rain or snow, as contrasted with water that has escaped from magma.

methane: a colorless, odorless gaseous hydrocarbon with the formula CH_4; also called marsh gas.

mica: a platy silicate mineral (one silicon atom surrounded by four oxygen atoms) that readily splits into thin, flexible sheets.

micas: a group of complex, hydrous sheet silicates crystallizing largely in the monoclinic system and possessing pearly, elastic sheets with perfect one-directional cleavage.

micronutrient: a substance that is needed by plants or animals in very small quantities.

midocean ridge: a large, undersea chain of volcanic mountains encircling the globe, branches of which are found in all the world's oceans.

migmatite: a rock exhibiting both igneous and metamorphic characteristics, which forms when light-colored silicate minerals melt and crystallize, while the dark silicate minerals remain solid.

millisievert (mSv): a measure of the biological effect of radiation; 1,000 mSv in a short time will result in radiation sickness, and 4,000 mSv will kill half of those exposed to it.

mineral: a naturally occurring solid with a specific chemical composition.

mineral species: a mineralogic division in which all the varieties in any one species have the same basic physical and chemical properties.

mineral variety: a division of a mineral species based upon color, type of optical phenomenon, or other distinguishing characteristics of appearance.

mobile belt: a linear belt of igneous and deformed metamorphic rocks produced by plate collision at a continental margin; relatively young mobile belts form major mountain ranges; synonymous with orogenic belt.

moderator: a material used in a nuclear reactor for slowing neutrons to increase their probability of causing fission.

Mohs hardness scale: a series of ten minerals arranged in order of increasing hardness, with talc as the softest mineral known (1) and diamond as the hardest (10).

moisture content: the weight of water in the soil divided by the dry weight of the soil, expressed as a percentage.

molecular weight: a measure of the mass of a molecule of a chemical compound, as determined by both the total number and the size of atoms in the molecule.

molecule: the smallest entity of an element or compound retaining chemical identity with the substance in mass.

monoclinic: a sulfuric type that is less dense than rhombic sulfur and that appears in the form of long needles.

monolith: a large soil profile created by dovetailing containers together and combining small, individual profiles.

montmorillonites: a group of clay minerals characterized by swelling in water; the primary agent in expansive soils.

moving dunes: collections of coarse soil materials that result from wind erosion and threaten marginal vegetation and settlements as they move across deserts.

mudflow: both the process and the landform characterized by very fluid movement of fine-grained material with a high water content.

mudrock: a rock composed of abundant clay minerals and extremely fine siliciclastic material.

multiple use: the simultaneous use of land for more than one purpose or activity.

multispectral scanning: a remote sensing technique that detects discrete bands of electromagnetic radiation.

naphthenic hydrocarbons: hydrocarbon molecules with a ring-shaped structure, in which any number of carbon atoms are all bonded to one another with single bonds.

natural gas: a flammable vapor found in sedimentary rocks, commonly but not always associated with crude oil; it is also known simply as gas or methane.

neutron: an uncharged, or electrically neutral, particle found in the nucleus of an atom.

nodule: a chemically precipitated and spherical to irregularly shaped mass of rock.

nonrenewable resource: a resource that is fixed in quantity and will not be renewed within a human lifetime.

norite: gabbro in which hypersthene is the principal pyroxene; it is commonly associated with anorthosites.

nuclear fission: the splitting of an atomic nucleus into two lighter nuclei, resulting in the release of neutrons and some of the binding energy that held the nucleus together.

nuclear fusion: the collision and combining of two nuclei to form a single nucleus with less mass than the original nuclei, with a release of energy equivalent to the mass reduction.

nucleation: the first stage of crystallization; a new crystal forms around a nucleus, initiated by a phase change.

nuée ardente: a type of dense pyroclastic flow that occurs after a volcanic dome collapse; also known as block and ash and Pelean flow.

nutrient: a substance required for the optimal functioning of a plant or animal; foods, vitamins, and minerals essential for life processes.

oceanic ridges: a system of mostly underwater rift mountains that bisect all the ocean basins; basalt is extruded along their central axes.

ocean thermal energy conversion: power derived from taking advantage of the significant temperature differences found in some tropical seas between the surface and deeper waters.

ocean wave power: the use of wind-generated ocean surface waves to propel various mechanical devices incorporated as an electrical generating system.

off-planet: pertaining to regions outside Earth in orbital or planetary space.

oil rights: the ownership of the oil and natural gas on another party's land, with the right to drill for and remove them.

oil shale: a sedimentary rock containing sufficient amounts of hydrocarbons that can be extracted by slow distillation to yield oil.

olivine: a silicate mineral found in the earth's mantle and some basalts, particularly the alkaline varieties.

ooze: seafloor sediments.

opal: a form of silica containing a varying proportion of water within the crystal structure.

open source: software in which the source code may be accessed by the public and modified without the application of copyright laws.

ophiolite complex: an assemblage of metamorphosed basaltic and ultramafic igneous rocks that originate at marine ridges and are subsequently emplaced in mobile belts by plate-collision tectonics.

optimization: aspect of a software program that enables the user to modify it to conform to the user's needs.

orbicule: formation within orbicular igneous rocks consisting of similar materials to the parent rock; organized in concentric layers of crystals surrounding a core comprising smaller crystal clusters, larger individual crystals, or fragments of other rock types.

orbitals: imaginary rings around atoms' nuclei in which electrons can be found; mathematical functions that describe the probable location of electrons relative to an atom's nucleus.

ore: a natural accumulation of mineral matter from which the owner expects to extract a metal at a profit; also called an ore deposit.

ore mineral: any mineral that can be mined and refined for its metal content at a profit.

organic molecules: molecules of carbon compounds produced in plants or animals, plus similar artificial compounds.

orphan lands: unreclaimed strip mines created prior to the passage of state or federal reclamation laws.

orthoquartzite: a sandstone in which more than 90 percent of the grains are quartz.

orthorhombic: referring to a crystal system possessing three axes of symmetry that are of unequal length and that intersect at right angles.

overburden: the material overlying the ore in a surface mine.

oxidation: a very common chemical reaction in which elements are combined with oxygen—for example, the burning of petroleum, wood, and coal; the rusting of metallic iron; and the metabolic respiration of organisms.

pahoehoe: rope-like structures formed when a flow of lava cools on the earth's surface.

panel: an area of underground coal excavation for production rather than development; the coal mine equivalent of a stope.

paraffin hydrocarbons: hydrocarbon compounds composed of carbon atoms connected with single bonds into straight chains; also known as n-alkanes.

partial melting: a process undergone by rocks as their temperature rises and metamorphism occurs; magmas are derived by the partial melting of pre-existing rock; also known as ultrametamorphism or anatexis.

pedologist: a soil scientist.

pedosphere: the soil, rocks, and organic matter that rest on the ground surface.

pegmatite: a very coarse-grained granitic rock, often enriched in rare minerals.

pelitic: an adjective for mudrocks and the metamorphic rocks derived from them.

pelitic rock: a rock whose protolith contained abundant clay or similar minerals.

penstock: the tube that carries water from a reservoir to a turbine.

peridotite: a class of ultramafic rocks made up principally of pyroxene and olivine, with subordinate amounts of other minerals; the most common rock type in the upper mantle, where basalt magma is produced.

periodic: a repeating pattern.

permafrost: a layer of soil and water ice frozen together.

permeability: the ability of a soil or rock to allow water to flow through it; sands and other materials with large pores have high permeabilities, whereas clays have very low permeabilities.

permeable formation: a rock formation that, through interconnected pore spaces or fractures, is capable of transmitting fluids.

petroleum: crude oil; a naturally occurring complex liquid hydrocarbon, which after distillation yields a range of combustible fuels, petrochemicals, and lubricants.

petrology: branch of geology that studies rocks and the conditions that lead to the formation of rocks.

pH: a term used to describe the hydrogen ion activity of a system; a solution of pH 0 to 7 is acid, pH of 7 is neutral, and pH 7 to 14 is alkaline.

phaneritic: a textural term that applies to an igneous rock composed of crystals that are macroscopic in size, ranging from about 1 to over 5 millimeters in diameter.

phenocryst: a large conspicuous crystal in a porphyritic rock.

phosphorite: sedimentary rock composed principally of phosphate minerals.

photolytics: the technology that makes use of sunlight's ability to alter chemical compounds in ways that can produce energy, fuels, or both.

photovoltaic cell: a device commonly made of layered silicon that produces electrical current in the presence of light; also, a solar cell.

photovoltaics: the technology employed to convert radiant solar energy directly into an electric current, using devices called solar cells.

phyllosilicate: a mineral with silica tetrahedra arranged in a sheet structure.

piezoelectricity: the property of some crystals (and some other solids) to create an electric charge from mechanical stress or pressure.

pillar: ore, coal, rock, or waste left in place underground to support the wall or roof of a mined opening.

pillow basalt: a submarine basaltic lava flow in which small cylindrical tongues of lava break through the surface, separate into pods, and accumulate downslope in a formation resembling a pile of sandbags.

placer deposit: a mass of sand, gravel, or soil resulting from the weathering of mineralized rocks that contains grains of gold, tin, platinum, or other valuable minerals derived from the original rock.

plagioclase: a silicate mineral found in many rocks; it is a member of the feldspar group.

plate tectonics: a theory that holds that the surface of the earth is divided into roughly one dozen rigid plates that move relative to one another, producing earthquakes, volcanoes, mountain belts, trenches, and many other large-scale features of the planet.

playa: dried salt lake environment known for concentrations of evaporite minerals.

Plinian: volcanic eruption that sends ash and debris high above the volcano in a column.

pluton: a body of igneous rock formed by intrusion.

plutonic rock: igneous rock formed at a great depth within the earth.

plutonium: a radioactive element with ninety-four protons in its nucleus; can be used as an energy source in the same way as uranium.

point group: a group of geometric symmetries with one or more fixed points.

polarized light: light whose waves vibrate or oscillate in a single plane.

pollution: a condition of air, soil, or water in which substances therein make it hazardous for human use.

polycrystalline: crystal structure consisting of crystallite fragments.

polymorph: minerals having the same chemical composition but a different crystal structure.

polypedons: bodies of individual kinds of soil in a geographic area.

pool: a continuous body of petroleum-saturated rock within a petroleum reservoir; a pool may be coextensive with a reservoir.

pore fluids: fluids, such as water (usually carrying dissolved minerals, gases, and hydrocarbons), in pore spaces in a rock.

porosity: a measure of the amount of open spaces capable of holding water or air in a rock or sediment.

porphyritic: a texture characteristic of an igneous rock in which macroscopic crystals are embedded in a fine phaneritic or aphanitic matrix.

porphyry: an igneous rock in which phenocrysts are set in a finer-grained groundmass.

Portland cement: common form of cement used in industry and construction; comprises hydrated, ground, and heated limestone, clay, and calcium silicates.

power: the rate at which energy is transferred or produced.

power flux: the amount of energy that can be obtained from a cross-sectional area of the wind at a given velocity.

power grid: the mechanical power distribution system of a social structure, and its independent power distribution systems.

Precambrian: The span of geologic time extending from early planetary origins to about 540 million years ago.

precipitation (chemistry): a process whereby a solid substance is separated from a solution.

precipitation (meteorology): rain, sleet, hail, snow, or other forms of liquid or solid water that fall from the atmosphere to the hydrosphere or lithosphere.

pressure-temperature regime: a sequence of metamorphic facies distinguished by the ratio of pressure to temperature, generally characteristic of a given geologic environment.

pressure vessel: a container designed to hold gases or liquids at a much higher pressure than the surrounding pressure.

primary: a term to describe minerals that crystallize at the time that the enclosing rock is formed; hydrothermal vein minerals are examples.

profile: a representative soil sample containing different layers of soil.

prograde: metamorphic changes that occur primarily because of increasing temperature conditions.

prograde metamorphism: recrystallization of solid rock masses induced by rising temperature; differs from metasomatism in that bulk rock composition is unchanged except for expelled fluids.

progressive metamorphism: mineralogical and textural changes that take place as temperature and pressures increase.

prospect: a limited geographic area identified as having all the characteristics of an oil or gas field but without a history of production.

protolith: the original igneous or sedimentary rock later affected by metamorphism.

proven reserve: a reserve supply of a valuable mineral substance that can be exploited at a future time.

pumice: light, frothy volcanic rock formed when lava is cooled above ground following a violent eruption and pyroclastic flow.

pumped hydro: a storage technique that utilizes surplus electricity to pump water into an elevated storage pond to be released later when more electricity is needed.

pyroclastic fall: the settling of debris under the influence of gravity from an explosively produced plume of material.

pyroclastic flow: a highly heated mixture of volcanic gases and ash that travels down the flank of a volcano; the relative concentration of particles is high; also called a pyroclastic surge.

pyroxene: a calcium, magnesium, and iron silicate mineral.

pyroxenes: a group of generally dark-colored, single-chain silicates crystallizing largely in the orthorhombic or monoclinic systems and possessing good cleavage in two directions intersecting at angles of about 87 and 93 degrees.

quad: 1 quadrillion Btu; equivalent to 8 billion barrels of gasoline.

quartz: crystalline form of silica that develops when silica is heated in a liquid or semiliquid environment and then slowly cooled.

radar: a remote sensing technique involving the emission of radio waves or microwaves.

radiation: unstable isotopes of an element decay in various ways; the emission of particles and energy from nuclei is known as radiation.

radioactive decay: process by which an unstable nucleus releases excess energy to reach a more stable state, transforming a radionuclide into a daughter nuclide.

radioactivity: the spontaneous release of energy accompanying the decay of a nucleus.

radionuclide: atom with an unstable nucleus due to a lack of balance among protons and neutrons within the atomic nucleus.

raise: a vertical or steeply inclined excavation of narrow dimensions that connects subsurface levels; unlike a winze, it is bored upward rather than sunk.

raster: cell-based, high-resolution image that relies on data on surface conditions.

rechargeable battery: a type of battery, in which the electrochemical reaction is reversible and the battery can be recharged; also called a secondary cell battery.

reclamation: all human efforts to improve conditions produced by mining wastes—mainly slope reshaping, revegetation, and erosion control.

recrystallization: a solid-state chemical reaction that eliminates unstable minerals in a rock and forms new stable minerals; the major process contributing to rock metamorphism.

reef: a provincial ore deposit term referring to a metalliferous mineral deposit, commonly of gold or platinum, which is usually in the form of a layer.

refractory: a term to describe minerals or manufactured materials that resist breakdown; most silicate minerals and furnace brick are examples.

regional geology: a study of the geologic characteristics of a geographic area.

regional metamorphic facies: the particular pressure and temperature conditions prevailing during metamorphism as recorded by the appearance of a new mineral assemblage.

regional metamorphism: metamorphism characterized by strong compression along one direction, usually affecting rocks over an extensive region or belt.

regional metasomatism: large-scale metasomatism related to regional metamorphism.

regolith: that layer of soil and rock fragments just above the planetary crust.

remote sensing: any number of techniques, such as aerial photography or satellite imagery, that can collect information by gathering energy reflected or emitted from a distant source.

rendering: manipulated GIS image that shows data overlain on the original map image.

renewable energy: energy source that does not depend on a finite, exhaustible resource.

reserves: the measured amount of petroleum present in a reservoir rock that can be profitably produced.

reservoir: a body of porous and permeable rock; petroleum reservoirs contain pools of oil or gas.

reservoir rock: sedimentary rock that holds oil or gas in the spaces between particles.

reservoir sand: a storage unit for various hydrocarbons; usually of sedimentary origin.

resistivity: a measure of the resistance offered by a cubic meter of a rock formation to the passage of an electric current.

resource: a naturally occurring substance, the form of which allows it to be extracted economically.

retort: a vessel used for the distillation or decomposition of substances using heat.

retrograde: metamorphic changes that occur primarily because of decreasing temperature conditions.

rhombic: a more common form of sulfur, with a dense and more stable composition.

rhombohedral: crystal structure based on the combination of different angular lengths and resulting in a combination of rectangular-shaped sections.

rills: small rivulets in channels.

rim syncline: a circular depression found at the base of many salt domes and diapirs.

rock: a mixture of minerals.

rock-forming mineral: the common minerals that compose the bulk of the earth's crust (outer layer).

rotary drilling: a fluid-circulating, rotating process that is the chief method of drilling oil and gas wells.

rotor: the bladed system on a wind-power machine that is set into motion by the wind and translated into power to a generator.

rough: gem mineral material of suitable quality to be used for fashioning gemstones.

salinity gradient energy conversion: power generated by the passage of water masses with different salinities through a special, semipermeable membrane, taking advantage of osmotic pressure to operate turbines.

salinization: the accumulation of salts in the soil.

salt: ionic compound formed from a reaction between an acidic and basic substance.

salt dome: an underground structure in the shape of a circular plug resulting from the upward movement of salt.

saltwater intrusion: aquifer contamination by salty waters that have migrated from deeper aquifers or from the sea.

sand: granular sedimentary silicate.

sandstone: a sedimentary rock that is known for its durability in resisting abrasive wear; it is likely to be used for paving stone.

saturated hydrocarbons: hydrocarbon compounds whose molecules are chemically stable, with carbon atoms fully bonded to other atoms.

saturated zone: that zone beneath the land surface where all the pores in the soil or rock are filled with water rather than air.

scale: the relationship between a distance on a map or diagram and the same distance on the earth.

scraper: a digging, hauling, and grading machine that has a cutting edge, a carrying bowl, a movable front wall, and a dumping or ejecting mechanism.

seal: a rock unit or bed that is impermeable and inhibits upward movement of oil or gas from the reservoir.

secondary: a mineral formed later than the enclosing rock, either by metamorphism or by weathering and transport; placers are examples.

sediment: rock fragments such as clay, silt, sand, gravel, and cobbles.

sedimentary: formed from material that has been deposited from solution or eroded from previously existing rocks.

sedimentary basin: lower area in the earth's crust in which sedimentary rocks and minerals accumulate.

sedimentary rock: a major group of rocks formed from the breakdown of preexisting rock material, or from the precipitation of minerals by organic or inorganic processes.

seep: an area where water, gas, petroleum, or another fluid emerges from subterranean chambers onto the surface of the earth.

segregation: the concentration of early-formed minerals in a magma by crystal settling or crystal floating.

seismic waves: low-frequency acoustic energy that travels through solid rock and other structures.

seismology: the application of the physics of elastic wave transmission and reflection to subsurface rock geometry.

selenium: an element of the sulfur family; several forms of selenium are known, including a metallic selenium.

semimetal: elements that have some properties of metals but are distinct because they are not malleable or ductile.

semipermeable membrane: a membrane with spaces large enough for the molecules, but not the ions, of a liquid to pass through.

sesquioxide: an oxide consisting of three oxygen atoms for every two atoms of other elements; the most common versions are aluminum (Al_2O_3) and iron (Fe_2O_3) sesquioxides.

shaft or winze: a vertical or steeply inclined excavation of narrow dimensions; shafts are sunk from the surface, and winzes are sunk from one subsurface level to another.

shale: a sedimentary rock with a high concentration of clays.

shale sheath: a variable thickness of ground-up sedimentary rock found along the flanks of many salt domes and diapirs.

shear (shearing) strength: the ability to withstand a lateral or tangential stress.

sheet silicates: a group of silicates characterized by the sharing of three of the four oxygen atoms in each silica tetrahedron with neighboring tetrahedra and the fourth oxygen atom with other atoms in adjacent structures to form flat sheets; also known as phyllosilicate or layer silicate.

shrinkage: an effect opposite to hydration, caused by evapotranspiration.

siderite: an iron carbonate that contains 48 percent iron.

silica: silicon dioxide; it reacts with lime and alkali oxides, and is a key component in cement.

silicate: mineral containing both silicon and oxygen, usually in combination with one or more other elements; includes mica, feldspar, and quartz.

silicate mineral: a mineral composed of silicon, oxygen, and other metals, such as iron, magnesium, potassium, and sodium.

silica tetrahedron: the fundamental molecular unit of silica; a silicon atom bonded to four adjacent oxygen atoms in a three-sided pyramid arrangement.

skeletal crystals: elongated mineral grains that may resemble chains, plates, or feathers.

skimmer: a vessel used to skim oil from the surface of water after an oil spill.

slate: a metamorphic rock that has a unique ability to be split into thin sheets; some slates are resistant to weathering and are thus good choices for exterior use.

slump: both the rotational slippage of material and the mass of material actually moved.

soil profile: a vertical section of a soil, extending through its horizons into the unweathered parent material.

soils and foundation engineering: the branch of civil engineering specializing in foundation and soil subgrade design and construction.

soil stabilization: engineering measures designed to minimize the opportunity and/or ability of expansive soils to shrink and swell.

solid solution: a solid that shows a continuous variation in composition in which two or more elements substitute for each other on the same position in the crystal structure.

solidus/liquidus temperature: the liquidus temperature marks the beginning of crystallization in magmas, and the solidus temperature marks the end; crystals and melt coexist only within the liquidus-solidus temperature interval.

solubility: the tendency for a solid to dissolve.

solute: a substance that is dissolved.

solution: a mixture in which a solute is dissolved in a solvent.

solvent: a substance that dissolves another.

sonar: a remote sensing technique for object detection and navigation that utilizes the propagation of sound waves.

sonde: the basic tool used in well logging; a long, slender instrument that is lowered into the borehole on an electrified cable and slowly withdrawn as it measures certain designated rock characteristics.

sorption: the process of removing a chemical from a fluid by either physical or chemical means.

source rock: rock containing organic materials that will be made into oil and gas.

spatial: pertaining to space and geography.

specific gravity: the ratio of the weight of any volume of a substance to the weight of an equal volume of water.

specific heat: the heat at which a liquid will transition to a gaseous state.

spectral signature: a unique identifier of a material based on how the material absorbs or reflects electromagnetic radiation wavelengths.

spontaneous fission: uninduced splitting of unstable atomic nuclei into two smaller nuclei; an energetic form of radioactive decay.

stable isotopes: atoms of the same element that differ by the number of neutrons in their nuclei yet are not radioactive.

star sapphire/ruby: a gem that has a starlike effect when viewed in reflected light because of the mineral's fibrous structure.

stellar nucleosynthesis: a nuclear reaction within a star that results in the production of elemental nuclei heavier than hydrogen.

step-up gearing: a gear system that increases the revolutions of the downstream shaft (generator) over the revolutions of the upstream shaft (rotor).

stock: a granite or diorite intrusion, smaller than a batholith, with an exposure area between 10 and 100 square kilometers.

stope: an underground excavation to remove ore, as opposed to development; the outlines of a stope are determined either by the limits of the ore body or by raises and levels.

strain: change in volume or size in response to stress.

strata: layers of sedimentary rock.

strategic resource: an Earth resource, such as manganese or oil, which would be essential to a nation's defense in wartime.

stratovolcano: a volcanic cone consisting of lava, mudflows, and pyroclastic rocks.

stress: force per unit of area.

strip mining: a method of mining that occurs on the earth's surface.

structure: the arrangement of primary soil particles into secondary units called peds.

S-type granitoid: granitic rock formed from magma generated by partial melting of sedimentary rocks within the crust.

subduction: geologic process whereby a tectonic plate is forced below another plate.

subduction zone: area in which two tectonic plates collide, forcing one plate to push beneath the other and driving the submerged plate into the earth's mantle.

subsurface features: land features or characteristics that are not visible or apparent, such as minerals, oil and gas, and structural features lying beneath the land surface.

sulfate: salt that forms from a chemical reaction between sulfuric acid and a basic material in the environment.

sulfide: a compound made with sulfur; an iron sulfide is a compound of iron and sulfur.

superconductors: materials that pass electrical current without exhibiting any electrical resistance.

surface tension: measure of the potential for the surface of a liquid to resist external force.

suspension: a condition in which particles are dispersed through a supporting medium.

synthetic mineral: a human-made reproduction of the structure, composition, and properties of a particular mineral.

syzygy: alignment of three astronomical bodies with regard to a certain direction of reference, such as when the sun, moon, and earth align twice each lunar cycle.

taconite: a low-grade ore that contains 20 to 30 percent iron.

tailings: waste from mining operations.

tar sand: a natural deposit that contains significant amounts of bitumen; also called oil sand.

taxation: a land-management tool that usually reflects the perceived best use of land.

tectonic movement: gradual movement of plates, which are portions of the earth's crust and lithosphere that are connected at deep sedimentary cores and that move as units.

tectonic plates: fragmented plates that make up the lithosphere and that are in constant motion.

tectonics: the study of the processes and products of large-scale movement and deformation within the earth.

tectonism: the formation of mountains because of the deformation of the crust of the earth on a large scale.

tenacity: the resistance of a mineral to bending, breakage, crushing, or tearing.

tephra: fragmentary volcanic rock materials ejected into the air during an eruption; also called pyroclasts.

tetrahedron (chemistry): a molecule in which the central atom forms four bonds.

tetrahedron (geometry): a four-sided pyramid made out of equilateral triangles.

texture: the size, shape, and arrangement of crystals or particles in a rock.

thermal gradient: the increase of temperature with depth below the earth's surface, expressed as degrees Celsius per kilometer; the average is 25 to 30 degrees Celsius per kilometer; also called geothermal gradient..

thermohaline conveyor belt: a system of oceanic circulation driven by the cooling and sinking of salty surface waters in the Nordic seas.

thorium: naturally occurring atomic element characterized by its mildly radioactive characteristics and a decay pattern that involves the release of alpha rays into the environment.

tidal (flow) power: power from turbines that are sited in coastal areas to take advantage of the tidal flow's rise and ebb; the kinetic energy created by moving tidal currents is used to generate electrical energy.

tidal force: secondary effect of the gravitational forces between two objects that cause elongation of the objects along the axis of the line connecting their respective center points.

tidal turbine: turbine housed within a tidal system that is used to generate electrical energy from the kinetic energy of tidal movement.

topography: the collective physical features of a region or area, such as hills, valleys, streams, cliffs, and plains.

topsoil: in reclamation, all soil that will support plant growth, but normally the 20 to 30 centimeters of the organically rich top layer.

toughness: the degree of resistance to fragmentation or resistance to plastic deformation.

toxicology: the study of the effects of toxic compounds on humans and environments.

transuranic: an isotope of an element that is heavier than uranium and formed in the processing and use of nuclear fuel and plutonium.

trap: a type of rock formation that causes hydrocarbons to accumulate (trapping them), forming a reservoir.

tree: a wellhead; a fixture to control the flow of oil and gas; sometimes called a Christmas tree.

trench: a long and narrow deep trough on the sea floor that forms where the ocean floor is pulled downward because of plate subduction.

troy ounce: a unit of weight equal to 31.1 grams; used in the United States for precious metals and gems.

tuff: a general term for all consolidated pyroclastic rocks.

turbine: a device used to convert the energy of flowing water into the spinning motion of the turbine's shaft; it does this by directing the flowing water against the blades mounted on the rotating shaft.

twinning: process in crystal formation in which two separate crystals share one or more points of connection between their respective lattices.

ultramafic: a term used to describe certain igneous rocks and most meteorites that contain less than 45 percent silica; they contain virtually no quartz or feldspar and are mainly of ferromagnesian silicates, metallic oxides and sulfides, and native metals.

understrata: material lying beneath the surface of the soil.

unit cell: an imaginary box that shows the three-dimensional pattern of the atoms, molecules, or ions of a crystal.

upper mantle: the region of the earth immediately below the crust, believed to be composed largely of periodotite (olivine and pyroxene rock), which is thought to melt to form basaltic liquids.

uranium: an element with ninety-two protons in the nucleus and with only radioactive isotopes.

uranium ore: uranium as it is typically found in nature, as shiny black rocks that may be found combined with a variety of other elements.

valence electron: an electron that is free to combine with the valence electrons of other atoms, resulting in a bond; in alkali metals and many other elements, only the outermost electrons can be valence electrons.

vector (geometry): a spatial point used in GIS to indicate surface features such as structures, bodies of water, and roadways.

vector (waste management): a term used in waste disposal when referring to rats, flies, mosquitoes, and other disease-carrying insects and animals that infest dumps.

vein: a mineral-filled fault or fracture in rock; veins represent late crystallization, most commonly in association with granite.

vesiculated: gas-infused rock within a pyroclastic flow.

viscosity: a property of fluids that measures their internal resistance to flowage; the inverse of fluidity or mobility.

volatiles: substances in coal that are capable of being gasified.

volcanic arc: a linear or arcuate belt of volcanoes that forms at a subduction zone because of rock melting near the top of the descending plate.

volcanic pipe: deposit of igneous rock formed when a deep magma pocket produces an explosion of gas and magma that rapidly rises to the surface and forms a cone-shaped deposit.

water table: the level below the earth's surface at which the ground becomes saturated with water.

water-to-rock ratio: the mass of free water in a given volume of rock divided by rock mass of the same volume; processes occurring at water-to-rock ratios less than 50 are "rock-dominated," while those greater than 50 are "fluid-dominated."

weathering: a geochemical process whereby rocks and minerals are broken down by water and wind.

well log: a graphic record of the physical and chemical characteristics of the rock units encountered in a drilled borehole.

xenobiotic: a chemical that is foreign to the natural environment; an artificial chemical that may not be biodegradable.

xenocrysts: minerals found as either crystals or fragments in some volcanic rocks; they are foreign to the body of the rock in which they occur.

xenoliths: various rock fragments that are foreign to the igneous body in which they are present

X ray: radiation that can be interpreted in terms of either very short electromagnetic waves or highly energetic photons (light particles).

yellowcake: uranium bonded with oxygen that has been milled from the uranium ore found in nature; has a yellow, powdery appearance.

zeolites: members of a mineral group with very complex compositions: aluminosilicates with variable amounts of calcium, sodium, and water and a very open atomic structure that permits atoms to move easily.

zone: an individual soil group within a horizon.

zoning: a land-management tool used to limit and define the conditions and extent of land use.

zoning ordinance: a legal method by which governments regulate private land by defining zones where specific activities are permitted.

BIBLIOGRAPHY

A Guide to Practical Management of Produced Water from Onshore Oil and Gas Operations in the United States. Interstate Oil and Gas Compact Commission and ALL Consulting, 2006.

Abandoned Mines and Mining Waste. Sacramento: California Environmental Protection Agency, 1996.

Agassi, Menachem, ed. *Soil Erosion, Conservation, and Rehabilitation.* New York: Marcel Dekker, 1996.

Ahr, Wayne M. *Geology of Carbonate Reservoirs.* Hoboken, N.J.: John Wiley & Sons, 2008.

Albarede, Francis. *Geochemistry: An Introduction.* 2d ed. Cambridge, Mass.: Cambridge University Press, 2009.

Albu, Marius, David Banks, Harriet Nash, et al., eds. *Mineral and Thermal Groundwater Resources.* London: Chapman and Hall, 1997.

Allan, J. A., ed. *The Sahara: Ecological Change and Early Economic History.* Outwell, England: Middle East and North American Studies Press, 1981.

Allaud, Louis A., and Maurice H. Martin. *Schlumberger: The History of a Technique.* New York: John Wiley & Sons, 1977.

Alloway, B. J. *Heavy Metals in Soils.* 2d ed. London: Blackie Academic and Professional, 1995.

Al-Rawas, Amer Ali, and Mattheus F. A. Goosen, eds. *Expansive Soils: Recent Advances in Characterization and Treatment.* New York: Taylor & Francis, 2006.

Alsop, G. Ian, Derek John Blundell, and Ian Davison, eds. *Salt Tectonics.* London: Geological Society, 1996.

Alward, Ron, Sherry Elisenbart, and John Volkman. *Micro-Hydro Power: Reviewing an Old Concept.* Butte, Mont.: National Center for Appropriate Technology, 1979.

American Petroleum Institute. *Primer of Oil and Gas Production.* 3d ed. Washington, D.C.: Author, 1976.

Anderson, Greg M., and David A. Crerar. *Thermodynamics in Geochemistry: The Equilibrium Model.* New York: Oxford University Press, 1993.

Andrews, John, and Nick Jelley. *Energy Science: Principles, Technologies, and Impact.* New York: Oxford University Press, 2007.

Armstead, H. Christopher. *Geothermal Energy.* 2d ed. Bristol, England: Arrowsmith, 1983.

_____, ed. *Geothermal Energy: Review of Research and Development.* Lanham, Md.: UNIPUB, 1973.

Arndt, N. T., D. Frances, and A. J. Hynes. "The Field Characteristics and Petrology of Archean and Proterozoic Komatiites." *Canadian Mineralogist* 17 (1985): 147-163.

Arndt, N., C. Michael Lesher, and Steven J. Barnes. *Komatiite.* New York: Cambridge University Press, 2008.

Arndt, N. T., and E. G. Nisbet. *Komatiites.* Winchester, Mass.: Allen & Unwin, 1982.

Arnopoulos, Paris. *Cosmopolitics: Public Policy of Outer Space.* Toronto: Guernica, 1998.

Atherton, Michael P., and J. Tarney, eds. *Origin of Granite Batholiths: Geochemical Evidence.* Orpington, England: Shiva Publishing, 1979.

Augustithis, S. S. *Atlas of Metamorphic-Metasomatic Textures and Processes.* Amsterdam: Elsevier, 1990.

Averitt, P. "Coal." In *United States Mineral Resources.* U.S. Geological Survey Professional Paper 820. Washington, D.C.: Government Printing Office, 1973.

Bagchi, Amalendu. *Design, Construction, and Monitoring of Landfills.* 2d ed. New York: John Wiley & Sons, 1994.

_____. *Design of Landfills and Integrated Solid Waste Management.* 3d ed. Hoboken, N.J.: John Wiley & Sons, 2004.

Baker, Ron. *A Primer of Oil Well Drilling.* 6th ed. Austin: University of Texas at Austin, 2000.

Ballard, Robert D. *Exploring Our Living Planet.* Rev. ed. Washington, D.C.: National Geographic Society, 1993.

Bar-Cohen, Yoseph, and Kris Zacny, eds. *Drilling in Extreme Environments: Penetration and Sampling on Earth and Other Planets.* Weinheim, Wiley-VCH. 2009.

Barker, A. J. *Introduction to Metamorphic Textures and Microstructures.* 2d ed. London: Routledge, 2004.

Barker, Colin. *Organic Geochemistry in Petroleum Exploration.* Tulsa, Okla.: American Association of Petroleum Geologists, 1979.

Barnard, Amanda. *The Diamond Formula.* Woburn, Mass.: Butterworth-Heinemann, 2000.

Barnes, H. L. *Geochemistry of Hydrothermal Ore Deposits.* 3d ed. New York: John Wiley & Sons, 1997.

Barry, Roger, and Thian Yew Gan. *The Global Cryosphere: Past, Present, and Future.* New York: Cambridge University Press, 2011.

Bartlett, Donald L., and James B. Steele. *Forevermore: Nuclear Waste in America.* New York: W. W. Norton, 1985.

Barton, William R. *Dimension Stone.* Information Circular 8391. Washington, D.C.: Government Printing Office, 1968.

Basaltic Volcanism Study Project. *Basaltic Volcanism on the Terrestrial Planets.* Elmsford, N.Y.: Pergamon Press, 1981.

Bascom, Willard. *Waves and Beaches: The Dynamics of the Ocean Surface.* Rev. ed. Garden City, N.Y.: Anchor Press/Doubleday, 1980.

Basson, M. S., et al., eds. *Probabilistic Management of Water Resources and Hydropower Systems.* Highlands Ranch, Colo.: Water Resources Publications, 1994.

Bates, Robert L. *Geology of the Industrial Rocks and Minerals.* Mineola, N.Y.: Dover, 1969.

_____. *Stone, Clay, Glass: How Building Materials Are Found and Used.* Hillside, N.J.: Enslow, 1987.

Bates, Robert L., and Julia A. Jackson. *Our Modern Stone Age.* Los Altos, Calif.: William Kaufmann, 1982.

Bathurst, Robin G. C. *Carbonate Sediments and Their Diagenesis.* 2d ed. New York: Elsevier, 1975.

Batie, Sandra S. *Soil Erosion: Crisis in America's Croplands?* Washington, D.C.: Conservation Foundation, 1983.

Batty, Lesley C., and Kevin B. Hallberg. *Ecology of Industrial Pollution.* New York: Cambridge University Press, 2010.

Baurer, Max. *Precious Stones.* 2 vols. Mineola, N.Y.: Dover, 1968.

Bear, Firman E. *Earth: The Stuff of Life.* 2d ed. Norman: University of Oklahoma Press, 1990.

Bebout, Gray E., et al., eds. *Subduction Top to Bottom.* Washington, D.C.: American Geophysical Union, 1996.

Bédard, Jean H. "Parental Magmas of the Nain Plutonic Suite Anorthosites and Mafic Cumulates: A Trace Element Modeling Approach." *Contributions to Mineralogy and Petrology* 141 (2001): 747-771.

Behrman, Daniel. *Solar Energy: The Awakening Science.* London: Routledge & Kegan Paul, 1979.

Bell, Keith, ed. *Carbonatites: Genesis and Evolution.* Boston: Unwin Hyman, 1989.

Bell, Keith, and J. Keller, eds. *Carbonatite Volcanism: Oldoinyo Lengai and the Petrogenesis of Natrocarbonatites.* New York: Springer-Verlag, 1995.

Bennett, Hugh Hammond. *Elements of Soil Conservation.* 2d ed. New York: McGraw-Hill, 1955.

Bennison, George M., Paul A. Olver, and A. Keith Moseley. *An Introduction to Geological Structures and Maps.* 8th ed. London: Hodder Education, 2011.

Berg, Robert R. *Reservoir Sandstones.* Englewood Cliffs, N.J.: Prentice-Hall, 1986.

Bergaya, F., B. K. G. Theng, and G. Lagaly, eds. *Handbook of Clay Science.* Amsterdam: Elsevier, 2006.

Berke, Philip R., and David R. Godschalk. *Urban Land Use Planning.* 5th ed. University of Illinois Press, 2006.

Berner, Elizabeth K., and Robert A. Berner. *The Global Water Cycle: Geochemistry and Environment.* Englewood Cliffs, N.J.: Prentice-Hall, 1987.

Berry, L. G., B. Mason, and R. V. Dietrich. *Mineralogy: Concepts, Descriptions, Determinations.* 2d ed. San Francisco: W. H. Freeman, 1983.

Best, Myron G. *Igneous and Metamorphic Petrology.* 2d ed. Malden, Mass.: Blackwell Science, 2003.

Beus, S. S., ed. *Rocky Mountain Section of the Geological Society of America: Centennial Field Guide.* Vol. 2. Denver, Colo.: Geological Society of America, 1987.

Birch, Francis, J. F. Schairer, and H. Cecil Spicer. *Handbook of Physical Constants.* Geological Society of America Special Paper 36. Baltimore: Waverly Press, 1942.

Birch, G. F. "Phosphatic Rocks on the Western Margin of South Africa." *Journal of Sedimentary Petrology* 49 (1979): 93-100.

Birkeland, P. W. *Soils and Geomorphology.* 3d ed. New York: Oxford University Press, 1999.

Black, Malcolm, and Joseph A. Mandarino. *Fleischer's Glossary of Mineral Species.* 10th ed. Tucson, Ariz.: The Mineralogical Record, 2008.

Blackburn, William H., and William H. Dennen. *Principles of Mineralogy.* 2d ed. Dubuque, Iowa: Wm. C. Brown, 1993.

Blair, Ian. *Taming the Atom: Facing the Future with Nuclear Power.* Bristol, England: Adam Hilger, 1983.

Blake, Alexander J., and Jacqueline M. Cole. *Crystal Structure Analysis: Principles and Practices.* 2d ed. New York: Oxford University Press, 2009.

Blanks, Robert F., and Henry L. Kennedy. *The Technology of Cement and Concrete.* New York: John Wiley & Sons, 1955.

Blatt, Harvey. *Sedimentary Petrology.* 2d ed. San Francisco: W. H. Freeman, 1992.

Blatt, Harvey, Gerard Middleton, and R. Murray. *Origin of Sedimentary Rocks.* 2d ed. Englewood Cliffs, N.J.: Prentice-Hall, 1980.

Blatt, Harvey, Robert J. Tracy, and Brent Owens. *Petrology: Igneous, Sedimentary, and Metamorphic.* 3d ed. New York: W. H. Freeman, 2005.

Blong, R. J. *Volcanic Hazards: A Sourcebook on the Effects of Eruptions.* Sydney, Australia: Academic Press, 1984.

Bloss, F. D. *An Introduction to the Methods of Optical Crystallography.* New York: Holt, Rinehart and Winston, 1961.

Boardman, John, and David Favis-Mortlock, eds. *Modeling Soil Erosion by Water.* New York: Springer-Verlag, 1998.

Bockheim, J. G., and A. N. Gannadiyev. "Soil-Factorial Models and Earth-System Science: A Review." *Geoderma* 159, nos. 3-4 (2010): 243-251.

Boeker, Egbert, and Rienk von Grondelle. *Environmental Physics: Sustainable Energy and Climate Change.* New York: John Wiley & Sons, 2011.

Boggs, Sam, Jr. *Petrology of Sedimentary Rocks.* New York: Cambridge University Press, 2009.

_____. *Principles of Sedimentology and Stratigraphy.* 5th ed. Upper Saddle River, N.J.: Prentice Hall, 2011.

Bohn, Heinrich L., Rick A. Myer, and George A. O'Connor. *Soil Chemistry.* 3d ed. New York: John Wiley & Sons, 2001.

Bolin, Cong, ed. *Ultrahigh-Pressure Metamorphic Rocks in the Dabieshan-Sulu Region of China.* New York: Springer, 1997.

Bolstad, Paul. *GIS Fundamentals: A First Text on Geographic Information Systems.* St. Paul, Minn.: Eider Press, 2007.

Borradaile, G. J., and D. Gauthier. "Magnetic Studies of Magma-Supply and Sea Floor Metamorphism: Troodos Ophiotile Dikes." *Tectonophysics* 418 (2006): 75-92.

Bowen, N. L. *The Evolution of Igneous Rocks.* Mineola, N.Y.: Dover, 1956.

Bowles, J. F. W., et al. *Rock-Forming Minerals: Non-Silicates; Oxides, Hydroxides and Sulfides.* 2d ed. London: Geological Society Publishing House, 2011.

Boyd, F. R., and H. O. A. Meyer, eds. *Kimberlites, Diatremes, and Diamonds: Their Geology, Petrology, and Geochemistry.* Washington, D.C.: American Geophysical Union, 1979.

Boyle, Godfrey, ed. *Renewable Energy.* 2d ed. New York: Oxford University Press, 2004.

Boyle, Robert W. *The Geochemistry of Silver and Its Deposits.* Geological Survey of Canada Bulletin 160. Ottawa: Queen's Printer, 1968.

_____. *Gold: History and Genesis of Deposits.* New York: Van Nostrand Reinhold, 1987.

Brady, Nyle C., and Ray R. Weil. *The Natures and Properties of Soils.* 14th ed. Upper Saddle River, N.J.: Prentice Hall, 2007.

Bragg, William Lawrence. *Atomic Structure of Minerals.* Ithaca, N.Y.: Cornell University Press, 1937.

Bragg, William Lawrence, G. F. Claringbull, and W. H. Taylor. *The Crystalline State.* The Crystal Structures of Minerals 4. Ithaca, N.Y.: Cornell University Press, 1965.

Braithwaite, C. J. R., G. Rizzi, and G. Darke, eds. "The Geometry and Petrogenesis of Dolomite Hydrocarbon Reservoirs: Introduction." In *The Geometry and Petrogenesis of Dolomite Hydrocarbon Reservoirs.* Bath, England: Geological Society, 2004.

Brandt, Daniel A., and Jairus C. Warner. *Metallurgy Fundamentals: Ferrous and Nonferrous.* Tinley Park, Ill.: Goodheart-Willcox, 2009.

Branney, Michael J., and Peter Kokelaar. *Pyroclastic Density Currents and the Sedimentation of Ignimbrites.* London: Geological Society of London, 2002.

Brill, Winston. "Agricultural Microbiology." *Scientific American* 245 (September, 1981): 198.

Brinkworth, B. J. *Solar Energy for Man.* New York: John Wiley & Sons, 1973.

Brock, Jim. *Analyzing Your Logs.* Vol. 1, *Fundamentals of Open Hole Log Interpretation.* 2d ed. Tyler, Tex.: Petro-Media, 1984.

Brooks, George E. "A Provisional Historical Schema for Western Africa Based on Seven Climate Periods, ca. 9000 b.c. to the Nineteenth Century." *Cahiers d'études africaines* 16 (1986): 46-62.

Brooks, Robert R., ed. *Noble Metals and Biological Systems: Their Role in Medicine, Mineral Exploration, and The Environment.* Boca Raton, Fla.: CRC Press, 1992.

Brown, G. C., C. J. Hawkesworth, and R. C. L. Wilson, eds. *Understanding the Earth: A New Synthesis.* Cambridge, England: Cambridge University Press, 1992.

Brown, Michael. *Laying Waste: The Poisoning of America by Toxic Chemicals.* New York: Pantheon, 1980.

Brumfiel, Geoff. "Iran's Nuclear Plan Revealed: Report Paints Detailed Picture of Nation's Intention to Build a Warhead." *Nature* 479 (2011): 282.

Brutsaert, Wilfried. *Hydrology: An Introduction.* New York: Cambridge University Press, 2005.

Bryson, Reid A., and Thomas J. Murray. *Climates of Hunger: Mankind and the World's Changing Weather.* Madison: University of Wisconsin Press, 1977.

Buchel, K. H., H.-H. Moretto, and P. Woditsch. *Industrial Inorganic Chemistry.* 2d ed. Weinheim: Wiley–VCH, 2000.

Bucher, Kurt, and Martin Frey. *Petrogenesis of Metamorphic Rocks.* 7th ed. New York: Springer-Verlag Berlin Heidelberg, 2002.

Bucher, Kurt, and Rodney Grapes. *Petrogenesis of Metamorphic Rocks.* 8th ed. New York: Springer, 2011.

Buol, S. W., F. D. Hole, and R. J. McCracken. *Soil Genesis and Classification.* 5th ed. Ames: Iowa State University Press, 2003.

Burby, Raymond J., ed. *Cooperating with Nature: Confronting Natural Hazards with Land-Use Planning for Sustainable Communities.* Washington D.C.: Joseph Henry Press, 1998.

Burger, Ana, and Slavko V. Šolar. "Mining Waste of Nonmetal Pits and Querries in Slovenia." In *Waste Management, Environmental Geotechnology and Global Sustainable Development.* Ljubljana, Slovenia: Geological Survey of Slovenia International Conference, 2007.

Burnham, C. W. "Contact Metamorphism of Magnesian Limestones at Crestmore, California." *Geological Society of America Bulletin* 70 (1959): 879-920.

Burns, Michael E., ed. *Low-Level Radioactive Waste Regulation: Science, Politics, and Fear.* Chelsea, Mich.: Lewis, 1987.

Butler, James Newton. *Carbon Dioxide Equilibria and Their Applications.* Chelsea, Mich.: Lewis, 1991.

Buzzi, O., S. Fityus, and D. Sheng, eds. *Unsaturated Soils.* New York: CRC Press, 2009.

Byrne, John, and Steven M. Hoffman, eds. *Governing the Atom: The Politics of Risk.* New Brunswick, N.J.: Transaction, 1996.

Cadogan, Peter. *The Moon: Our Sister Planet.* New York: Cambridge University Press, 1981.

CAFTA DR and U.S. Country EIA. *EIA Technical Review Guideline: Non-metal and Metal Mining.* Washington, D.C.: U.S. Environmental Protection Agency, 2011.

California Integrated Waste Management Board, Permitting and Enforcement Division. *Active Landfills.* Publication Number 251-96-001. Sacramento: Author, 1998.

Camacho, E. F., Manuel Berenguel, and Francisco R. Rubio. *Advanced Control of Solar Plants.* New York: Springer, 1997.

Cameron, Eugene N., R. H. Jahns, A. H. McNair, and L. R. Page. *Internal Structure of Granitic Pegmatites.* Urbana, Ill.: Economic Geology Publishing, 1949.

Cameron, I. R. *Nuclear Fission Reactors.* New York: Plenum Press, 1983.

Cann, J. R., H. Elderfield, A. Laughton, et al., eds. *Mid-Ocean Ridges: Dynamics of Processes Associated with Creation of New Ocean Crust.* Cambridge, England: Cambridge University Press, 1999.

Carlson, Diane H., Charles C. Plummer, and Lisa Hammersley. *Physical Geology: Earth Revealed.* 9th ed. New York: McGraw-Hill Science/Engineering/Math, 2010.

Carlson, R. W., ed. *The Mantle and Core.* Amsterdam: Elsevier Science, 2005.

Carmichael, Ian S. E., F. J. Turner, and John Verhoogan. *Igneous Petrology.* New York: McGraw-Hill, 1974.

Carmichael, Robert S., et al. *Practical Handbook of Physical Properties of Rocks and Minerals.* Boca Raton, Fla.: CRC Press, 1989.

Carr, Donald E. *Energy and the Earth Machine.* New York: W. W. Norton, 1976.

Carr, Margaret, and Paul Zwick. *Smart Land-Use Analysis: The LUCIS Model.* Redlands, Calif.: ESRI Press, 2007.

Carrigy, M. A. "New Production Techniques for Alberta Oil Sands." *Science* 234 (December, 1986): 1515-1518.

Carter, Luther J. *Nuclear Imperatives and Public Trust: Dealing with Radioactive Waste.* Washington, D.C.: Resources for the Future, 1989.

Cartwright, David Edgar. *Tides: A Scientific History.* New York: Cambridge University Press, 2001.

Carver, Robert E., ed. *Procedures in Sedimentary Petrology.* New York: John Wiley & Sons, 1971.

Cas, R. A. F., and J. V. Wright. *Volcanic Successions: Modern and Ancient.* Winchester, Mass.: Unwin Hyman, 1987.

Casale, Riccardo, and Claudio Margottini, eds. *Floods and Landslides: Integrated Risk Assessment.* New York: Springer, 1999.

Cattermole, Peter John. *Planetary Volcanism: A Study of Volcanic Activity in the Solar System.* 2d ed. New York: John Wiley & Sons, 1996.

Cavey, Christopher. *Gems and Jewels.* London: Studio, 1992.

Cepeda, Joseph C. *Introduction to Minerals and Rocks.* New York: Macmillan, 1994.

Chacon, Mark A. *Architectural Stone: Fabrication, Installation, and Selection.* New York: Wiley, 1999.

Chang, Raymond. *Chemistry.* Boston: McGraw-Hill Higher Education, 2007.

Chang, Raymond, and Jason Overby. *General Chemistry: The Essential Concepts.* 6th ed. Columbus, Ohio: McGraw-Hill, 2010.

Chapman, R. E. *Petroleum Geology.* New York: Elsevier, 1983.

Chappell, B. W. and A. J. R. White. "Two Contrasting Granite Types: 25 Years Later." *Australian Journal of Earth Sciences* 48 (2001): 489-499.

Chen, Ren-Xu, Yong-Fei Zheng, and Bing Gong. "Mineral Hydrogen Isotopes and Water Contents in Ultrahigh-Pressure Metabasite and Metagranite: Constraints on Fluid Flow During Continental Subduction-Zone Metamorphism." *Chemical Geology* 281, nos. 1/2 (2011): 103-124.

Chernicoff, Stanley, and Donna Whitney. *Geology: An Introduction to Physical Geology.* 4th ed. Upper Saddle River, N.J.: Prentice Hall, 2006.

Chesterman, C. W., and K. E. Lowe. *The Audubon Society Field Guide to North American Rocks and Minerals.* New York: Alfred A. Knopf, 1988.

Chiras, Daniel D., and John P. Reganold. *Natural Resource Conservation: Management for a Sustainable Future.* 10th ed. Boston: Addison Wesley, 2009.

Chironis, N. P., ed. *Coal Age Operating Handbook of Coal Surface Mining and Reclamation.* New York: McGraw-Hill, 1978.

Chouhan, T. S. *Desertification in the World and Its Control.* Jodhpur, India: Scientific Publishers, 1992.

Christodouleas, John P., et al. "Short-Term and Long-Term Health Risks of Nuclear-Power-Plant Accidents." *New England Journal of Medicine* 364 (2011): 2334-2341.

Clark, Wilson. *Energy for Survival.* Garden City, N.Y.: Doubleday, 1974.

Clarke, G. R. *The Study of Soil in the Field.* 5th ed. Oxford, England: Clarendon Press, 1971.

Clawson, Marion. *The Federal Lands Revisited.* Washington, D.C.: Resources for the Future, 1983.

_____. *Forests for Whom and for What?* Baltimore: Johns Hopkins University Press, 1975.

_____. *Man, Land, and the Forest Environment.* Seattle: University of Washington Press, 1977.

Climate, Drought, and Desertification. Geneva, Switzerland: World Meteorological Organization, 1997.

Close, Upton, and Elsie McCormick. "Where the Mountains Walked." *National Geographic* 41 (May, 1922): 445-464.

Coates, George R., Lizhi Xiao, and Manfred G. Prammer. *NMR Logging Principles and Applications.* Houston: Halliburton Energy Services Publication, 1999.

Coch, Nicholas K. *Geohazards: Natural and Human.* Englewood Cliffs, N.J.: Prentice Hall, 1995.

Coe, Angela L., ed. *The Sedimentary Record of Sea-Level Change.* New York: Cambridge University Press, 2003.

Cohen, Bernard L. *Before It's Too Late: A Scientist's Case for Nuclear Energy.* New York: Plenum Press, 1982.

Collier, C. R., et al., eds. *Influences of Strip Mining on the Hydrologic Environment of Parts of Beaver Creek Basin, Kentucky, 1955-66.* U.S. Geological Survey Professional Paper 427-C. Washington, D.C.: Government Printing Office, 1970.

Coltorti, M., and M. Gregoire. *Metasomatism in Oceanic and Continental Lithospheric Mantle.* Special Publication no. 293. London: Geological Society of London, 2008.

Colwell, J. D. *Estimating Fertilizer Requirements: A Quantitative Approach.* Wallingford, England: CAB International, 1994.

Compton, Robert. *Geology in the Field.* New York: John Wiley & Sons, 1985.

Condie, Kent C. *Plate Tectonics and Crustal Evolution.* 4th ed. Oxford: Butterworth-Heinemann, 1997.

Conservation Foundation. *State of the Environment: An Assessment at Mid-Decade.* Washington, D.C.: Author, 1984.

Constans, Jacques A. *Marine Sources of Energy.* Elmsford, N.Y.: Pergamon Press, 1980.

Constantinescu, Robert, Jean-Claude Thouret, and Ioan-Aurel Irimus. "Computer Modeling as a Tool for Volcanic Hazards Assessment: An Example of Pyroclastic Flow Modeling at El Misti Volcano, Southern Peru." *Geographia Technica* 11, no. 2 (2011): 1-14.

Constantopoulos, James T. *Earth Resources Laboratory Investigations.* Upper Saddle River, N.J.: Prentice Hall, 1997.

Cook, James. "Not in Anybody's Back Yard." *Forbes* 142 (November 28, 1988): 172-177.

Cooper, Henry S. F., Jr. *The Search for Life on Mars: Evolution of an Idea.* New York: Holt, Rinehart and Winston, 1980.

Cornforth, Derek H. *Landslides in Practice.* Hoboken, N.J.: John Wiley & Sons, 2005.

Costa, J. E., and V. R. Baker. *Surficial Geology: Building with the Earth.* New York: John Wiley & Sons, 1981.

Costa, J. E., and G. F. Wieczorek, eds. *Debris Flows/ Avalanches: Process, Recognition, and Mitigation.* Reviews in Engineering Geology 7. Boulder, Colo.: Geological Society of America, 1987.

Cotton, Simon. *Chemistry of Precious Metals.* London: Blackie Academic and Professional, 1997.

Cox, Allan, ed. *Plate Tectonics and Geomagnetic Reversals.* San Francisco: W. H. Freeman, 1973.

Cox, K. G. "Kimberlite Pipes." In *Volcanoes and the Earth's Interior,* edited by Robert Decker and Barbara Decker. San Francisco: W. H. Freeman, 1982.

CPM Group. *The CPM Platinum Group Metals Yearbook 2011.* Hoboken, N.J.: John Wiley & Sons, 2011.

Craig, James R., David J. Vaughan, and Brian J. Skinner. *Earth Resources and the Environment.* 4th ed. Upper Saddle River, N.J.: Prentice Hall, 2010.

_____. *Resources of the Earth.* 3d ed. Upper Saddle River, N.J.: Prentice Hall, 2001.

Crawford, Clifford S., and James R. Gosz. "Desert Ecosystems: Their Resources in Space and Time." *Environmental Conservation* 9 (1982): 181-195.

Crickmer, D. F., and D. A. Zegeer, eds. *Elements of Practical Coal Mining.* 2d ed. New York: Society of Mining Engineers of the American Institute of Mining, Metallurgical, and Petroleum Engineers, 1981.

Crittenden, John C. *Water Treatment Principles and Design.* Hoboken, N.J.: Wiley, 2005.

Cronan, D. S. *Handbook of Marine Mineral Deposits.* Boca Raton, Fla.: CRC Press, 2000.

Crundwell, Frank K., Michael S. Moats, Venkoba Ramachandran, et al. *Extractive Metallurgy of Nickel, Cobalt, and Platinum-Group Metals.* Amsterdam: Elsevier, 2011.

Culver, David C., and William B. White, eds. *Encyclopedia of Caves.* Academic Press, 2004.

Cummans, J. *Mudflows Resulting from the May 18, 1980, Eruption of Mount St. Helens.* U.S. Geological Survey Circular 850-B. Washington, D.C.: Government Printing Office, 1981.

Cutter, Susan L., and William H. Renwick. *Exploitation Conservation Preservation: A Geographic Perspective on Natural Resource Use.* 4th ed. Hoboken, N.J.: John Wiley & Sons, 2003.

Dadd, Debra L. *The Nontoxic Home and Office: Protecting Yourself and Your Family from Everyday Toxics and Health Hazards.* Los Angeles: Jeremy P. Tarcher, 1992.

Dahl, P. S., and D. F. Palmer. "The Petrology and Origin of Orbicular Tonalite from Western Taylor Valley, Southern Victoria Land, Antarctica." In *Antarctic Earth Science,* edited by R. L. Oliver, P. R. James, and J. B. Jago. New York: Cambridge University Press, 2011.

Dahlin, Greger R., and Kalle E. Strøm. *Lithium Batteries: Research, Technology, and Applications.* New York: Nova Science, 2010.

Dale, Virginia H., et al., eds. *Ecological Responses to the 1980 Eruption of Mount St. Helens.* New York: Springer, 2005.

D'Alessandro, Bill. "Dark Days for Solar." *Sierra Magazine* (July/August, 1987): 34.

Dalverny, Louis E. *Pyrite Leaching from Coal and Coal Waste.* Pittsburgh, Penn.: U.S. Department of Energy, 1996.

Dandekar, Abhijit Y. *Petroleum Reservoir Rock and Fluid Properties.* Boca Raton, Fla.: CRC Press, 2006.

Da Rosa, Aldo Viera. *Fundamentals of Renewable Energy Processes.* New York: Academic Press, 2009.

Dasmann, Raymond F. *No Further Retreat.* New York: Macmillan, 1971.

Davenport, William G., and Matthew J. King. *Sulfuric Acid Manufacture.* Boston: Elsevier Science, 2006.

Davidson, Donald A. *Soils and Land Use Planning.* New York: Longman, 1980.

Davidson, Jon P., Walter E. Reed, and Paul M. Davis. *Exploring Earth: An Introduction to Physical Geology.* 2d ed. Upper Saddle River, N.J.: Prentice Hall, 2001.

Davis, George H. *Structural Geology of Rocks and Regions.* 2d ed. New York: John Wiley & Sons, 1996.

Davis, Kenneth P. *Land Use.* New York: McGraw-Hill, 1976.

Dawson, J. B. *Kimberlites and Their Xenoliths.* New York: Springer-Verlag, 1980.

Decker, Robert, and Barbara Decker. *Volcanoes.* 4th ed. New York: W. H. Freeman, 2005.

Deer, W. A., R. A. Howie, and J. Zussman. *Rock-Forming Minerals.* 2d ed. London: Pearson Education Limited, 1992.

Delamare, Francois, and Bernard Guineau. *Colors: The Story of Dyes and Pigments.* New York: H. N. Abrams, 2000.

Demirbas, Ayhan. *Methane Gas Hydrates*. New York: Springer, 2010.

De Nevers, Noel. "Tar Sands and Oil Shales." *Scientific American* 214 (February, 1966): 21-29.

Desautels, P. E. *The Gem Kingdom*. New York: Random House, 1970.

_____. *The Mineral Kingdom*. New York: Madison Square Press, 1972.

Di Pippo, Ronald. *Geothermal Power Plants*. 2d ed. Burlington, Mass.: Butterworth-Heinemann, 2008.

Dietrich, R. V., and B. J. Skinner. *Rocks and Rock Minerals*. New York: John Wiley & Sons, 1979.

Dixon, Dougal. *The Practical Geologist*. New York: Simon & Schuster, 1992.

Dluhy, Milan J., and Kan Chen, eds. *Interdisciplinary Planning: A Perspective for the Future*. New Brunswick, N.J.: Center for Urban Policy Research, 1986.

Dobrzhinetskaya, Larissa, et al., eds. *Ultrahigh-Pressure Metamorphism*. Burlington, Mass.: Elsevier, 2011.

Dodds, Walter K., and Matt R. Whiles. *Freshwater Ecology: Concepts and Environmental Applications of Limnology*. 2d ed. Burlington: Academic Press, 2010.

Dodson, Ronald P., and Samuel P. Hammar, eds. *Asbestos: Risk Assessment, Epidemiology, and Health Effects*. Boca Raton, Fla.: Taylor & Francis, 2006.

Dolan, Robert. "Barrier Islands: Natural and Controlled." In *Coastal Geomorphology*, edited by Donald R. Coates. Binghamton: State University of New York, 1972.

Dolgoff, Anatole. *Physical Geology*. Boston: Houghton Mifflin, 1999.

Donahue, Roy L., Raymond W. Miller, and John C. Shickluna. *Soils: An Introduction to Soils and Plant Growth*. 6th ed. Upper Saddle River, N.J.: Prentice-Hall, 1990.

Dorn, Ronald I. *Rock Coatings*. New York: Elsevier, 1998.

Dorton, Peter, Saman Alavi, and T. K. Woo. "Free Energies of Carbon Dioxide Sequestration and Methane Recovery in Clathrate Hydrates." *Journal of Chemical Physics* 127, no. 12 (2007).

Drevnick, Paul E., et al. "Increased Accumulation of Sulfur in Lake Sediments of the High Arctic." *Environmental Science and Technology* 44, no. 22 (2010): 8415-8421.

Duchesne, J. C., and A. Korneliussen, eds. *Ilmenite Deposits and Their Geological Environment. Norges Geologiski Undersokelse*. Special publication 9. Trondheim, Norway: Geological Survey of Norway, 2003.

Duff, D. *Holmes' Principles of Physical Geology*. 4th ed. London: Chapman & Hall, 1998.

Duffield, Wendell, and John Sas. *Geothermal Energy: Clean Power from the Earth's Heat*. USGS Circular 1249. 2003.

Duffy, Simon, E. Ohtani, and D. C. Rubie, eds. *New Developments in High-Pressure Mineral Physics and Applications to the Earth's Interior*. Atlanta: Elsevier, 2005.

Duncan, Donald C., and Vernon E. Swanson. *Organic-Rich Shale of the United States and World Land Area*. U.S. Geological Survey Circular 523. Washington, D.C.: Government Printing Office, 1965.

Dunmur, David, and Timothy J. Sluckin. *Soap, Science, and Flat-Screen TVs: A History of Liquid Crystals*. New York: Oxford University Press, 2011.

Durand, Bernard, ed. *Kerogen: Insoluble Organic Matter from Sedimentary Rocks*. Paris: Éditions Technip, 1980.

Eckholm, Erik, and Lester R. Brown. *Spreading Deserts: The Hand of Man*. New York: Worldwatch Institute, 1971.

Economic Commission for Europe. *Land Administration Guidelines: With Special Reference to Countries in Transition*. New York: United Nations, 1996.

Edwards, Richard, and Keith Atkinson. *Ore Deposit Geology and Its Influence on Mineral Exploration*. London: Chapman and Hall, 1986.

Edzwald, James K. *Water Quality and Treatment: A Handbook on Drinking Water*. New York: McGraw-Hill, 2011.

Einaudi, M. T., L. D. Meinert, and R. J. Newberry. "Skarn Deposits." In *Economic Geology: Seventy-fifth Anniversary Volume*. El Paso, Tex.: Economic Geology Publishing, 1981.

Eisenbud, Merril, and Thomas F. Gesell. *Environmental Radioactivity: From Natural, Industrial, and Military Sources*. Waltham, Mass.: Academic Press, 1997.

Ellers, F. S. "Advanced Offshore Oil Platforms." *Scientific American* 246 (April, 1982): 39-49.

Engel, Leonard. *The Sea*. Boston: Time-Life Books, 1967.

Ericsson, Magnus, Anton Löf, and Olle Östensson. "Iron Ore Review 2010/2011." *Nordic Steel and Mining Review* 195 (2011): 14-19.

Eriksson, P. G., et al. *The Precambrian Earth: Tempos and Events*. Edited by K. C. Condie. Amsterdam: Elsevier, 2004.

Ernst, W. G. *Earth Materials.* Englewood Cliffs, N.J.: Prentice-Hall, 1969.

_____. *Subduction Zone Metamorphism.* Stroudsburg, Pa.: Dowden, Hutchinson and Ross, 1975.

Eslinger, Eric, and David Pevear. *Clay Minerals for Petroleum Geologists and Engineers.* Tulsa, Okla.: Society of Economic Paleontologists and Mineralogists, 1988.

Evangelou, V. P. *Environmental Soil and Water Chemistry: Principles and Applications.* New York: Wiley, 1998.

Evans, Anthony M. *An Introduction to Economic Geology and Its Environmental Impact.* Malden, Mass.: Blackwell Science, 1997.

Evans, Bernard W., and Edwin H. Brown, eds. *Blueschists and Eclogites.* Boulder, Colo.: Geological Society of America, 1986.

Evans, Robert Crispin. *An Introduction to Crystal Chemistry.* 2d ed. New York: Cambridge University Press, 1964.

Evenari, Michael, Leslie Shanan, and Nephtali Tadmor. *The Negev: The Challenge of a Desert.* 2d ed. Cambridge, Mass.: Harvard University Press, 1982.

Fabos, Julius Gy. *Land-Use Planning.* New York: Chapman and Hall, 1985.

Fairbridge, Rhodes W., and Joanne Burgeois, eds. *The Encyclopedia of Sedimentology.* New York: Springer, 1978.

Fall, J.A., et al. *Long-Term Consequences of the Exxon Valdez Oil Spill for Coastal Communities of Southcentral Alaska.* Technical Report 163. Alaska Department of Fish and Game, Division of Subsistence, Anchorage, Alaska. 2001.

Fanchi, John R., and Christopher J. Fanchi. *Energy in the Twenty-First Century.* 2d ed. Hackensack, N.J.: World Scientific, 2011.

Faulkner, Edwin B., and Russell J. Schwartz. *High Performance Pigments.* Hoboken, N.J.: Wiley, 2009.

Faure, Gunter. *Origin of Igneous Rocks: The Isotopic Evidence.* New York: Springer-Verlag, 2010.

Feldman, V. I., L. V. Sazonova, and E. A. Kozlov. "High-Pressure Polymorphic Modifications in Minerals in the Products of Impact Metamorphism of Polymineral Rocks." *AIP Conference Proceedings* 849, no. 1 (2006): 444-450.

Fenner, Janis L., Debora J. Hamberg, and John D. Nelson. *Building on Expansive Soils.* Fort Collins: Colorado State University, 1983.

Fenton, Carol Lane, and Mildred Adams Fenton. *The Rock Book.* Mineola, N.Y.: Dover, 2003.

Ferraris, Giovanni, Emil Makovicky, and Stefano Merlino. *Crystallography of Modular Materials.* New York: Oxford University Press, 2004.

Ferrey, Steven, and R. Anil Cabral. *Renewable Power in Developing Countries: Winning the War on Global Warming.* Tulsa, Okla.: Pennwell Books, 2006.

Fettes, Douglas, and Jacqueline Desmons, eds. *Metamorphic Rocks: A Classification and Glossary of Terms.* New York: Cambridge University Press, 2007.

Finley, Fred N., Younkeong Nam, and John Oughton. "Earth Systems Science: An Analytic Framework." *Science Education* 95, no. 6 (2011): 1066-1085.

Fischetti, Mark. "The Human Cost of Energy." *Scientific American* 305, no. 3 (2011): 96.

Fisher, P. J. *The Science of Gems.* New York: Charles Scribner's Sons, 1966.

Fisher, R. V., and H. U. Schmincke. *Pyroclastic Rocks.* New York: Springer-Verlag, 1984.

Fletcher, Seth. *Bottled Lightning: Superbatteries, Electric Cars, and the New Lithium Economy.* New York: Hill and Wang, 2011.

Flugel, Erik. *Microfacies of Carbonate Rocks: Analysis, Interpretation, and Application.* 2d ed. New York: Springer, 2010.

Foreman, T. L., and N. L. Ziemba. "Cleanup on a Large Scale." *Civil Engineering* (August, 1987): 46-48.

Foth, Henry D., and Boyd G. Ellis. *Soil Fertility.* 2d ed. Boca Raton, Fla.: CRC Lewis, 1997.

Francis, Peter, and Clive Oppenheimer. *Volcanoes.* 2d ed. New York: Oxford University Press. 2004.

Francis, Wilfrid. *Coal: Its Formation and Composition.* London: Edward Arnold, 1961.

Freese, Barbara. *Coal: A Human History.* Cambridge, Mass.: Penguin, 2004.

Freeze, R. Allan, and John A. Cherry. *Groundwater.* Englewood Cliffs, N.J.: Prentice-Hall, 1979.

Freudenburg, William, and Robert Gramling. *Blowout in the Gulf: The BP Oil Spill Disaster and the Future of Energy in America.* Cambridge, Mass.: MIT Press, 2011.

Fridleifsson, Gudmundur O., and Wilfred A. Elders. "The Iceland Deep Drilling Project: A Search for Deep Unconventional Geothermal Resources." *Geothermics* 34 (2005): 269-285.

Frisch, Wolfgang, Martin Meschede, and Ronald Blakey. *Plate Tectonics: Continental Drift and Mountain Building.* New York: Springer, 2011.

Frost, B. Ronald, et al. "A Geochemical Classification for Granitic Rocks." *Journal of Petrology* 42 (2001): 2033-2048.

Frye, Keith. *Mineral Science: An Introductory Survey.* New York: Macmillan, 1993.

_____. *Modern Mineralogy.* Englewood Cliffs, N.J.: Prentice-Hall, 1974.

Fyfe, W. S., N. J. Price, and A. B. Thompson. *Fluids in the Earth's Crust.* New York: Elsevier, 1978.

Gage, Thomas E., and Richard Merrill, eds. *Energy Primer, Solar, Water, Wind, and Biofuels.* 2d ed. New York: Dell Publishing, 1978.

Gait, R. I. *Exploring Minerals and Crystals.* Toronto: McGraw-Hill Ryerson, 1972.

Galarraga, Ibon, Mikel Gonzalez-Eguino, and Anil Markandya. *Handbook of Sustainable Energy.* Northampton, Mass.: Edward Elgar, 2011.

Galloway, W. E., and D. K. Hobday. *Terrigenous Clastic Depositional Systems.* 2d ed. New York: Springer-Verlag, 1996.

Gani, Mary S. J. *Cement and Concrete.* London: Chapman and Hall, 1997.

Gartner, E. M., and H. Uchikawa, eds. *Cement Technology.* Westerville, Ohio: American Ceramic Society, 1994.

Gashus, O. K., and T. J. Gray, eds. *Tidal Power.* New York: Plenum Press, 1972.

Gasparrini, Claudia. *Gold and Other Precious Metals: From Ore to Market.* New York: Springer-Verlag, 1993.

Gavezzotti, Angelo. *Molecular Aggregation: Structural Analysis and Molecular Simulation of Crystals and Liquids.* New York: Oxford University Press, 2007.

Gay, Kathlyn. *Silent Killers: Radon and Other Hazards.* New York: Franklin Watts, 1988.

Gayer, Rodney A., and Jierai Peesek, eds. *European Coal Geology and Technology.* London: Geological Society, 1997.

Gee, George E. *Recovering Precious Metals.* Palm Springs, Calif.: Wexford College Press, 2002.

Geeson, N. A., C. J. Brandt, and J. B. Thornes, eds. *Mediterranean Desertification: A Mosaic of Processes and Responses.* Hoboken, N.J.: John Wiley & Sons, 2002.

Gerard, A. J. *Soils and Landforms.* London: Allen & Unwin, 1981.

Gerber, Michele Stenehjem. *On the Home Front: The Cold War Legacy of the Hanford Nuclear Site.* 2d ed. Lincoln: University of Nebraska Press, 2002.

Gerritsen, Margot G., and Louis J. Durlofsky. "Modeling Fluid Flow in Oil Reservoirs." *Annual Review of Fluid Mechanics* 37 (2005): 211-238.

Ghose, Ajoy K., ed. *Mining on a Small and Medium Scale: A Global Perspective.* London: Intermediate Technology, 1997.

Giavarini, Carlo, and Keith Hester. *Gas Hydrates: Immense Energy Potential and Environmental Challenges.* New York: Springer, 2011.

Gill, J. B. *Orogenic Andesite and Plate Tectonics.* New York: Springer-Verlag, 1981.

Gill, Robin. *Igneous Rocks and Processes: A Practical Guide.* Hoboken, N.J.: Wiley-Blackwell, 2011.

Gillen, Cornelius. *Metamorphic Geology: An Introduction to Tectonic and Metamorphic Processes.* 2d ed. Berlin: Springer, 2011.

Gipe, Paul. *Wind Energy Comes of Age.* New York: John Wiley, 1995.

Glantz, Michael H., ed. *Desertification: Environmental Degradation in and Around Arid Lands.* Boulder, Colo.: Westview Press, 1977.

Glaser, Peter E., Frank P. Davidson, Katinka I. Csigi, et al., eds. *Solar Power Satellites: The Emerging Energy Option.* New York: E. Howard, 1993.

Gluyas, Jon, and Richard Swarbrick. *Petroleum Geoscience.* Malden, Mass.: Blackwell Science, 2004.

Gorbachev, Yury I. *Well Logging: Fundamentals of Methods.* New York: John Wiley & Sons, 1995.

Gorman, D. G., and June Neilson, eds. *Decommissioning Offshore Structures.* New York: Springer, 1997.

Gosnell, Mariana. *Ice: The Nature, the History, and the Uses of an Astonishing Substance.* New York: Alfred A. Knopf, 2005.

Govett, G. J. S., and M. H. Govett, eds. *World Mineral Supplies: Assessment and Perspective.* New York: Elsevier, 1976.

Gowariker, Vasant, et al. *The Fertilizer Encyclopedia.* Hoboken, N.J.: John Wiley & Sons, 2009.

Graham, Bob, et al. *Deep Water: The Gulf Oil Disaster and the Future of Offshore Drilling.* National Commission on the BP Deepwater Horizon Oil Spill and Offshore Drilling, 2011.

Grapes, Rodney. *Pyrometamorphism.* 2d ed. New York: Springer, 2010.

Gray, Forest. *Petroleum Production for the Nontechnical Person.* Tulsa, Okla.: PennWell, 1986.

Gray, G. Ronald. "Oil Sand." In *McGraw-Hill Encyclopedia of the Geological Sciences,* edited by S. P. Parker. 2d ed. New York: McGraw-Hill, 1988.

Gray, Theodore. *The Elements: A Visual Exploration of Every Known Atom in the Universe.* New York: Black Dog & Leventhal Publishers, 2009.

Gregory, Snyder A., Clive R. Neal, and W. Gary Ernst, eds. *Planetary Petrology and Geochemistry.* Columbia, Md.: Geological Society of North America, 1999.

Griggs, Gary, and Lauret Savoy, eds. *Living with the California Coast.* Durham, N.C.: Duke University Press, 1985.

Grossman, Dan, and Seth Shulman. "A Nuclear Dump: The Experiment Begins (Beneath Yucca Mountain, Nevada)." *Discover* 10 (March, 1989): 48.

Grotzinger, J. P. "New Views of Old Carbonate Sediments." *Geotimes* 38 (September, 1993): 12-15.

Grotzinger, John, Thomas H. Jordan, Frank Press, and Raymond Siever. *Understanding Earth.* 5th ed. New York: W. H. Freeman, 2009.

Gubelin, E. J. *Internal World of Gemstones.* 3d ed. Santa Monica, Calif.: Gemological Institute of America, 1983.

Guilbert, John M., and Charles F. Park, Jr. *The Geology of Ore Deposits.* 4th ed. Long Grove, Ill.: Waveland Press, Inc., 1986.

Gulkis, Samuel, D. S. Stetson, Ellen Renee Stofan, et al., eds. *Mission to the Solar System: Exploration and Discovery.* Pasadena, Calif.: National Aeronautics and Space Administration, Jet Propulsion Laboratory, 1998.

Gupta, Chiranjib, and Harvinderpal Singh. *Uranium Resource Processing: Secondary Resources.* Berlin: Springer-Verlag, 2010.

Hacker, Bradley R., Ju G. Liou, et al., eds. *When Continents Collide: Geodynamics and Geochemistry of Ultra-High Pressure Rocks.* Boston: Kluwer, 2010.

Hackman, Christian L., E. Ellsworth Hackman III, and Matthew E. Hackman. *Hazardous Waste Operations and Emergency Response Manual and Desk Reference.* New York: McGraw-Hill, 2002.

Halacy, D. S. *Earth, Water, Wind, and Sun: Our Energy Alternatives.* New York: Harper & Row, 1977.

Halbouty, M. T. *Salt Domes: Gulf Region, United States and Mexico.* 2d ed. Houston, Tex.: Gulf Publishing, 1981.

Halfar, Jochen, and Rodney M. Fujita. *Science* 316, no. 5827 (2007): 987.

Halka, Monica, and Brian Nordstrom. *Metals and Metalloids.* New York: Facts on File, 2011.

Hall, Anthony. *Igneous Petrology.* 2d ed. New York: John Wiley & Sons, 1996.

Hall, Cally. *Gemstones.* 2d ed. London: Darling Kindersley, 2002.

Hall, R. Stewart, ed. *Drilling and Producing Offshore.* Tulsa, Okla.: PennWell Books, 1983.

Hallam, A. *A Revolution in the Earth Sciences.* New York: Oxford University Press, 1973.

Ham, W. E., ed. *Classification of Carbonate Rocks: A Symposium.* Tulsa, Okla.: American Association of Petroleum Geologists, 1962.

Hamblin, Jacob Darwin. *Poison in the Well.* Piscataway, N.J.: Rutgers University Press, 2008.

Hamblin, Kenneth W., and Eric H. Christiansen. *Earth's Dynamic Systems.* 10th ed. Upper Saddle River, N.J.: Prentice Hall, 2003.

Hamilton, Warren B., and W. Bradley Meyers. *The Nature of Batholiths.* Professional Paper 554-C. Denver, Colo.: U.S. Geological Survey, 1967.

_____. "Nature of the Boulder Batholith of Montana." *Geological Society of America Bulletin* 85 (1974): 365-378.

Hammond, Christopher. *The Basics of Crystallography and Diffraction.* 2d ed. New York: Oxford University Press, 2001.

Hanson, David T. *Waste Land: Meditations on a Ravaged Landscape.* New York: Aperture, 1997.

Harlow, George E., ed. *The Nature of Diamonds.* Cambridge, England: Cambridge University Press, 1998.

Harpstead, Milo I., Thomas J. Sauer, and William F. Bennett. *Soil Science Simplified.* 4th ed. Ames: Iowa State University Press, 2001.

Harsh, K., and Sukanta Roy Gupta. *Geothermal Energy: An Alternative Resource for the 21st Century.* Boston: Elsevier Science, 2006.

Hartley, Frank R., ed. *Chemistry of the Platinum Group Metals: Recent Developments.* New York: Elsevier, 1991.

Hatch, F. H., R. H. Rastall, and J. T. Greensmith. *Petrology of the Sedimentary Rocks.* 7th ed. New York: Springer, 1988.

Haun, John D., and L. W. LeRoy, eds. *Subsurface Geology in Petroleum Exploration.* Golden: Colorado School of Mines, 1958.

Haussühl, Siegfried. *Physical Properties of Crystals: An Introduction.* Weinheim: Wiley-VCH, 2007.

Havlin, John L., et al. *Soil Fertility and Fertilizers.* 7th ed. Upper Saddle River, N.J.: Prentice Hall, 2004.

Hawkes, Nigel. *Toxic Waste and Recycling.* New York: Gloucester, 1988.

Haymon, R. M. "Growth History of Hydrothermal Black Smoker Chimneys." *Nature* 301 (1983): 695-698.

Hays, W. W. *Facing Geologic and Hydrologic Hazards.* U.S. Geological Survey Professional Paper 1240-B. Washington, D.C.: Government Printing Office, 1981.

Hazen, Robert M. *The Breakthrough: The Race for the Superconductor.* New York: Summit Books, 1988.

Healy, Robert G. *Competition for Land in the American South.* Washington, D.C.: The Conservation Foundation, 1985.

Heinberg, Richard. *Blackout: Coal, Climate and the Last Energy Crisis.* Gabriola Island, B.C.: 2009.

Heinrich, E. William. *The Geology of Carbonatites.* Skokie, Ill.: Rand McNally, 1966.

Heppenheimer, T. A. *Colonies in Space.* Harrisburg, Pa.: Stackpole Books, 1977.

Hewitt, Paul G., John Suchocki, and Leslie A. Hewitt. *Conceptual Physical Science.* San Francisco: Pearson/ Addison Wesley, 2008.

Hewlette, Peter C., ed. *Lea's Chemistry of Cement and Concrete.* 4th ed. Burlington, Mass.: Butterworth-Heinemann, 2004.

Hibbard, Malcolm J. *Mineralogy: A Geologist's Point of View.* New York: McGraw-Hill, 2002.

Hill, John W., and Doris K. Kolb. *Chemistry for Changing Times.* Upper Saddle River, N.J.: Prentice Hall, 2001.

Hill, Mary. *Geology of the Sierra Nevada.* Rev. ed. Berkeley: University of California Press, 2006.

Hillebrand, W. F. *The Analysis of Silicate and Carbonate Rocks.* Ann Arbor: University of Michigan Library, 2009.

Hobson, G. D., and E. N. Tiratsoo. *Introduction to Petroleum Geology.* 2d ed. Houston: Gulf Publishing, 1979.

Hodgson, Susan F. *A Geysers Album: Five Eras of Geothermal History.* Sacramento: California Department of Conservation, Division of Oil, Gas and Geothermal Resources, 1997.

Hoek, E., and J. W. Bray. *Rock Slope Engineering.* 3d ed. Brookfield, Vt.: IMM/North American Publications Center, 1981.

Hofmann, W. "2.03: Sampling Mantle Heterogeneity Through Oceanic Basalts: Isotopes and Trace Elements." In *Treatise on Geochemistry.* Volume 2. Edited by K. K. Turekian and H. D. Holland. San Diego: Elsevier, 2003.

Holtz, Wesley G., and Stephen S. Hart. *Home Construction on Shrinking and Swelling Soils.* Denver: Colorado Geological Survey, 1978.

Houseknecht, David W., and Edward D. Pittman, eds. *Origin, Diagenesis, and Petrophysics of Clay Minerals in Sandstones.* Tulsa, Okla.: Society for Sedimentary Geology, 1992.

Howie, Frank M., ed. *The Care and Conservation of Geological Material: Mineral Rocks, Meteorites, and Lunar Finds.* Oxford: Butterworth-Heinemann, 1992.

Hoyle, Brian, ed. *Cityports, Coastal Zones, and Regional Change: International Perspectives on Planning and Management.* New York: Wiley, 1996.

Hsu, Kenneth J. *Physics of Sedimentology.* 2d ed. New York: Springer, 2010.

Hudson, Norman. *Soil Conservation.* 3d ed. Ames: Iowa State University Press, 1995.

Hughes, Richard V. "Oil Shale." In *The Encyclopedia of Sedimentology,* edited by R. W. Fairbridge and J. Bourgeois. New York: Springer, 1978.

Humphris, S. E., and G. Thompson. "Hydrothermal Alteration of Oceanic Basalts by Seawater." *Geochimica et Cosmothimica Acta* 42 (January, 1978): 107-125.

Hunt, John Meacham. *Petroleum Geochemistry and Geology.* 2d ed. New York: W. H. Freeman, 1996.

Huo, Hu, et al. "Mechanical and Thermal Properties of Methane Clathrate Hydrates as an Alternative Energy Resource." *Journal of Renewable and Sustainable Energy* 3, no. 6 (2011).

Hurlbut, C. S., Jr., and Robert C. Kammerling. *Gemology.* 2d ed. New York: John Wiley & Sons, 1991.

Hutchison, Charles S. *Economic Deposits and Their Tectonic Setting.* New York: John Wiley & Sons, 1983.

_____. *Laboratory Handbook of Petrographic Techniques.* New York: John Wiley & Sons, 1974.

Hyndman, Donald W. *Petrology of Igneous and Metamorphic Rocks.* 2d ed. New York: McGraw-Hill, 1985.

Hyne, Norman J. *Dictionary of Petroleum Exploration, Drilling, and Production.* Tulsa, Okla.: PennWell, 1991.

_____. *Nontechnical Guide to Petroleum Geology, Exploration, Drilling, and Production.* 2d ed. Tulsa, Okla.: PennWell, 2001.

Icon Group International. *The 2009 World Forecasts of Silicates and Commercial Alkalai Metal Silicates Export Supplies.* San Diego, Calif.: Author, 2009.

Idriss, I. M., and R. W. Boulanger. *Soil Liquefaction During Earthquakes.* Oakland, Calif.: Earthquake Engineering Research Institute, 2008.

Illinois State Geological Survey. *How to Read Illinois Topographic Maps.* Champaigne, Ill.: Illinois Department of Natural Resources. 2005.

Imeson, Anton. *Desertification, Land Degradation and Sustainability.* Hoboken, N.J.: John Wiley & Sons, 2012.

Inglis, David R. *Nuclear Energy: Its Physics and Its Social Challenge.* Reading, Mass.: Addison-Wesley, 1973.

Institution of Mining and Metallurgy. *Surface Mining and Quarrying.* Brookfield, Vt.: IMM/North American Publications Center, 1983.

Isachsen, Yngvar W., ed. *Origin of Anorthosite and Related Rocks.* Memoir 18. Albany, N.Y.: State Department of Education of New York, 1968.

Jackson, M. P. A., D. G. Roberts, and S. Snelson, eds. *Salt Tectonics: A Global Perspective.* Tulsa, Okla.: American Association of Petroleum Geologists, 1996.

Jackson, M. P. A., ed. *Salt Diapirs of the Great Kavir, Central Iran.* Boulder, Colo.: Geological Society of America, 1991.

Jacobson, Michael, et al. *From Biogeochemical Cycles to Global Changes.* Vol. 7 in *Earth System Science.* Maryland Heights, Mo.: Academic Press, 2000.

Jacobson, Michael C., et al. *Earth System Science: From Biogeochemical Cycles to Global Change.* San Diego, Calif.: Academic Press, 2006.

James, P. *The Future of Coal.* 2d ed. London: Macmillan Press, 1984.

Jenny, H. *Factors of Soil Formation: A System of Quantitative Pedology.* New York: Dover, 1994.

_____. *The Soil Resource: Origin and Behavior.* New York: Springer-Verlag, 1980.

Jensen, John R. *Remote Sensing of the Environment: An Earth Resource Perspective.* 2d ed. Upper Saddle River, N.J.: Prentice Hall, 2006.

Jensen, M. L., and A. M. Bateman. *Economic Mineral Deposits.* 3d ed. New York: John Wiley & Sons, 1981.

Jerram, Dougal, and Nick Petford. *The Field Description of Igneous Rocks.* 2d ed. Hoboken, N.J.: Wiley-Blackwell, 2011.

Jez, Joseph. *Sulfur: A Missing Link Between Soils, Crops, and Nutrition.* Madison, Wis.: American Society of Agronomy, 2008.

Jiandong, Tong, et al. *Mini Hydropower.* New York: John Wiley & Sons, 1997.

Jochim, Candace L. *Home Landscaping and Maintenance on Swelling Soil.* Special Publication 14. Denver: Colorado Geological Survey, 1981.

Johnson, T. "Hot-Water Power from the Earth." *Popular Science* 222 (January, 1983): 70.

Jonas, E. C., and E. McBride. *Diagenesis of Sandstone and Shale: Application to Exploration for Hydrocarbons.* Austin: Department of Geological Sciences, University of Texas, Continuing Education Program, 1977.

Jones, Adrian P., Frances Wall, and C. Terry Williams, eds. *Rare Earth Minerals: Chemistry, Origin, and Ore Deposits.* New York: Chapman and Hall, 1996.

Jones, David S. J., and Peter P. Pujadó. *Handbook of Petroleum Processing.* Dordrecht: Springer, 2006.

Judson, S., and M. E. Kauffman. *Physical Geology.* 8th ed. Englewood Cliffs, N.J.: Prentice-Hall, 1990.

Kalvoda, Jan, and Charles L. Rosenfeld, eds. *Geomorphological Hazards in High Mountain Areas.* Boston: Kluwer Academic, 1998.

Kapur, Selim, and George Stoops, eds. *New Trends in Soil Micromorphology.* New York: Springer, 2010.

Kapustin, Yuri L. *Mineralogy of Carbonatites.* Washington, D.C.: Amerind, 1980.

Karl, Herman A., et al., eds. *Restoring Lands; Coordinating Science, Politics, and Action.* New York: Springer, 2012.

Katti, R. K. *Behaviour of Saturated Expansive Soil and Control Methods.* Rev. ed. Oxford: Taylor & Francis, 2002.

Kaufman, Wallace, and Orrin H. Pilkey, Jr. *The Beaches Are Moving: The Drowning of America's Shorelines.* Durham, N.C.: Duke University Press, 1983.

Keefer, D. K. "Landslides Caused by Earthquakes." *Geological Society of America Bulletin* 95 (April, 1984): 406-421.

Keefer, Robert F., and Kenneth S. Sajwan. *Trace Elements in Coal and Coal Combustion Residues.* Boca Raton, Fla.: Lewis Publishers, 1993.

Kelland, Malcolm A. *Production Chemicals for the Oil and Gas Industry.* Boca Raton, Fla.: CRC Press, 2009.

Keller, Edward A. *Environmental Geology.* 9th ed. Upper Saddle River, N.J.: Prentice Hall, 2010.

_____. *Introduction to Environmental Geology.* 4th ed. Upper Saddle River, N.J.: Pearson Prentice Hall, 2008.

Kellogg, Charles E. *Our Garden Soils.* New York: Macmillan, 1952.

_____. *The Soils That Support Us.* New York: Macmillan, 1956.

Kennedy, John L. *Fundamentals of Drilling.* Tulsa, Okla.: PennWell, 1982.

Kennedy, Nathaniel T. "California's Trial by Mud and Water." *National Geographic* 136 (October, 1969): 552-573.

Kerr, P. F. *Optical Mineralogy.* New York: McGraw-Hill, 1977.

Kerr, Richard A. "Hot Dry Rock: Problems, Promise." *Science* 238 (November, 1987): 1226-1229.

Kesler, S. E. *Our Finite Mineral Resources.* New York: McGraw-Hill, 1976.

Khoo, Iam-Choon. *Liquid Crystals.* Hoboken, N.J.: Wiley-Interscience, 2007.

Kiersch, G. A. "Vaiont Reservoir Disaster." *Civil Engineering* 34 (1964): 32-39.

King, Robert E., ed. *Stratigraphic Oil and Gas Fields.* Tulsa, Okla.: American Association of Petroleum Geologists, 1972.

King, Vandall T. *A Collector's Guide to the Granite Pegmatite.* Atglen, Pa.: Schiffer Publishing, 2010.

Klein, C., and C. S. Hurlbut, Jr. *Manual of Mineralogy.* 23d ed. New York: John Wiley & Sons, 2008.

Kluge, P. F. "The Man Who Is Diamond's Best Friend." *Smithsonian* 19 (May, 1988): 72.

Knight, Richard L., and Peter B. Landres, eds. *Stewardship Across Boundaries.* Washington D.C.: Island Press, 1998.

Koerner, Robert M. *Construction and Geotechnical Methods in Foundation Engineering.* New York: McGraw-Hill, 1984.

Konishi, Hiromi, Reijo Alviola, and Peter R. Buseck. "2111 Biopyribole Intermediate Between Pyroxene and Amphibole: Artifact or Natural Product?" *American Mineralogist* 89 (2004): 15-19.

Koslow, J. A. *The Silent Deep: The Discovery, Ecology, and Conservation of the Deep Sea.* Chicago: University of Chicago Press, 2007.

Kourkoulis, Stavros K., ed. *Fracture and Failure of Natural Building Stones.* Dordrecht: Springer, 2006.

Kraus, Nicholas C. *Shoreline Changed and Storm-Induced Beach Erosion Modeling.* Springfield, Va.: National Technical Information Service, 1990.

Kruger, Paul, and Carel Otte, eds. *Geothermal Energy: Resources, Production, Stimulation.* Stanford, Calif.: Stanford University Press, 1973.

Krygowski, Daniel, George B. Asquith, and Charles R. Gibson. *Basic Well Log Analysis.* 2d ed. Tulsa, Okla.: American Association of Petroleum Geologists, 2004.

Kunz, George. *The Curious Love of Precious Stones.* Mineola, N.Y.: Dover, 1971.

Kupfer, D. H., ed. *Geology and Technology of Gulf Coast Salt Domes: A Symposium.* Baton Rouge, La.: School of Geoscience, Louisiana State University, 1970.

Kurten, T., et al. "Large Methane Releases Lead to Strong Aerosol Forcing and Reduced Cloudiness." *Atmospheric Chemistry and Physics Discussions* 11, no. 3 (2011): 9057-9081.

LaGrega, Michael D., Philip L. Buckingham, and Jeffrey C. Evans. *Hazardous Waste Management.* 2d ed. Long Grove, Ill.: Waveland Press, 2001.

Lahti, Seppo L., Paula Raivio, and Ilka Laitakari, eds. *Orbicular Rocks in Finland.* Espoo: Geological Survey of Finland, 2005.

Lal, Rattan, ed. *Soil Quality and Soil Erosion.* Boca Raton, Fla.: CRC Press, 1999.

Lambert, David. *The Field Guide to Geology.* 2d ed. New York: Checkmark Books, 2007.

Lamey, C. A. *Metallic and Industrial Mineral Deposits.* New York: McGraw-Hill, 1966.

Lancaster, David E., ed. *Production from Fractured Shales.* Richardson, Tex.: Society of Petroleum Engineers, 1996.

Landa, Edward R., and Christian Feller, eds. *Soil and Culture.* New York: Springer, 2010.

Landsberg, Hans H., et al. *Energy: The Next Twenty Years.* Cambridge, Mass.: Ballinger, 1979.

Langenkamp, Robert D. *Oil Business Fundamentals.* Tulsa, Okla.: PennWell Books, 1982.

Larsen, Esper Signius. *Alkalic Rocks of Iron Hill, Gunnison County, Colorado.* Geological Survey Professional Paper 197-A. Washington, D.C.: Government Printing Office, 1942.

Larsen, Gunnar, and George V. Chilingar, eds. *Diagenesis in Sediments and Sedimentary Rocks.* New York: Elsevier, 1979.

Lauf, Robert. *The Collector's Guide to Silicate Crystal Structures.* Atglen, Pa.: Schiffer, 2010.

_____. *Introduction to Radioactive Minerals.* Atglen, Pa.: Schiffer, 2007.

Law, Dennis L. *Mined-Land Rehabilitation.* New York: Van Nostrand Reinhold, 1984.

Laznicka, Peter. *Giant Metallic Deposits.* 2d ed. Berlin: Springer-Verlag, 2010.

Le Roex, Anton P., David R. Bell, and Peter Davis. "Petrogenesis of Group I Kimberlites from Kimberley, South Africa: Evidence from Bulk-Rock Geochemistry." *Journal of Petrology* 44 (2003): 2261-2286.

League of Women Voters Educational Fund Staff. *The Nuclear Waste Primer: A Handbook for Citizens.* New York: Lyons, 1987.

Leake, Bernard E. "Nomenclature of Amphiboles." *American Mineralogist* 63 (November, 1978): 1023-1052.

LeBas, Michael John. *Carbonatite-Nephelinite Volcanism: An African Case History.* New York: John Wiley & Sons, 1977.

Lee, G. Fred, and Anne Jones-Lee. *Overview of Subtitle D Landfill Design, Operation, Closure and Postclosure Care Relative to Providing Public Health and Environmental Protections for as Long as the Wastes in the Landfill Will Be a Threat.* El Macero, Calif.: G. Fred Lee & Associates, 2004.

Leeder, Mike. *Sedimentology and Sedimentary Basins: From Turbulence to Tectonics.* 2d ed. Malden, Mass.: Wiley-Blackwell, 2011.

Lefond, S. J. *Handbook of World Salt Resources.* New York: Plenum Press, 1969.

Legget, Robert F. *Cities and Geology.* New York: McGraw-Hill, 1973.

Legrand, Jacques. *Diamonds: Myth, Magic, and Reality.* Rev. ed. New York: Crown, 1985.

Lennox, Jethro, ed. *The Times Comprehensive Atlas of the World.* 13th ed. New York: Times Books, 2001.

Lerche, Ian, and Kenneth Peters. *Salt and Sediment Dynamics.* Boca Raton, Fla.: CRC Press, 1995.

Lerner, Steve. *Sacrifice Zones: The Front Lines of Toxic Chemical Exposure in the United States.* Cambridge, Mass.: MIT Press, 2010.

LeRoy, L. W., and D. O. LeRoy, eds. *Subsurface Geology.* 4th ed. Golden: Colorado School of Mines, 1977.

Lesnov, Felix P. *Rare Earth Elements in Ultramafic and Mafic Rocks and their Minerals.* New York: CRC Press, 2010.

Letcher, Trevor M., ed. *Future Energy: Improved, Sustainable and Clean.* Amsterdam: Elsevier, 2008.

Levorsen, A. I. *Geology of Petroleum.* 2d ed. Tulsa, Okla.: American Associations of Petroleum Geologists, 2001.

Lew, Kristi. *The Alkali Metals: Lithium, Sodium, Potassium, Rubidium, Cesium, Francium.* New York: Rosen Central, 2010.

Lewis, Thomas A., ed. *Volcano.* Alexandria, Va.: Time-Life Books, 1982.

Li, Guoyu. *World Atlas of Oil and Gas Basins.* Hoboken, N.J.: John Wiley & Sons, 2011.

Libes, Susan M. *Introduction to Marine Biogeochemistry.* Burlington, Mass.: Academic Press, 2009.

Liddicoat, Richard T., Jr. *Handbook of Gem Identification.* 12th ed. Santa Monica, Calif.: Gemological Institute of America, 1993.

Lillesand, Thomas M., Ralph W. Kiefer, and Jonathan W. Chipman. *Remote Sensing and Image Interpretation.* Hoboken, N.J.: John Wiley & Sons, 2008.

Lima-de-Faria, José. *Structural Mineralogy: An Introduction.* Dordrecht: Kluwer, 1994.

Lindsley, Donald H., ed. *Oxide Minerals: Petrologic and Magnetic Significance.* Washington, D.C.: Mineralogical Society of America, 1991.

Link, Peter K. *Basic Petroleum Geology.* 3d ed. Tulsa, Okla.: Oil and Gas Consultants International, 2007.

Lipson, Henry S. *Crystals and X-Rays.* New York: Springer-Verlag, 1970.

Liu, Jian Guo, and Philippa J. Mason. *Essential Image Processing and GIS for Remote Sensing.* Hoboken, N.J.: Wiley-Blackwell, 2009.

Liu, Yi-Can, et al. "Ultrahigh-Pressure Metamorphism and Multistage Exhumation of Eclogite of the Luotian Dome, North Dabie Complex Zone (Central China): Evidence from Mineral Inclusions and Decompression Textures." *Journal of Asian Earth Sciences* 42, no. 4 (2011): 607-617.

Lloyd, G. B. *Don't Call It Dirt.* Ontario, Calif.: Bookworm Publishing, 1976.

Loferski, Patricia J. "Platinum-Group Metals." In *Minerals Yearbook.* Washington, D.C.: Department of the Interior, issued annually.

Longley, Paul A., et al. *Geographic Information Systems and Science.* 3d ed. Hoboken, N.J.: Wiley, 2010.

Longstaffe, F. J. *Short Course in Clays and the Resource Geologist.* Toronto: Mineralogical Association of Canada, 1981.

Loomis, John B. *Integrated Public Lands Management.* 2d ed. New York: Columbia University Press, 2002.

Lottermoser, Bernd G. *Mine Wastes: Characterization, Treatment, Environmental Impacts.* 2d ed. New York: Springer, 2007.

Loucks, Robert G., and J. Frederick Sarg, eds. *Carbonate Sequence Stratigraphy: Recent Developments and Applications.* Tulsa, Okla.: American Association of Petroleum Geologists, 1993.

Loucks, Robert. *Shale Oil.* Bloomington, Ind.: Xlibris Corporation, 2002.

Lundgren, Lawrence W. *Environmental Geology.* 2d ed. Upper Saddle River, N.J.: Prentice Hall, 2010.

Lutgens, Frederick K., Edward J. Tarbuck, and Dennis Tasa. *Earth: An Introduction to Physical Geology.* 10th ed. Upper Saddle River, N.J.: Prentice Hall, 2010.

_____. *Essentials of Geology.* 11th ed. Upper Saddle River, N.J.: Prentice Hall, 2011.

Luthi, S. *Geological Well Logs.* New York: Springer-Verlag, 2001.

Macdonald, Gordon A. *Volcanoes.* Englewood Cliffs,

N.J.: Prentice-Hall, 1972.

MacDougall, Douglas. *Frozen Earth*. Berkeley: University of California Press, 2006.

Mackenzie, F. T., ed. *Our Changing Planet: An Introduction to Earth System Science and Global Environmental Change*. 4th ed. Upper Saddle River, N.J.: Prentice Hall, 1998.

_____. *Sediments, Diagenesis, and Sedimentary Rocks*. Amsterdam: Elsevier, 2005.

MacLachlan, Malcolm. *An Introduction to Marine Drilling*. Herefordshire: Dayton's Oilfield Publications, 1987.

MacLeod, P. A. Tyler, and C. L. Walker, eds. *Tectonic, Magmatic, Hydrothermal, and Biological Segmentation of Mid-Ocean Ridges*. London: Geological Society, 1996.

Magner, Mike. *Poisoned Legacy: The Human Cost of BP's Rise to Power*. New York: St. Martin's Press, 2011.

Mairota, Poala, John B. Thornes, and Nichola Geeson, ed. *Atlas of Mediterranean Environments in Europe: The Desertification Context*. New York: John Wiley & Sons, 1996.

Malley, Marjorie C. *Radioactivity: A History of a Mysterious Science*. New York: Oxford University Press, 2011.

Manutchehr-Danai, Mohsen. *Dictionary of Gems and Gemology*. 3d ed. Germany: Springer-Verlag, 2005.

Marier, Donald. *Wind Power for the Homeowner*. Emmaus, Pa.: Rodale Press, 1981.

Marino, Maurizio, and Massimo Santantonio. "Understanding the Geological Record of Carbonate Platform Drowning Across Rifted Tethyan Margins: Examples from the Lower Jurassic of the Apennines and Sicily (Italy)." *Sedimentary Geology* 225 (2010): 116-137.

Marion, J. B., and M. L. Roush. *Energy in Perspective*. 2d ed. New York: Academic Press, 1982.

Marsden, John O., and C. Iain House. *The Chemistry of Gold Extraction*. 2d ed. Littleton, Colo.: Society for Mining, Metallurgy and Exploration, 2006.

Marsh, William M. *Landscape Planning: Environmental Applications*. 5th ed. New York: John Wiley & Sons, 2010.

Mason, Brian, and L. G. Berry. *Elements of Mineralogy*. San Francisco: W. H. Freeman, 1968.

Mason, R. *Petrology of the Metamorphic Rocks*. 2d ed. Berlin: Springer, 2011.

Mathews, Jay. "Solar Energy Complex Hailed as Beacon for Utility Innovation." *The Washington Post* (March 2, 1989): A25.

Maugeri, Leonardo. *Beyond the Age of Oil: The Myths, Realities, and Future of Fossil Fuels and Their Alternatives*. Santa Barbara, Calif.: Praeger, 2010.

Maurizio, Galimberti, ed. *Rubber-Clay Nanocomposites*. Hoboken, N.J.: John Wiley & Sons, 2011.

Maynard, J. B., E. R. Force, and J. J. Eidel, eds. *Sedimentary and Diagenetic Mineral Deposits: A Basin Analysis Approach to Exploration*. Chelsea, Mich.: Society of Economic Geologists, 1991.

McBride, E. F. *Silica in Sediments: Nodular and Bedded Chert*. Tulsa, Okla.: Society of Economic Paleontologists and Mineralogists, 1979.

McBride, Murray B. *Environmental Chemistry of Soils*. New York: Oxford University Press, 1994.

McDonald, D. A., and R. C. Surdam, eds. *Clastic Diagenesis*. Memoir 37. Tulsa, Okla.: American Association of Petroleum Geologists, 1984.

McDowell, Bart. "Avalanche!" *National Geographic* 121 (June, 1962): 855-880.

McGuigan, Dermott. *Harnessing Water Power for Home Energy*. Pownal, Vt.: Garden Way Publishing, 1978.

McHarg, Ian L. *Design with Nature*. Hoboken, N.J.: John Wiley & Sons, 1992.

McMillan, Nancy J., et al. "Laser-Induced Breakdown Spectroscopy Analysis of Complex Silicate Materials: Beryl." *Analytical and Bioanalytical Chemistry* 385, no. 2 (2006): 263-271.

Meade, Lance P. "Defining a Commercial Dimension Stone Marble Property." In *Twelfth Forum on the Geology of Industrial Minerals*. Atlanta: Georgia Department of Natural Resources, 1976.

Meador, Roy. *Future Energy Alternatives*. Ann Arbor, Mich.: Ann Arbor Science Publishers, 1979.

Mehta, P. Kumar. *Concrete: Structure, Properties, and Materials*. 2d ed. Englewood Cliffs, N.J.: Prentice-Hall, 1992.

Menzie, W. D. *The Global Flow of Aluminum from 2006 through 2025*. Reston, Va.: U.S. Geological Survey, 2010.

Menzies, L. A. D., and J. M. Martins. "The Jacupiranga Mine, São Paulo, Brazil." *Mineralogical Record* 15 (1984): 261-270.

Merritts, Dorothy, Andrew De Wet, and Kirsten Menking. *Environmental Geology: An Earth System Science Approach*. Cranbury, N.J.: W. H. Freeman, 1998.

Mertie, J. B., Jr. *Economic Geology of the Platinum Metals*. U.S. Geological Survey Professional Paper 630. Washington, D.C.: Government Printing Office, 1969.

Meunier, Alain. *Clays*. Berlin: Springer, 2010.

Meyers, J. S. "Cauldron Subsidence and Fluidization: Mechanisms of Intrusion of the Coastal Batholith of Peru into Its Own Volcanic Ejecta." *Geological Society of America Bulletin* 86 (1975): 1209-1220.

Middleton, Gerard V., ed. *Encyclopedia of Sediments and Sedimentary Rocks*. Dordrecht: Springer, 2003.

Miles, Frank, and Nicholas Booth. *Race to Mars: The Harper and Row Mars Flight Atlas*. New York: Harper & Row, 1988.

Miller, G. Tyler, Jr. *Living in the Environment*. 17th ed. Belmont, Calif.: Brooks/Cole, Cengage Learning, 2012.

Millot, Georges. "Clay." *Scientific American* 240 (April, 1979): 108.

Mindess, Sidney, J. Francis Young, and David Darwin. *Concrete*. 2d ed. Upper Saddle River, N.J.: Prentice-Hall, 2002.

"Mineral Resource of the Month: Sulfur." *Earth* 55, no. 6 (2010): 26-27.

Mitchell, R. S. "Metamict Minerals: A Review, Part 1. Chemical and Physical Characteristics, Occurrence." *The Mineralogical Record* 4 (July/August, 1973): 177.

_____. "Metamict Minerals: A Review, Part 2. Origin of Metamictization, Methods of Analysis, Miscellaneous Topics." *The Mineralogical Record* 4 (September/October, 1973): 214.

Mitchell, Roger H. *Kimberlites, Orangeites, and Related Rocks*. New York: Plenum Press, 1995.

_____. *Kimberlites, Orangeites, Lamproites, Melilitites, and Minettes: A Petrographic Atlas*. Thunder Bay, Ontario: Almaz, 1997.

Miyashiro, A. *Metamorphism and Metamorphic Belts*. New York: Springer, 1978.

Moeller, Dade W. *Environmental Health*. 3d ed. Boston: President and Fellows of Harvard College, 2005.

Monroe, James S., Reed Wicander, and Richard Hazlett. *Physical Geology*. 6th ed. Belmont, Calif.: Thompson Higher Education, 2007.

Montgomery, Carla W. *Environmental Geology*. 4th ed. Columbus, Ohio: McGraw-Hill, 2008.

Mooney, Chris. "The Truth About Fracking." *Scientific American* 305, no. 5 (2011): 80-85.

Moore, Carl A. *Handbook of Subsurface Geology*. New York: Harper & Row, 1963.

Morgan, Lisa A. *Integrated Geoscience Studies in the Greater Yellowstone Area: Volcanic, Tectonic, and Hydrothermal Processes in the Yellowstone Geoecosystem*. Reston, Va.: U.S. Geological Survey, 2007.

Morgan, Royston Philip Charles. *Soil Erosion and Conservation*. 3d ed. Malden, Mass.: Blackwell Science, 2005.

Morris, E. M., and J. D. Pasteris. *Mantle Metasomatism and Alkaline Magmatism*. Boulder, Colo.: Geological Society of America, 1987.

Mortimore, Michael. *Adapting to Drought: Farmers, Famines and Desertification in West Africa*. New York: Cambridge University Press, 2009.

Mottl, M. J. "Metabasalts, Axial Hot Springs, and the Structure of Hydrothermal Systems at Mid-Ocean Ridges." *Geological Society of America Bulletin* 94 (1983): 161-180.

Mulamoottil, George, Edward A. McBean, Frank Rovers, et al., eds. *Constructed Wetland for the Treatment of Landfill Leachates*. Boca Raton, Fla.: Lewis Publishers, 1999.

Murck, Barbara W., Brian J. Skinner, and Stephen C. Porter. *Dangerous Earth: An Introduction to Geologic Hazards*. New York: John Wiley & Sons, 1997.

Murillo-Amador, B., et al. "Influence of Calcium Silicate on Growth, Physiological Parameters, and Mineral Nutrition in Two Legume Species Under Salt Stress." *Journal of Agronomy and Crop Science* 193, no. 6 (2007): 413-421.

Murray, G. E. *Geology of the Atlantic and Gulf Coastal Province of North America*. New York: Harper & Brothers, 1961.

Myers, A., D. Edmonds, and K. Donegani. *Offshore Information Guide*. Tulsa, Okla.: PennWell Books, 1988.

Naisbitt, John. *Megatrends: Ten New Directions Transforming Our Lives*. New York: Warner Books, 1984.

Nardone, Paul J. *Well Testing Project Management: Onshore and Offshore Operations*. Burlington, Mass.: Gulf Professional Publishing, 2009.

Narten, Perry F., et al. *Reclamation of Mined Lands in the Western Coal Region*. U.S. Geological Survey Circular 872. Alexandria, Va.: U.S. Geological Survey, 1983.

National Commission on the BP Deepwater Horizon Oil Spill and Offshore Drilling. *Deep Water: The Gulf Oil Disaster and the Future of Offshore Drilling*. Washington, D.C.: Author, 2011.

National Research Council Committee on Medical Isotope Production Without Highly Enriched Uranium. *Medical Isotope Production Without Highly*

Enriched Uranium. Washington, D.C.: National Academies Press, 2009.

Nazri, Gholam-Abbas, and Gianfranco Pistoia. *Lithium Batteries.* New York: Springer, 2009.

Nebel, Bernard J., and Richard T. Wright. *Environmental Science: Towards a Sustainable Future.* Englewood Cliffs, N.J.: Prentice Hall, 2008.

Nelson, John D, and Debora J. Miller. *Expansive Soils: Problems and Practice in Foundation and Pavement Engineering.* New York: John Wiley and Sons, 1997.

Nespolo, Massimo, and Giovanni Ferraris. "A Survey of Hybrid Twins in Non-Silicate Minerals." *European Journal of Mineralogy* 21 (2009): 673-690.

Neumann, A. Conrad. "Scenery for Sale." In *Coastal Development and Areas of Environmental Concern.* Raleigh: North Carolina State University Press, 1975.

Neville, A. M. *Properties of Concrete.* 4th ed. New York: John Wiley & Sons, 1996.

Neville, A. M., and J. J. Brooks. *Concrete Technology.* 2d ed. Upper Saddle River, N.J.: Prentice Hall, 2010.

Newman, Cathy, and Pierre Boulat. "Carrara Marble: Touchstone of Eternity." *National Geographic* 162 (July, 1982): 42-58.

Newman, Renée. *Diamond Handbook.* 2d ed. Los Angeles: International Jewelry Publications, 2008.

Newson, Malcolm. *Land, Water and Development: Sustainable and Adaptive Management of Rivers.* 3d ed. London: Routledge, 2008.

Newton, Robert C. "The Three Partners of Metamorphic Petrology." *American Mineralogist* 96 (2011): 457-469.

Nikiforuk, Andrew. *Tar Sands: Dirty Oil and the Future of a Continent.* Vancouver: Greystone Books, 2010.

Nixon, Peter H., ed. *Mantle Xenoliths.* New York: John Wiley & Sons, 1987.

Nolon, John R., and Patricia E. Salkin. *Land Use in a Nutshell.* St. Paul, Minn.: Thomson/West, 2006.

Norman, M. D., L. E. Borg, L. E. Nyquist, and D. D. Bogard. "Chronology, Geochemistry, and Petrology of a Ferroan Noritic Anorthosite Clast from Descartes Breccia 67215: Clues to the Age, Origin, Structure, and Impact History of the Lunar Crust." *Meteoritics and Planetary Science* 38 (2003): 645-661.

Norris, Joanne E., et al., eds. *Slope Stability and Erosion Control: Ecotechnological Solutions.* New York: Springer, 2010.

North, F. K. *Petroleum Geology.* Boston: Allen & Unwin, 1985.

Norton, James J. "Sequence of Minerals Assemblages in Differentiated Granitic Pegmatites." *Economic Geology* 78 (August, 1983): 854-874.

Noyes, Robert. *Nuclear Waste Cleanup Technology and Opportunities.* Park Ridge, N.J.: Noyes Publications, 1995.

Oates, Joseph A. H. *Lime and Limestone: Chemistry and Technology, Production and Uses.* New York: Wiley-VCH, 1998.

O'Donoghue, Michael. *Gems.* 6th ed. London: Robert Hale, 2008.

_____. *The Encyclopedia of Minerals and Gemstones.* London: Orbis, 1976.

Oldershaw, Cally. *Rocks and Minerals.* New York: DK, 1999.

O'Leary, Philip R., Patrick W. Walsh, and Robert K. Ham. "Managing Solid Waste." *Scientific American* (December, 1988): 36.

Oliver, F. W. "Dust-Storms in Egypt and Their Relation to the War Period, as Noted in Maryut, 1939-1945." *Geographical Journal* 106-108 (1945-1946): 26-49, 221-226.

Olmsted, John A., and Gregory M. Williams. *Chemistry.* 4th ed. Hoboken, N.J.: Wiley, 2006.

Olsen, J. C., D. R. Shawe, L. C. Pray, and W. N. Sharp. *Rare Earth Mineral Deposits of the Mountain Pass District, San Bernardino County, California.* U.S. Geological Survey Professional Paper 261. Washington, D.C.: Government Printing Office, 1954.

Olson, Gerald W. *Soils and the Environment: A Guide to Soil Surveys and Their Application.* New York: Chapman & Hall, 1981.

O'Neil, Paul. *Gemstones.* Alexandria, Va.: Time-Life Books, 1983.

O'Neill, Gerard K. *2081: A Hopeful View of the Human Future.* New York: Simon & Schuster, 1981.

Oppenheimer, Clive. *Eruptions That Shook the World.* New York: Cambridge University Press, 2011.

Orchard, Dennis Frank. *Concrete Technology.* Vol. 1, Properties of Materials. 4th ed. London: Applied Science Publishers, 1979.

Ormsby, Tim, et al. *Getting to Know ArcGIS Desktop.* 2d ed. Redlands, Calif.: ESRI Press, 2010.

Orszulik, Stefan T., ed. *Environmental Technology in the Oil Industry.* 2d ed. New York: Springer Science, 2010.

O'Very, David P., Christopher E. Paine, and Dan W. Reicher, eds. *Controlling the Atom in the 21st Century.* Boulder, Colo.: Westview Press, 1994.

Owen, E. W. *Trek of the Oil Finders: A History of Exploration for Petroleum.* Tulsa, Okla.: American Association of Petroleum Geologists, 1975.

Ozawa, Kazunori. *Lithium Ion Rechargeable Batteries.* Hoboken, N.J.: Wiley, 2009.

Paddock, Joe, Nancy Paddock, and Carol Bly. *Soil and Survival.* San Francisco: Sierra Club Books, 1986.

Palmer, Tim. *Endangered Rivers and the Conservation Movement.* Berkeley: University of California Press, 1986.

Paquet, Hélène, and Norbert Clauer, eds. *Soils and Sediments: Mineralogy and Geochemistry.* New York: Springer, 1997.

Parker, Raymond L., and William N. Sharp. *Mafic-Ultramafic Igneous Rocks and Associated Carbonatites of the Gem Park Complex, Custer and Fremont Counties, Colorado.* U.S. Geological Survey Professional Paper 649. Washington, D.C.: Government Printing Office, 1970.

Parker, Sybil P., ed. *McGraw-Hill Encyclopedia of the Geological Sciences.* 2d ed. New York: McGraw-Hill, 1988.

Parsons, Ian, ed. *Feldspars and Their Reactions.* Dordrecht, Netherlands: Kluwer, 1994.

Passchier, Cees W., and Rudolph A. J. Trouw. *Microtectonics.* Berlin: Springer, 2005.

Patel, Mukund R. *Wind and Solar Power Systems.* Boca Raton, Fla.: CRC Press, 1999.

Paton, T. R., G. S. Humphreys, and P. B. Mitchell. *Soils: A New Global View.* New Haven, Conn.: Yale University Press, 1995.

Pauling, Linus. *The Nature of the Chemical Bond and the Structure of Molecules and Crystals: An Introduction to Modern Structural Chemistry.* 3d ed. Ithaca, N.Y.: Cornell University Press, 1960.

Pellant, Chris. *Smithsonian Handbooks: Rocks and Minerals.* New York: Dorling Kindersley, 2002.

Perchuk, L. L., ed. *Progress in Metamorphic and Magmatic Petrology.* New York: Cambridge University Press, 1991.

Perkins, Dexter. *Mineralogy.* 3d ed. Upper Saddle River, N.J.: Prentice Hall, 2010.

Pernetta, John, ed. *The Rand McNally Atlas of the Oceans.* Skokie, Ill.: Rand McNally, 1994.

Peters, K. E. "Guidelines for Evaluating Petroleum Source Rock Using Programmed Pyrolysis." *American Association of Petroleum Geologists Bulletin* 70 (March, 1986): 318-329.

Peters, W. C. *Exploration and Mining Geology.* 2d ed. New York: John Wiley & Sons, 1987.

Petersen, R. E. "Geological Hazards to Petroleum Exploration and Development in Lower Cook Inlet." *Environmental Assessment of the Alaskan Continental Shelf: Lower Cook Inlet Interim Synthesis Report* (1979): 51-70.

Peterson, Jocelyn A. *Platinum-Group Elements in Sedimentary Environments in the Conterminous United States.* Washington, D.C.: Government Printing Office, 1994.

Pettijohn, F. J. *Sedimentary Rocks.* 3d ed. New York: Harper & Row, 1975.

Pettijohn, F. J., P. E. Potter, and R. Siever. *Sand and Sandstone.* 2d ed. New York: Springer-Verlag, 1987.

Phillips, William Revell. *Mineral Optics: Principles and Techniques.* San Francisco: W. H. Freeman, 1971.

Philpotts, Anthony, and Jay Aque. *Principles of Igneous and Metamorphic Petrology.* 2d ed. New York: Cambridge University Press, 2009.

Pichtel, John. *Waste Management Practices: Municipal, Hazardous, and Industrial.* Boca Raton, Fla.: CRC Press, 2005.

Pick, James B. "Geographic Information Systems: A Tutorial and Introduction." *Communications of AIS* 14 (2004): 307-331.

Pirson, Sylvain J. *Geologic Well Log Analysis.* 3d ed. Houston: Gulf Publishing, 1983.

Plate, Erich J., et al., eds. *Buoyant Convention in Geophysical Flows.* Boston: Kluwer Academic Publishers, 1998.

Plummer, Charles C., Diane H. Carlson, and David McGeary. *Physical Geology.* 13th ed. Columbus, Ohio: McGraw-Hill Higher Education, 2009.

Pohl, Walter L. *Economic Geology: Principles and Practice.* Hoboken, N.J.: Wiley-Blackwell, 2011.

Popovics, Sandor. *Strength and Related Properties of Concrete: A Quantitative Approach.* New York: John Wiley & Sons, 1998.

Portland Cement Association. *Principles of Quality Concrete.* New York: John Wiley & Sons, 1975.

Potts, P. J., A. G. Tindle, and P. C. Webb. *Geochemical Reference Material Compositions: Rocks, Minerals, Sediments, Soils, Carbonates, Refractories, and Ore Used in Research and Industry.* Boca Raton, Fla.: CRC Press, 1992.

Pough, F. H. *A Field Guide to Rocks and Minerals.* 5th ed. Boston: Houghton Mifflin, 1998.

Press, F., and R. Siever. *Earth.* 4th ed. New York: W. H. Freeman, 1986.

Priest, Joseph. *Energy: Principles, Problems, Alternatives.* 6th ed. Dubuque, Iowa: Kendal Hunt Publishing, 2006.

Prikryl, R., and B. J. Smith, eds. *Building Stone Decay: from Diagnosis to Conservation.* London: The Geological Society, 2007.

Prikryl, R., ed. *Dimension Stone.* New York: A. A. Balkema Publishers, 2004.

Prinz, Martin, George Harlow, and Joseph Peters, eds. *Simon and Schuster's Guide to Rocks and Minerals.* New York: Simon & Schuster, 1978.

Prothero, Donald R., and Fred Schwab. *Sedimentary Geology: An Introduction to Sedimentary Rocks and Stratigraphy.* 2d ed. New York: W. H. Freeman, 2003.

Putnam, William C. *Geology.* 2d ed. New York: Oxford University Press, 1971.

Qian, Xuede, Robert M. Koerner, and Donald H Gray. *Geotechnical Aspects of Landfill Design and Construction.* Upper Saddle River, N.J.: Prentice Hall, 2001.

Radbruch-Hall, Dorothy H. *Landslide Overview Map of the Conterminous United States.* U.S. Geological Survey Professional Paper 1183. Washington, D.C.: Government Printing Office, 1981.

Rahn, P. H. *Engineering Geology: An Environmental Approach.* 2d ed. New York: Elsevier, 1996.

Rajapakse, Ruwan. *Geotechnical Engineering Calculations and Rules of Thumb.* Burlington, Mass.: Butterworth-Heinemann, 2008.

Ramo, O. T. *Granitic Systems.* Amsterdam: Elsevier, 2005.

Randolph, John. *Environmental Land Use Planning and Management.* Washington, D.C.: Island Press, 2004.

Ransom, Jay E. *Gems and Minerals of America.* New York: Harper & Row, 1974.

Raymond, L. A. *Petrology: The Study of Igneous, Sedimentary, and Metamorphic Rocks.* 2d ed. Long Grove, Ill.: Waveland Press, 2007.

Raymond, Martin S., and William L. Leffler. *Oil and Gas Production in Nontechnical Language.* Tulsa, Okla.: PennWell, 2005.

Reading, H. G., ed. *Sedimentary Environments: Processes, Facies, and Stratigraphy.* Oxford: Blackwell Science, 1996.

Reece, Erik. *Lost Mountain: A Year in the Vanishing Wilderness: Radical Strip Mining and the Devastation of Appalachia.* New York: Riverhead Books, 2007.

Reiche, Ina, et al. "Development of a Nondestructive Method for Underglaze Painted Tiles—Demonstrated by the Analysis of Persian Objects from the Nineteenth Century." *Analytical and Bioanalytical Chemistry* 393, no. 3 (2009): 1025-1041.

Rencz, Andrew N., ed. *Remote Sensing for the Earth Sciences.* New York: John Wiley & Sons, 1999.

Rencz, Andrew N., and Susan L. Ustin. *Manual of Remote Sensing.* Hoboken, N.J.: John Wiley & Sons, 2004.

Retallack, G. J. *Soils of the Past.* 2d ed. New York: Blackwell Science, 2001.

Rhind, D., and R. Hudson. *Land Use.* London: Methuen, 1980.

Ribbe, P. H., ed. *Orthosilicates: Reviews in Mineralogy.* Vol. 5. Washington, D.C.: Mineralogical Society of America, 1980.

Rider, Malcolm H. *The Geological Interpretation of Well Logs.* 2d ed. Houston: Gulf, 1999.

Ridge, J. D. *Economic Geology: Seventy-fifth Anniversary Volume.* El Paso, Tex.: Economic Geology Publishing, 1981.

_____. *Ore Deposits of the United States, 1933-1967.* 2 vols. New York: American Institute of Mining, Metallurgical, and Petroleum Engineers, 1968.

Riding, Robert E., and Stanley M. Awramik. *Microbial Sediments.* Berlin: Springer-Verlag, 2010.

Robb, Laurence. *Introduction to Ore-Forming Processes.* Malden, Mass.: Blackwell, 2005.

Roberts, Willard Lincoln, Thomas J. Campbell, and George Robert Rapp, Jr. *Encyclopedia of Minerals.* 2d ed. New York: Van Nostrand Reinhold, 1990.

Robinson, George. "Amphiboles: A Closer Look." *Rocks and Minerals* (November/December, 1981).

Rona, Peter A. "Resources of the Seafloor." *Science* 299, no. 5607, n.s. (2003): 673-674.

Rosenfeld, Paul E., and Lydia G. H. Feng. *Risk of Hazardous Wastes.* Burlington, Mass.: Elsevier, 2011.

Ross, David. *Power from the Waves.* New York: Oxford University Press, 1995.

Ross, Pierre-Simon, et al. "Basaltic to Andesitic Volcaniclastic Rocks in the Blake River Group, Abitibi Greenstone Belt: 2. Origin, Geochemistry, and Geochronology." *Canadian Journal of Earth Sciences.* 48 (2011): 757-777.

Rygle, Kathy J., and Stephen F. Pedersen. *The Treasure Hunter's Gem and Mineral Guides to the U.S.A.* 5th ed. Woodstock, Vt.: GemStone Press, 2008.

Sacher, Hubert; and Rene Schiemann. "When Do Deep Drilling Geothermal Projects Make Good Economic Sense?" *Renewable Energy Focus* 11 (2010): 30-31.

Sage, R. P., and T. Gareau. *A Compilation of References for Kimberlite, Diamond and Related Topics: Ontario Geological Survey, Open File Report 6067.* Ontario: Queen's Printer for Ontario, 2001.

Salomons, Willem, Ulrich Feorstner, and Pavel Mader, eds. *Heavy Metals: Problems and Solutions.* New York: Springer-Verlag, 1995.

Sasowsky, Ian D., and John Mylroie, eds. *Studies of Cave Sediments: Physical and Chemical Records of Paleoclimate.* Rev. ed. Dordrecht: Springer, 2007.

Sassa, Kyoji, and Paolo Canuti, eds. *Landslides: Disaster Risk Reduction.* New York: Springer, 2009.

Saucedo, R., et al. "Modeling of Pyroclastic Flows of Colima Volcano, Mexico: Implications for Hazard Assessment." *Journal of Volcanology and Geothermal Research* 139, nos. 1/2 (2005): 103-115.

Scarth, Alwyn. *La Catastrophe: The Eruption of Mount Pelée, the Worst Volcanic Disaster of the Twentieth Century.* New York: Oxford University Press, 2002.

Schaetzl, Randall J., and Sharon Anderson. *Soils: Genesis and Geomorphology.* New York: Cambridge University Press, 2005.

Schlager, Wolfgang. "Earth's Layers, Their Cycles, and Earth System Science." *Austrian Journal of Earth Sciences* 101 (2008): 4-16.

Schlesinger, Mark E. *Aluminum Recycling.* Boca Raton, Fla.: CRC Press, 2007.

Schmidt, Victor. *Planet Earth and the New Geoscience.* 4th ed. Dubuque, Iowa: Kendall/Hunt, 2003.

Scholle, Peter A. *Color Illustrated Guide to Constituents, Textures, Cements, and Porosities of Sandstones and Associated Rocks.* Memoir 28. Tulsa, Okla.: American Association of Petroleum Geologists, 1979.

Scholle, Peter A., D. G. Bebout, and C. H. Moore. *Carbonate Depositional Environments.* Memoir 33. Tulsa: American Association of Petroleum Geologists, 1983.

Scholle, Peter A., and P. R. Schluger, eds. *Aspects of Diagenesis.* Special Publication 26. Tulsa, Okla.: Society of Economic Paleontologists and Mineralogists, 1979.

Scholle, Peter A., and D. R. Spearing, eds. *Sandstone Depositional Environments.* Memoir 31. Tulsa, Okla.: American Association of Petroleum Geologists, 1982.

Schon, J. H. *Physical Properties of Rocks.* Amsterdam: Elsevier, 2004.

Schott, John R. *Remote Sensing: The Image Chain Approach.* New York: Oxford University Press, 2007.

Schowengerdt, Robert A. *Remote Sensing: Models and Methods for Image Processing.* Burlington, Mass.: Academic Press, 2007.

Schreiber, B. C., S. Lugli, and M. Babel, eds. *Evaporites Through Space and Time.* Geological Society Special Publication 285. Bath, England: Geological Society, 2007.

Schumann, Walter. *Minerals of the World.* 2d ed. New York: Sterling, 2008.

Schuster, Robert L., ed. *Landslide Dams: Processes, Risk, and Mitigation.* New York: American Society of Civil Engineers, 1986.

Schuster, Robert L., and Keith Turne, eds. *Landslide: Investigation and Mitigation.* Washington, D.C.: National Academy Press, 1996.

Science Mate: Plate Tectonic Cycle. Fremont, Calif.: Math/Science Nucleus, 1990.

Seabrooke, W., and Charles William Noel Miles. *Recreational Land Management.* 2d ed. New York: E. and F. N. Spon, 1993.

Segar, Douglas. *An Introduction to Ocean Sciences.* 2d ed. New York: W. W. Norton & Company, 2007.

Sei-Ichi, Saitoh, et al. "Some Operational Uses of Satellite Remote Sensing and Marine GIS for Sustainable Fisheries and Aquaculture." *ICES Journal of Marine Science/Journal du Conseil* 68, no. 4 (2011): 687-695.

Selley, Richard C. *Elements of Petroleum Geology.* 2d ed. San Diego: Academic Press, 1998.

Sengupta, Supiya. *Introduction to Sedimentology.* Rotterdam, Vt.: A. A. Balkema, 1994.

Serra, Lorenzo. *Well Logging: Data Acquisition and Applications.* Paris: Technip Editions, 2004.

Seyfried, W. E., Jr. "Experimental and Theoretical Constraints on Hydrothermal Alteration Processes at Mid-Ocean Ridges." *Annual Review of Earth and Planetary Sciences* 15 (1987): 317-335.

Shadmon, Asher. *Stone: An Introduction.* 2d ed. London: Intermediate Technology, 1996.

Shepherd, Mike. *Oil Field Production Geology.* Tulsa, Okla.: American Association of Petroleum Geologists, 2009.

Sherwood, Dennis, and Jon Cooper. *Crystals, X-rays, and Proteins: Comprehensive Protein Crystallography.* New York: Oxford University Press, 2011.

Shipley, Robert M. *Dictionary of Gems and Gemology.* 6th ed. Santa Monica, Calif.: Gemological Institute of America, 1974.

Sides, Susan. "Grow Powder." *Mother Earth News* 114 (November/December, 1988).

Sigurdsson, Haraldur, ed. *Encyclopedia of Volcanoes.* San Diego, Calif.: Academic Press, 2000.

Simkin, Tom, L. Siebert, L. McClelland, et al. *Volcanoes of the World: A Regional Directory, Gazetteer, and Chronology of Volcanism During the Last Ten Thousand Years.* Berkeley: University of California Press, 2011.

Simmons, William, et al. *Pegmatology.* Rubellite Press, 2003.

Simms, O. Harper. *The Soil Conservation Service.* New York: Praeger, 1970.

Simonson, R. W. "Outline of a Generalized Theory of Soil Genesis." *Soil Science Society of America Proceedings* 23 (1959): 152-156.

_____. "What Soils Are." In *Soil.* Edited by the U.S. Department of Agriculture. Washington, D.C.: Government Printing Office, 1957.

Sinding-Larsen, Richard, and Friedrich-W. Wellmer, eds. *Non-renewable Resource Issues: Geoscientific and Societal Challenges.* New York: Springer, 2012.

Singer, M. J., and D. N. Munns. *Soils.* 6th ed. Upper Saddle River, N.J.: Prentice Hall, 2005.

Sinkankas, John. *Gemstones of North America.* Tucson, Ariz.: Geoscience Press, 1997.

_____. *Mineralogy for Amateurs.* New York: Van Nostrand Reinhold, 1964.

Skidmore, Andrew. *Environmental Modeling with GIS and Remote Sensing.* New York: Taylor & Francis, 2003.

Skinner, Brian J., and Barbara W. Murck. *The Blue Planet: An Introduction to Earth System Science.* 3d ed. Hoboken, N.J.: Wiley, 2011.

Sloan, E. Dendy, Jr., and Carolyn Koh. *Clathrate Hydrates of Natural Gases.* 3d ed. London: CRC Press, 2007.

Smith, David G., ed. *The Cambridge Encyclopedia of Earth Sciences.* New York: Cambridge University Press, 1981.

Smith, George D. *From Monopoly to Competition: The Transformations of Alcoa, 1888-1986.* New York: Cambridge University Press, 2003.

Smith, John W. "Synfuels: Oil Shale and Tar Sands." In *Perspectives on Energy, Issues, Ideas, and Environmental Dilemmas,* edited by L. C. Ruedisili and M.

W. Firebaugh. 3d ed. New York: Oxford University Press, 1982.

Smith, John W., and Howard B. Jensen. "Oil Shale." In *McGraw-Hill Encyclopedia of the Geological Sciences,* edited by S. P. Parker. 2d ed. New York: McGraw-Hill, 1988.

Smith, Mike R., ed. *Stone: Building Stone, Rock Fill, and Armourstone in Construction.* London: Geological Society, 1999.

Snape, Colin. *Composition, Geochemistry, and Conversion of Oil Shales.* Boston: Kluwer Academic Publishers, 1995.

So, Frank S., ed. *The Practice of Local Government Planning.* 3d ed. Washington, D.C.: International City/County Management Association, 2000.

So, Frank S., Irving Hand, and Bruce McDowell. *The Practice of State and Regional Planning.* Chicago: Planners Press, 1986.

Sobolev, A. V., et al. "The Amount of Recycled Crust in Sources of Mantle-Derived Melts." *Science* 316 (2007): 412-417.

Sobolevsky, Dmitry Yu. *Strength of Dilating Soil and Load-Holding Capacity of Deep Foundations: Introduction to Theory and Practical Application.* Rotterdam, Netherlands: A. A. Balkema, 1995.

Soil Science Society of America. *Glossary of Soil Science Terms.* Madison, Wis.: Author, 2001.

Sopher, Charles D., and Jack V. Baird. *Soils and Soil Management.* 3d ed. Reston, Va.: Reston Publishing, 1986.

Sorensen, Bent. *Renewable Energy: Its Physics, Engineering, Environmental Impacts, Economics, and Planning.* 4th ed. Burlington, Mass.: Academic Press, 2011.

Sorrella, C. A. *Minerals of the World.* Racine, Wis.: Western Publishing, 1973.

Spangle, William, and Associates; F. Beach Leighton and Associates; and Baxter, McDonald and Company. *Earth-Science Information in Land-Use Planning: Guidelines for Earth Scientists and Planners.* Geological Survey Circular 721. Arlington, Va.: U.S. Department of the Interior, Geological Survey, 1976.

Sparks, Donald S., ed. *Soil Physical Chemistry.* 2d ed. Boca Raton, Fla.: CRC Press, 1999.

Speight, James G. *An Introduction to Petroleum Technology, Economics, and Politics.* Hoboken, N.J.: John Wiley & Sons, 2011.

Spiess, F. N., and RISE Project Group. "East Pacific Rise: Hot Springs and Geophysical Experiments." *Science* 207 (March 28, 1980): 1421-1432.

Spooner, E. T. C., and W. S. Fyfe. "Sub-Sea-Floor Metamorphism, Heat, and Mass Transfer." *Contributions to Mineralogy and Petrology* 42, no. 4 (1973): 287-304.

Sposito, Garrison. *The Chemistry of Soils.* 2d ed. New York: Oxford University Press, 2008.

Spry, A. *Metamorphic Textures.* Oxford, England: Pergamon Press, 1969.

Stach, E. *Stach's Textbook of Coal Petrography.* 3d ed. Berlin-Stuttgart: Gebruder Borntraeger, 1983.

Stallings, J. H. *Soil: Use and Improvement.* Englewood Cliffs, N.J.: Prentice-Hall, 1957.

Stanley, Steven M. *Earth System History.* 2d ed. New York: W. H. Freeman, 2004.

Steel, Brent S., ed. *Public Lands Management in the West: Citizens, Interest Groups, and Values.* Westport, Conn.: Praeger, 1997.

Stefanko, Robert. *Coal Mining Technology Theory and Practice.* New York: Society of Mining Engineers, 1983.

Stewart, S. A. "Salt Tectonics in the North Sea Basin: A Structural Style Template for Seismic Interpreters." *Special Publication of the Geological Society* 272 (2007): 361-396.

Stewart, W. N. *Paleobotany and the Evolution of Plants.* 2d ed. New York: Cambridge University Press, 2010.

St. John, Jeffrey. *Noble Metals.* Alexandria, Va.: Time-Life Books, 1984.

Stockton, Peter, and Ingrid Drake. "From Danger to Dollars: What the US Should Do with Its Highly Enriched Uranium." *Bulletin of the Atomic Scientists* 66, no. 6 (2010): 43-55.

Storer, Donald A. *The Chemistry of Soil Analysis.* Middletown, Ohio: Terrific Science Press, 2005.

Stout, K. S. *Mining Methods and Equipment.* Chicago: Maclean-Hunter, 1989.

Strahler, Arthur N. *Physical Geology.* New York: Harper & Row, 1981.

Strohmeier, Brian R., et al. "What Is Asbestos and Why Is It Important? Challenges of Defining and Characterizing Asbestos." *International Geology Review* 52, nos. 7/8 (2010): 801-872.

Sullivan, Charles W. *Small-Hydropower Development: The Process, Pitfalls, and Experience.* 4 vols. Palo Alto,

Calif.: Electric Power Research Institute, 1985-1986.

Suppe, John. *Principles of Structural Geology.* Englewood Cliffs, N.J.: Prentice-Hall, 1985.

Sutherland, Lin. *The Volcanic Earth: Volcanoes and Plate Tectonics, Past, Present, and Future.* Sydney, Australia: University of New South Wales Press, 1995.

Sutton, Gerard K.; and Joseph A. Cassalli, eds. *Catastrophe in Japan: The Earthquake and Tsunami of 2011.* Nova Science Publishers, 2011.

Sverdrup, Keith A., Alyn C. Duxbury, and Alison B. Duxbury. *An Introduction to the World's Oceans.* 8th ed. Columbus, Ohio: McGraw-Hill Science, 2004.

Swaine, Dalway J. *Trace Elements in Coal.* Boston: Butterworth, 1990.

Swart, Peter K., Gregor Eberli, and Judith A. McKenzie, eds. *Perspectives in Carbonate Geology.* Hoboken, N.J.: Wiley-Blackwell, 2009.

Tagare, Digamber M. *Electricity Power Generation: The Changing Dimensions.* Hoboken, N.J.: Wiley-Blackwell, 2011.

Tamura, Y., et al. "Andesites and Dactites from Daisen Volcano, Japan: Partial-to-Total Remelting of an Andesite Magma Body." *Journal of Petrology* 44 (2003): 2243-2260.

Tan, Kim Howard. *Principles of Soil Chemistry.* 4th ed. Boca Raton, Fla.: CRC Press, 2011.

Tappert, Ralf, and Michelle C. Tappert. *Diamonds in Nature: A Guide to Rough Diamonds.* New York: Springer, 2011.

Tarbuck, Edward J., Frederick K. Lutgens, and Dennis Tasa. *Earth: An Introduction to Physical Geology.* 10th ed. Upper Saddle River, N.J.: Prentice Hall, 2010.

Taylor, G. H., et al., eds. *Organic Petrology.* Berlin: Gebreuder Borntraeger, 1998.

Taylor, G. Jeffrey. *A Close Look at the Moon.* New York: Dodd, Mead, 1980.

Taylor, Harry F. W. *Cement Chemistry.* 2d ed. London: T. Telford, 1997.

Taylor, Roger G. *Geology of Tin Deposits.* New York: Elsevier, 1979.

Teller, Edward. *Energy from Heaven and Earth.* San Francisco: W. H. Freeman, 1979.

Tennissen, A. C. *Nature of Earth Materials.* 2d ed. Englewood Cliffs, N.J.: Prentice-Hall, 1983.

Terabayashi, Masaru, et al. "Silicification of Politic Schist in the Ryoke Low-Pressure/Temperature Metamorphic Belt, Southwest Japan: Origin of

Competent Layers in the Middle Crust." *Island Arc* 19 (2010): 17-29.

Thomas, L. J. *An Introduction to Mining: Exploration, Feasibility, Extraction, Rock Mechanics.* Rev. ed. Sydney: Methuen of Australia, 1978.

Thomas, Larry. *Coal Geology.* Hoboken, N.J.: John Wiley & Sons, 2002.

Thomas, William L., Jr., ed. *Man's Role in Changing the Face of the Earth.* Chicago: University of Chicago Press, 1956.

Thompson, Alan Bruce, and Jo Laird. "Calibrations of Modal Space for Metamorphism of Mafic Schist." *American Mineralogist* 90 (2005): 843-856.

Thompson, Graham R., and Jonathan Turk. *Introduction to Physical Geology.* Belmont, Calif.: Brooks/Cole, Cengage Learning, 1997.

Thompson, James B., Jr. "Modal Spaces for Pelitic Schists." *Reviews in Mineralogy and Geochemistry* 46 (2002): 449-462.

Thrall, Grant Ian. "MapInfo Professional 8.5: Internet-Enabled GIS." *Geospatial Solutions* 16, no. 9 (2006): 36-39.

Tilley, Richard J. D. *Crystals and Crystal Structures.* Hoboken, N.J.: John Wiley & Sons, 2006.

Tissot, Bernard P., Bernard Durand, J. Espitalié, and A. Combaz. "Influence of Nature and Diagenesis of Organic Matter in Formation of Petroleum." *American Association of Petroleum Geologists Bulletin* 58 (March, 1974): 499-506.

Tissot, Bernard P., and D. H. Welte. *Petroleum Formation and Occurrence.* 2d ed. Berlin: Springer-Verlag, 1984.

Torrey, Volta. *Wind-Catchers: American Windmills of Yesterday and Tomorrow.* Brattleboro, Vt.: Stephen Greene Press, 1981.

Totten, George E., and D. Scott MacKenzie, eds. *Handbook of Aluminum.* New York: Marcel Dekker, 2003.

_____. *Mineral Commodity Summaries.* Washington, D.C.: Government Printing Office, issued annually.

_____. *Minerals Yearbook.* Vol. 1, *Metals and Minerals.* Washington, D.C.: Government Printing Office, issued annually.

Troeh, Frederick R., and Louis M. Thompson. *Soils and Soil Fertility.* 5th ed. New York: Oxford University Press, 1993.

Troxell, George Earl, Harmer E. Davis, and Joe W. Kelly. *Composition and Properties of Concrete.* 2d ed. New York: McGraw-Hill, 1968.

Tucker, M., ed. *Sedimentary Petrology: An Introduction to the Origin of Sedimentary Rocks.* 3d ed. Malden, Mass.: Blackwell Science, 2001.

_____, ed. *Sedimentary Rocks in the Field.* 4th ed. New York: John Wiley & Sons, 2011.

_____. ed. *Techniques in Sedimentology.* Boston: Blackwell Scientific Publications, 1988.

Turnberg, Wayne L. *Biohazardous Waste: Risk Assessment, Policy, and Management.* New York: John Wiley & Sons, 1996.

Tuttle, O. F., and J. Gittins. *Carbonatites.* New York: Wiley-Interscience, 1966.

United Nations Conference on Desertification, Nairobi, Kenya. *Desertification: Its Causes and Consequences.* Elmsford, N.Y.: Pergamon Press, 1977.

U.S. Bonneville Power Administration. *Columbia River Power for the People: History of Policies of the Bonneville Power Administration.* Portland, Ore.: Author, 1980.

U.S. Bureau of Mines. *Minerals Yearbook.* Washington, D.C.: Government Printing Office, issued annually.

U.S. Department of Agriculture, Soil Survey Staff. *Keys to Soil Taxonomy.* Honolulu: University Press of the Pacific, 2005.

U.S. Department of Labor, Mine Safety and Health Administration. *Coal Mining.* Washington, D.C.: Government Printing Office, 1997.

U.S. Department of the Interior. *Surface Mining and Our Environment: A Special Report to the Nation.* Washington, D.C.: Government Printing Office, 1967.

U.S. Department of the Interior, Minerals Management Service (USDOI MMS). *Programmatic Environmental Assessment: Arctic Ocean Outer Continental Shelf Seismic Surveys.* U.S. Department of the Interior Minerals Management Service Alaska OCS Region, 2006.

Usdowski, Eberhard, and Martin Dietzel. *Atlas and Data of Solid-Solution Equilibria of Marine Evaporites.* New York: Springer, 1998.

U.S. Environmental Protection Agency. *Erosion and Sediment Control: Surface Mining in the Eastern U.S.* EPA Technology Transfer Seminar Publication. Washington, D.C.: Government Printing Office, 1976.

_____. *Profile of the Non-metal, Non-fuel Mining Industry*. Washington, D.C.: U.S. Environmental Protection Agency, 1995.

_____. *RCRA Subtitle D Study*. Washington, D.C.: Government Printing Office, 1986, 1988.

U.S. Federal Power Commission. *Hydroelectric Power Resources of the United States, Developed and Undeveloped*. Washington, D.C.: Government Printing Office, 1992.

U.S. Forest Service. *Final Environmental Impact Statement: Changes Proposed to Existing Yellowstone Pipeline Between Thompson Falls and Kingston*. Missoula, Mont.: Author, 2000.

U.S. Geological Survey. *Mineral Facts and Problems*. Bulletin 675. Washington, D.C.: Government Printing Office, 1985.

_____. *United States Mineral Resources*. Professional Paper 820. Washington, D.C.: Government Printing Office, 1973.

_____. *Water Fact Sheet: Toxic Waste—Groundwater Contamination Program*. Washington, D.C.: Government Printing Office, 1988.

_____. *Yearbook*. Washington, D.C.: Government Printing Office, issued annually.

U.S. Soil Conservation Series. *Our American Land: Use the Land, Save the Soil*. Washington, D.C.: Government Printing Office, 1967.

Van Breemen, Nico, and Peter Buurman. *Soil Formation*. 2d ed. Boston: Kluwer Academic Publishers, 2002.

Van Krevelen, D. W. *Coal: Typology-Chemistry-Physics-Constitution*. Amsterdam: Elsevier, 1961.

Veblen, D. R. "Biopyriboles." In *McGraw-Hill Encyclopedia of Science and Technology*. 10th ed. New York: McGraw-Hill, 2007.

Velde, B. *Introduction to Clay Minerals: Chemistry, Origins, Uses, and Environmental Significance*. New York: Chapman and Hall, 1992.

_____, ed. *Origin and Mineralogy of Clays: Clays and the Environment*. London: Springer-Verlag, 1995.

Vermaak, C. Frank. *The Platinum-Group Metals: A Global Perspective*. Randburg, South Africa: Mintek, 1995.

Vernon, Ron H. *A Practical Guide to Rock Microstructure*. New York: Cambridge University Press, 2004.

Vernon, Ron H., and Geoffrey Clarke. *Principles of Metamorphic Petrology*. New York: Cambridge University Press, 2008.

Vogel, Shawna. "Wind Power." *Discover* 10 (May, 1989): 46-49.

Vogel, Willis G. *A Guide for Revegetating Coal Minesoils in the Eastern United States*. USDA Forest Service General Technical Report NE-68. Broomall, Pa.: Northeast Forest Experiment Station, 1981.

Voll, Gerhard, ed. *Equilibrium and Kinetics in Contact Metamorphism: The Ballachulish Igneous Complex and Its Aureole*. New York: Springer-Verlag, 1991.

Wade, Timothy G., et al. "A Comparison of Vector and Raster GIS Methods for Calculating Landscape Metrics Used in Environmental Assessments." *Photogrammetric Engineering and Remote Sensing* 69, no. 12 (2003): 1399-1405.

Wagner, Kathryn D., et al., eds. *Environmental Management in Healthcare Facilities*. Philadelphia: W. B. Saunders, 1998.

Walker, John F., and Nicholas Jenkins. *Wind Energy Technology*. New York: John Wiley & Sons, 1997.

Wallace, John M., and Peter V. Hobbs. *Atmospheric Science: An Introductory Survey*. 2d ed. Burlington, Mass.: Academic Press, 2006.

Walls, James. *Land, Man, and Sand: Desertification and Its Solution*. New York: Macmillan, 1980.

Waltham, Tony. *Great Caves of the World*. Buffalo, N.Y.: Firefly Books, 2008.

Walther, John Victor. *Essentials of Geochemistry*. 2d ed. Burlington, Mass.: Jones & Bartlett Publishers, 2008.

Waples, Douglas W. *Geochemistry in Petroleum Exploration*. Boston: International Human Resources Development Corporation, 1985.

Ward, Peter L. "Sulfur Dioxide Initiates Global Climate Change in Four Ways." *Thin Solid Films* 517, no. 11 (2009): 3188-3203.

Warren, John K. *Evaporites: Sediment, Resources and Hydrocarbons*. Berlin: Springer-Verlag, 2006.

_____. *Evaporites: Their Evolution and Economics*. Malden, Mass.: Blackwell Science, 1999.

Watkins, Tom H. *Gold and Silver in the West: The Illustrated History of an American Dream*. Palo Alto, Calif.: American West, 1971.

Weast, Robert C., ed. *CRC Handbook of Chemistry and Physics*. 92d ed. Boca Raton, Fla.: CRC Press, 2011.

Webster, J. D. *Melt Inclusions in Plutonic Rocks: Short Course Series 36*. Ottawa: Mineralogical Associations of Canada, 2006.

Webster, Robert F. G. A. *Gems: Their Sources, Descriptions, and Identification.* 5th ed. Boston: Butterworth-Heinemann, 1994.

Weiner, Jonathan. *Planet Earth.* New York: Bantam Books, 1986.

Welker, Anthony J. *The Oil and Gas Book.* Tulsa, Okla.: PennWell, 1985.

Welte, Dietrich H., Brian Horsfield, and Donald R. Baker, eds. *Petroleum and Basin Evolution: Insights from Petroleum Geochemistry, Geology, and Basin Modeling.* New York: Springer, 1997.

Welton, J. E. *SEM Petrology Atlas.* Tulsa, Okla.: American Association of Petroleum Geologists, 1984.

Wenk, Hans-Rudolf, and Andrei Bulakh. *Minerals: Their Constitution and Origin.* New York: Cambridge University Press, 2004.

Wessel, G. R., ed. *Native Sulfur: Developments in Geology and Exploration.* Englewood, Colo.: Society for Mining Metallurgy, 1992.

West Texas Geological Society. *Geological Examples in West Texas and South-eastern New Mexico (the Permian Basin) Basic to the Proposed National Energy Act.* Midland: West Texas Geological Society, 1979.

White, Benjamin. *Silver: Its History and Romance.* Detroit, Mich.: Tower Books, 1971.

Wicander, Reed, and James S. Monroe, eds. *The Changing Earth: Exploring Geology and Evolution.* 5th ed. Belmont, Calif.: Brooks/Cole, 2009.

_____. eds. *Historical Geology: Evolution of Earth and Life Through Time.* 6th ed. Belmont, Calif.: Brooks/Cole, 2010.

Wilhelm, Helmut, Walter Zuern, Hans-Georg Wenzel, et al., eds. *Tidal Phenomena.* Berlin: Springer, 1997.

Will, Thomas M. *Phase Equilibria in Metamorphic Rocks: Thermodynamic Background and Petrological Applications.* New York: Springer, 1998.

Williams, H., and A. R. McBirney. *Volcanology.* San Francisco: Freeman, Cooper, 1979.

Williams, Howel, F. J. Turner, and C. M. Gilbert. *Petrography: An Introduction to the Study of Rocks in Thin Sections.* 2d ed. San Francisco: W. H. Freeman, 1982.

Wilson, Alan D., and John W. Nicholson. *Acid-Base Cements: Their Biomedical and Industrial Applications.* New York: University of Cambridge, 2005.

Wilson, B. M. *Igneous Petrogenesis: A Global Tectonic Approach.* Dordrecht: Springer, 2007.

Wilson, J. L. *Carbonate Facies in Geologic History.* New York: Springer-Verlag, 1975.

Wilson, Marjorie. *Igneous Petrogenesis: A Global Tectonic Approach.* New York: Springer, 2007.

Wilson, Mitchell, et al., eds. *Life Science Library: Energy.* New York: Time, 1963.

Windley, B. F. *The Evolving Continents.* 3d ed. New York: John Wiley & Sons, 1995.

Winkler, H. G. F. *Petrogenesis of Metamorphic Rocks.* 5th ed. New York: Springer-Verlag, 1979.

Winter, J. D. *An Introduction to Igneous and Metamorphic Petrology.* Upper Saddle River, N.J.: Prentice Hall, 2001.

_____. *Principles of Igneous and Metamorphic Petrology.* 2d ed. New York: Pearson Education, Inc, 2010.

Woltman, Scott J., Gregory Philip Crawford, and Gregory D. Jay. *Liquid Crystals: Frontiers in Biomedical Applications.* Hackensack, N.J.: World Scientific, 2007.

Wood, Elizabeth A. *Crystals and Light.* 2d ed. Princeton, N.J.: Van Nostrand Reinhold, 1977.

Woodhead, James A., ed. *Geology.* 2 vols. Pasadena, Calif.: Salem Press, 1999.

Woolley, A. R. *Alkaline Rocks and Carbonatites of the World.* Tulsa, Okla.: Geological Society, 2001.

World Bank. *Sustainable Land Management: Challenges, Opportunities, and Trade-offs.* Washington, D.C.: International Bank for Reconstruction and Development, 2006.

Wyllie, P. J., ed. *Ultramafic and Related Rocks.* Malabar, Fla.: Kieger Publishing Company, 1979.

Yardley, B. W. D. *An Introduction to Metamorphic Petrology.* New York: John Wiley & Sons, 1989.

Yen, T. F., and G. V. Chilingarian, eds. *Oil Shale.* New York: Elsevier, 1976.

Young, Gordon. "The Miracle Metal: Platinum." *National Geographic* 164 (November, 1983): 686-706.

Young, Warren. *Atomic Energy Costing.* Boston: Kluwer Academic Publishers, 1998.

Youngquist, Walter Lewellyn. *GeoDestinies: The Inevitable Control of the Earth Resources Over Nations and Individuals.* Portland, Ore.: National Book Company, 1997.

Yuan, Xianxia, Hansan Liu, and Jiujun Zhang. *Lithium-Ion Batteries: Advanced Materials and Technologies.* Boca Raton, Fla.: CRC Press, 2012.

Zamel, Bernard. *Tracers in the Oil Field.* New York: Elsevier, 1995.

Zdruli, Pandi, et al., eds. *Land Degradation and desertification: Assessment, Mitigation and Remediation.* New York: Springer, 2010.

Zektser, Igor S., et al., eds. *Geology and Ecosystems.* New York: Springer, 2010.

Zhang, Ming, et al. "Metamictization of Zircon: Raman Spectroscopic Study." *Journal of Physics: Condensed Matter* 12 (2000): 1915-1925.

Zoellner, Tom. *Uranium: War, Energy, and the Rock That Shaped the World.* New York: Viking Penguin, 2009.

Zussman, J., ed. *Physical Methods in Determinative Mineralogy.* 2d ed. New York: Academic Press, 1978.

Periodic Table of Elements

1	2	3	4	5	6	7	8	9	10	11	12	13	14	15	16	17	18
1 H Hydrogen 1																	1 He Helium 4
3 Li Lithium 7	4 Be Beryllium 9											5 B Boron 11	6 C Carbon 12	7 N Nitrogen 14	8 O Oxygen 16	9 F Fluoride 19	10 Ne Neon 20
11 Na Sodium 28	12 Mg Magnesium 24											13 Al Aluminum 27	14 Si Silicon 28	15 P Phosphorus 31	16 S Sulfur 32	17 Cl Chloride 35	18 Ar Argon 40
19 K Potassium 39	20 Ca Calcium 40	21 Sc Scandium 45	22 Ti Titanium 48	23 Vi Vanadium 51	24 Cr Chromium 52	25 Ma Manganese 55	26 Fe Iron 56	27 Co Cobalt 59	28 Ni Nickel 59	29 Cu Copper 64	30 Zn Zinc 65	31 Ga Gallium 70	32 Ge Germanium 73	33 As Arsenic 75	34 Se Selenium 79	35 Br Bromine 80	36 Kr Krypton 84
37 Rb Rubidium 85	38 Sr Strontium 88	39 Y Yttrium 89	40 Zr Zirconium 91	41 Nb Niobium 93	42 Mo Molybdenum 96	43 Tc Technium 98	44 Ru Ruthenium 101	45 Rh Rhodium 103	46 Pd Palladium 106	47 Ag Silver 108	48 Cd Cadmium 112	49 In Indium 115	50 Sn Tin 119	51 Sb Indium 122	52 Te Tellurium 128	53 I Iodine 127	54 Xe Xenon 131
55 Cs Cesium 133	56 Ba Barium 137	57 La Lanthanum 139	72 Hf Hafnium 178	73 Ta Tantalum 181	74 W Tungsten 184	75 Re Rhenium 186	76 Os Osmium 190	77 Ir Iridium 192	78 Pt Platinum 195	79 Au Gold 197	80 Hg Mercury 201	81 Ti Thallium 204	82 Pb Lead 207	83 Bi Bismuth 209	84 Po Polonium 209	85 At Asatine 210	86 Rn Radon 222
87 Fa Francium 223	88 Ra Radium 226	89 Ac Actinium 227	116 Rf Rutherfordium 267	105 Db Dubnium 268	106 Sg Seaborgium 271	107 Bh Bohrium 272	108 Hs Hassium 277	109 Mt Unnilhexium 276	110 Ds Cobalt 281	111 Rg Roentgenium 280	112 Uub Ununbium 285	113 Uut Unutrium 284	114 Lv Livermorium 289	115 Uup Unnilpentium 288	116 Fl Flerovium 293	117 Uus Unnilseptium 294	118 Uuo Unniloctium 294

57 La Lanthanum 139	58 Ce Cerium 140	59 Pr Praseodymium 141	60 Nd Neodymium 144	61 Pm Promethium 145	62 Sm Samarium 150	63 Eu Europium 152	64 Gd Gadolinium 157	65 Tb Terbium 159	66 Dy Dysprosium 163	67 Ho Holmium 165	68 Er Erbium 167	69 Tm Thulium 160	70 Yb Ytterbium 173	71 Lu Lutetium 175
89 Ac Actinium 227	90 Th Thorium 232	91 Pa Protactinium 231	92 U Uranium 238	93 Np Neptunium 237	94 Pu Plutonium 244	95 Am Americium 243	96 Cm Curium 247	97 Bk Berkelium 247	98 Cf Californium 251	99 Es Einsteinium 252	100 Fm Fermium 257	101 Md Mendelevium 258	102 No Nobelium 259	103 Lr Lawrencium 262

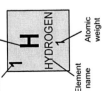

Atomic number — Element symbol — 1 H HYDROGEN — Atomic weight — Element name

609

Element Groups

Nonmetals	Aliali Metals	Alkali Earth Metals

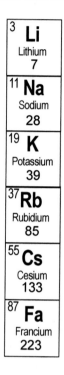

Nonmetals

1 **H** Hydrogen 1

Aliali Metals

3 **Li** Lithium 7
11 **Na** Sodium 28
19 **K** Potassium 39
37 **Rb** Rubidium 85
55 **Cs** Cesium 133
87 **Fa** Francium 223

Alkali Earth Metals

4 **Be** Beryllium 9
12 **Mg** Magnesium 24
20 **Ca** Calcium 40
38 **Sr** Strontium 88
56 **Ba** Barium 137
88 **Ra** Radium 226

Transition Metals

21 **Sc** Scandium 45	22 **Ti** Titanium 48	23 **Vi** Vanadium 51	24 **Cr** Chromium 52	25 **Ma** Manganese 55	26 **Fe** Iron 56	27 **Co** Cobalt 59	28 **Ni** Nickel 59	29 **Cu** Copper 64	30 **Zn** Zinc 65
39 **Y** Yttrium 89	40 **Zr** Zirconium 91	41 **Nb** Niobium 93	42 **Mo** Molybdenium 96	43 **Tc** Technitium 98	44 **Ru** Ruthenium 101	45 **Rh** Rhodium 103	46 **Pd** Palladium 106	47 **Ag** Silver 108	48 **Cd** Cadmium 112
72 **Hf** Hafnium 178	73 **Ta** Tantalum 181	74 **W** Tungsten 184	75 **Re** Rhenium 186	76 **Os** Osmium 190	77 **Ir** Iridium 192	78 **Pt** Platinum 195	79 **Au** Gold 197	80 **Hg** Mercury 201	
116 **Rf** Rutherfordium 267	105 **Db** Dubnium 268	106 **Sg** Seaborgium 271	107 **Bh** Bohrium 272	108 **Hs** Hassium 277	109 **Mt** Unnilhexium 276	110 **Ds** Cobalt 281	111 **Rg** Roentgenium 280	112 **Uub** Ununbium 285	

Post-transition Metals

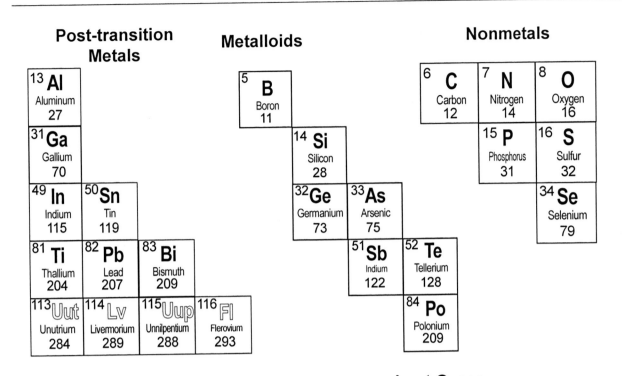

13 **Al** Aluminum 27		
31 **Ga** Gallium 70		
49 **In** Indium 115	50 **Sn** Tin 119	
81 **Ti** Thallium 204	82 **Pb** Lead 207	83 **Bi** Bismuth 209
113 Uut Unutrium 284	114 Lv Livermorium 289	115 Uup Unnilpentium 288
			116 Fl Flerovium 293

Metalloids

5 **B** Boron 11	
14 **Si** Silicon 28	
32 **Ge** Germanium 73	33 **As** Arsenic 75
51 **Sb** Indium 122	52 **Te** Tellerium 128
	84 **Po** Polonium 209

Nonmetals

6 **C** Carbon 12	7 **N** Nitrogen 14	8 **O** Oxygen 16
	15 **P** Phosphorus 31	16 **S** Sulfur 32
		34 **Se** Selenium 79

Halogens

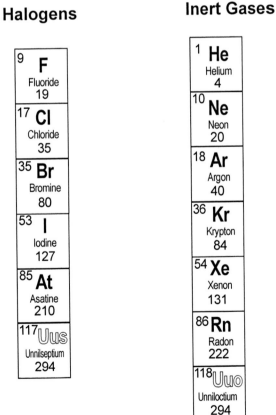

9 **F** Fluoride 19
17 **Cl** Chloride 35
35 **Br** Bromine 80
53 **I** Iodine 127
85 **At** Asatine 210
117 Uus Unnilseptium 294

Inert Gases

1 **He** Helium 4
10 **Ne** Neon 20
18 **Ar** Argon 40
36 **Kr** Krypton 84
54 **Xe** Xenon 131
86 **Rn** Radon 222
118 Uuo Unniloctium 294

Rare Earth Metals

^{57}La	^{58}Ce	^{59}Pr	^{60}Nd	^{61}Pm	^{62}Sm	^{63}Eu	^{64}Gd	^{65}Tb	^{66}Dy
Lanthanium 139	Cerium 140	Praseodynium 141	Neodymium 144	Promethium 145	Samarium 150	Europium 152	Gadolimium 157	Terbium 159	Dysprosium 163
^{89}Ac	^{90}Th	^{91}Pa	^{92}U	^{93}Np	^{94}Pu	^{95}Am	^{96}Cm	^{97}Bk	^{98}Cf
Actinium 227	Thorium 232	Protactinium 231	Uranium 238	Neptunium 237	Plutonium 244	Americium 243	Curium 247	Berkelium 247	Californium 251

^{67}Ho	^{68}Er	^{69}Tm	^{70}Yb	^{71}Lu
Holmium 165	Erbium 167	Thulium 160	Ytterbium 173	Lutetium 175
^{99}Es	^{100}Fm	^{101}Md	^{102}No	^{103}Lr
Einsteinium 252	Fermium 257	Mendelevium 258	Nobelium 259	Lawrencium 262

MOHS SCALE OF MINERAL HARDNESS

In 1812 the Mohs scale of mineral hardness was devised by the German mineralogist Friedrich Mohs (1773–1839), who selected the ten minerals because they were common or readily available. The scale is not a linear scale, but somewhat arbitrary.

Hardness	Mineral	Associations and Uses	Hardness of some other items:
1	Talc	Talcum powder.	
2	Gypsum	Plaster of Paris. Gypsum is formed when sea-water evaporates from the ground surface.	Fingernail 2.5 Gold/silver 2.5–3 Copper penny 3
3	Calcite	Limestone and most shells contain calcite.	
4	Fluorite	Fluorine in fluorite prevents tooth decay.	Platinum 4–4.5 Iron 4–5 Knife blade 5.5
5	Apatite	Apatite is the primary component of bones and teeth.	
6	Orthoclase	Orthoclase is a feldspar, and in German, "feld" means "field."	Glass 6–7 Iron pyrite 6.5 Hardened steel 7+
7	Quartz		
8	Topaz	The November birthstone. Emerald and aquamarine are varieties of beryl with a hardness of 8.	
9	Corundum	Sapphire and ruby are varieties of corundum. Twice as hard as topaz.	
10	Diamond	Used in jewelry and cutting tools. Four times as hard as corundum.	

ENERGY PRODUCTION AND CONSUMPTION

The charts below provide a comparison of the major producers and consumers in each energy resource industry. The top five countries are identified for production and consumption of various energy sources. All other countries (varying, dependent on the factor) are combined into a single "Other" category. Electricity resource charts are treated differently, as all electricity produced runs through a combined grid regardless of the source. Therefore, the amount of electricity consumed by each country cannot be categorized based on its original source and only one value is available for consumption of all electricity sources.

FOSSIL FUEL PRODUCTIONS AND CONSUMPTION

Coal Production

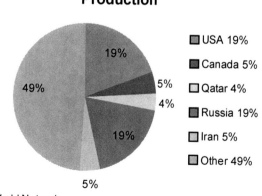

- USA 14%
- India 8%
- Indonesia 5%
- China 44%
- Australia 6%
- Other 24%

World Coal
Production: 7,984,900.0 Thousand Short Tons

Coal Consumption

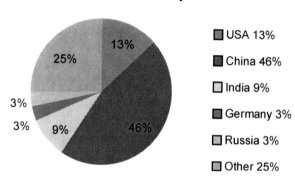

- USA 13%
- China 46%
- India 9%
- Germany 3%
- Russia 3%
- Other 25%

World Coal
Consumption: 7,994,703 Thousand Short Tons

Natural Gas Production

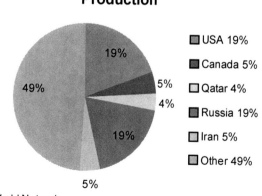

- USA 19%
- Canada 5%
- Qatar 4%
- Russia 19%
- Iran 5%
- Other 49%

World Natural
Gas Production: 112,090 Billion Cubic Feet

Natural Gas Consumption

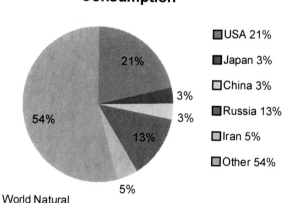

- USA 21%
- Japan 3%
- China 3%
- Russia 13%
- Iran 5%
- Other 54%

World Natural
Gas Consumption: 112,920.0 Billion Cubic Feet

FOSSIL FUEL PRODUCTIONS AND CONSUMPTION

Petroleum Production

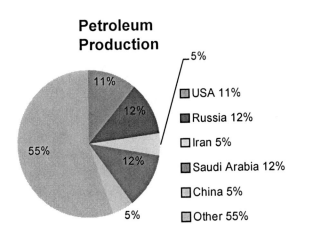

- ☐ USA 11%
- ■ Russia 12%
- ☐ Iran 5%
- ■ Saudi Arabia 12%
- ☐ China 5%
- ☐ Other 55%

World Petroleum
Production: 86,838.8 Thousand Barrels per Day

Petroleum Consumption

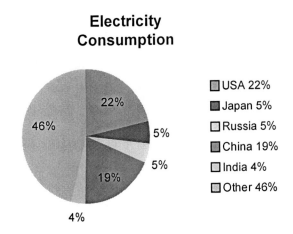

- ☐ USA 22%
- ■ Russia 4%
- ☐ India 4%
- ■ Japan 5%
- ☐ China 11%
- ☐ Other 55%

World Petroleum
Consumption: 85,714.0 Thousand Barrels per Day

ELECTICITY PRODUCTION AND CONSUMPTION

Electricity Production

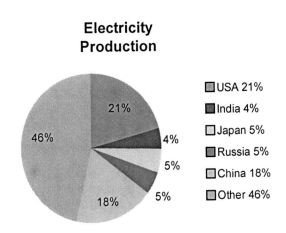

- ☐ USA 21%
- ■ India 4%
- ☐ Japan 5%
- ■ Russia 5%
- ☐ China 18%
- ☐ Other 46%

World Electricity
Production: 18,979.9 Billion Kilowatt-Hours

Electricity Consumption

- ☐ USA 22%
- ■ Japan 5%
- ☐ Russia 5%
- ■ China 19%
- ☐ India 4%
- ☐ Other 46%

World Electricity
Consumption: 17,313.6 Billion Kilowatt-Hours

RENEWABLE RESOURCE PRODUCTION AND CONSUMPTION

Biofuel Production

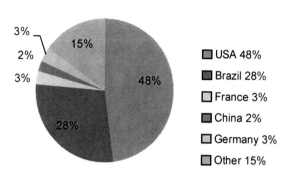

3%
2%
3%
15%
48%
28%

- USA 48%
- Brazil 28%
- France 3%
- China 2%
- Germany 3%
- Other 15%

World Biofuel
Production: 1,855.6 Thousand Barrels per Day

Biofuel Consumption

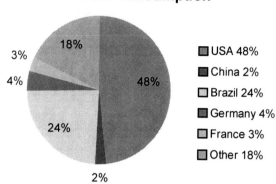

3%
4%
18%
48%
24%
2%

- USA 48%
- China 2%
- Brazil 24%
- Germany 4%
- France 3%
- Other 18%

World Biofuel
Consumption: 1,768.3 Thousand Barrels per Day

RENEWABLE RESOURCE PRODUCTION

Biomass and Waste Electricity Production

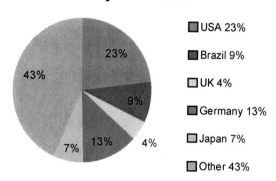

23%
43%
9%
7%
13%
4%

- USA 23%
- Brazil 9%
- UK 4%
- Germany 13%
- Japan 7%
- Other 43%

World Biomass and Waste Electricity
Production: 294.2 Billion Kilowatt-Hours

Geothermal Electricity Production

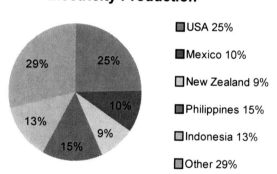

29%
25%
10%
9%
15%
13%

- USA 25%
- Mexico 10%
- New Zealand 9%
- Philippines 15%
- Indonesia 13%
- Other 29%

World Geothermal Electricity
Production: 63.9 Billion Kilowatt-Hours

RENEWABLE RESOURCE PRODUCTION

Hydroelectricity Production

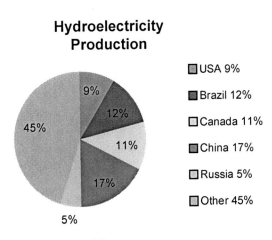

- USA 9%
- Brazil 12%
- Canada 11%
- China 17%
- Russia 5%
- Other 45%

World Hydroelectricity
Production: 3,145.2 Billion Kilowatt-Hours

Nuclear Electricity Production

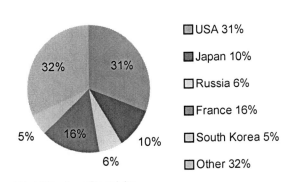

- USA 31%
- Japan 10%
- Russia 6%
- France 16%
- South Korea 5%
- Other 32%

World Nuclear Electricity
Production: 2,619.9 Billion Kilowatt-Hours

Solar, Tide, and Wave Electricity Production

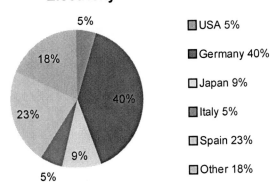

- USA 5%
- Germany 40%
- Japan 9%
- Italy 5%
- Spain 23%
- Other 18%

World Solar, Tide, and Wave Electricity
Production: 27.9 Billion Kilowatt-Hours

Wind Electricity Production

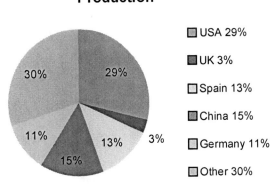

- USA 29%
- UK 3%
- Spain 13%
- China 15%
- Germany 11%
- Other 30%

World Wind Electricity
Production: 327.9 Billion Kilowatt-Hours

Source: US Energy Information Administration (EIA) www.eia.gov

SUBJECT INDEX

Note: Page numbers in bold indicate main discussion.

X-ray defined, 268, 297

X-ray diffraction: biopyriboles, 34; blueschists, 40; carbonates, 51; clays and clay minerals, 73–74; diagenesis, 98; diamonds, 104; dolomite, 108; kimberlite, 221; metamictization, 268, 270, 271; metamorphic rock classification, 277; minerals: structure, 299–303; nesosilicates and sorosilicates, 315; oxides, 378; regional metamorphism, 428; sedimentary rock classification, 459; siliciclastic rocks, 470

X-ray flourescence: blueschists, 40–41; kimberlite, 221

X-ray flourescence spectrometers, 135

X-ray fluorescence: feldspars, 141

X-ray fluorescence spectroscopy, 432

X-ray spectroscopy, 464

Y

yardangs, 477

yellowcake: radioactive minerals, 428; uranium deposits, 536, 538

Yellowstone National Park, 194, 195

Yosemite National Park, California batholiths, 21

Young's modulus, 442

Yucca Mountain, Nevada, nuclear waste disposal, 330

Z

Zechstein Basin, Europe, salt domes, 447

zeolites, 430, 431, 432

zicron: nesosilicates and sorosilicates, 315

zinc: industrial metal, 203, 204, 205; metasomatism, 288; plutonic rocks, 410; toxic minerals, 520

zircon: metamictization, 268–269, 270, 271; plutonic rocks, 409

zirconium, 204, 206

zone-refining hypothesis, 220

zones of soil types, 491, 493

zoning: kimberlite, 219; metasomatism, 287, 288; pegmatites, 383, 384; regional metamorphism, 431

zoning ordinances: land management, 234, 235; land-use planning, 245, 246, 247